Precision Photometry: Astrophysics of the Galaxy

Sponsored by

Union College
Schenectady, New York

Van Vleck Observatory
Middletown, Connecticut

October 3 – 5, 1990
Meeting held in Schenectady, New York at Union College

Edited by

A. G. Davis Philip
Union College and Van Vleck Observatory

Arthur R. Upgren
Van Vleck Observatory

Kenneth A. Janes
Boston University

Van Vleck Observatory Contribution No. 11

L. Davis Press
Schenectady, N. Y.
1991

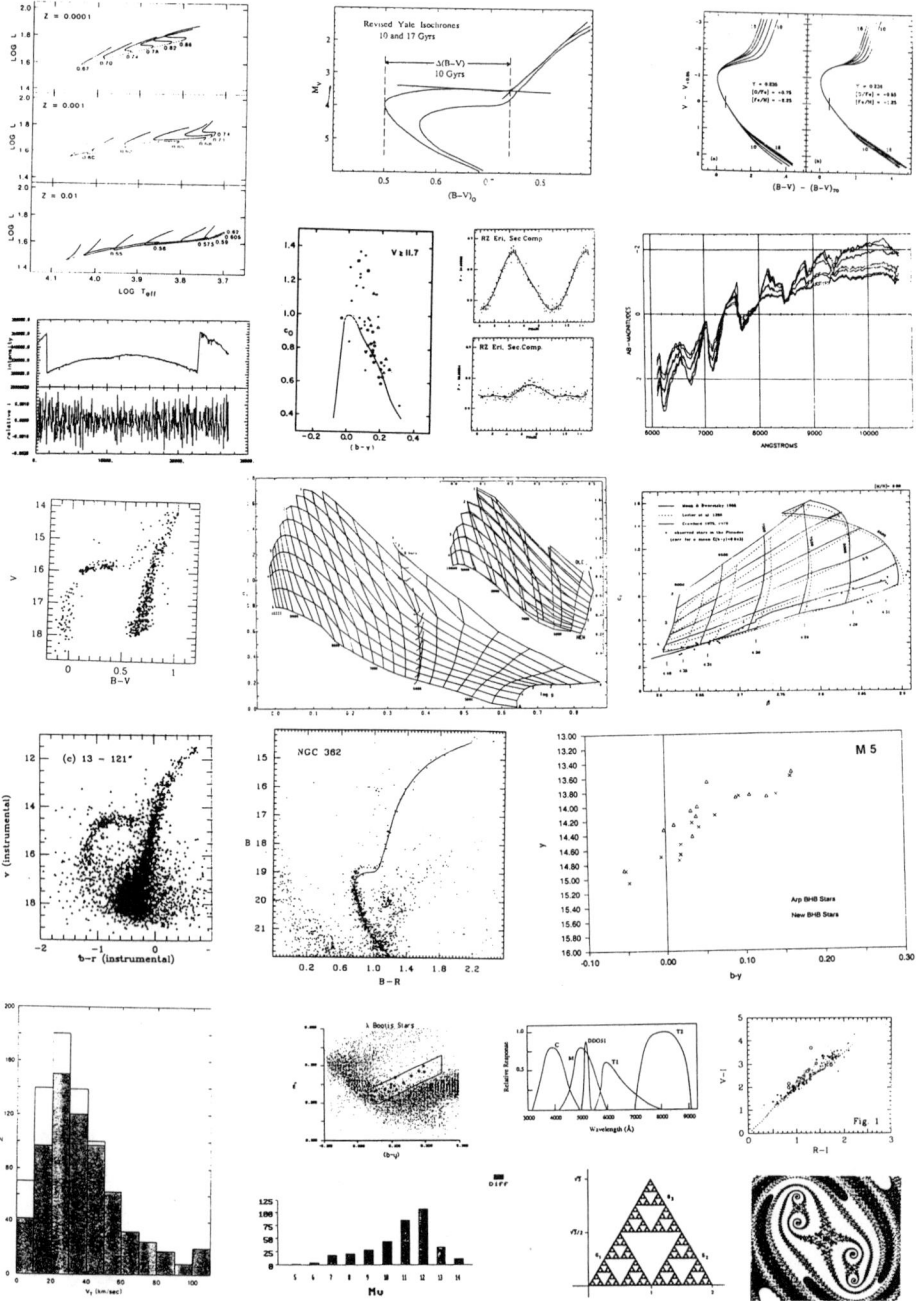

An official picture of Bengt Strömgren, taken on the occasion of his 70th birthday in 1978.

Table of Contents

Preface

List of Participants

Section 1 - Review Papers

ON BENGT STROMGREN'S WORK IN PHOTOMETRY

                                            Jens Knude   3

HORIZONTAL-BRANCH MODELS

                                        Allen V. Sweigart   13

NEW LINES, NEW MODELS, NEW COLORS

                                        Robert L. Kurucz   27

THE AGE SPREAD AMONG GALACTIC GLOBULAR CLUSTERS

                                        Pierre Demarque   45

GLOBULAR CLUSTER PHOTOMETRY NEAR THE TURN-OFF: RELATIVE AGES AND THE HORIZONTAL BRANCH

                                        Ata Sarajedini   55

ON THE COLOR AND POPULATION GRADIENTS IN THE CENTRAL-CUSP GLOBULAR CLUSTER M 15

                                        Peter Stetson   69

CCD DETECTORS APPLIED TO GLOBULAR CLUSTER RESEARCH

                                        Harvey B. Richer and
                                        Gregory C. Fahlman   89

THE A(16)-λ(9) PHOTOMETRIC SYSTEM REVISITED

                                        Eugenio E. Mendoza V   99

V, (B-V) AND DDO COLORS EXTRACTED FROM GRISM SPECTRA

                                        David J. Bell and
                                        Kenneth M. Yoss   111

PHOTOMETRY OF STARS IN SEVERAL SOURCE CATALOGUES

        A. R. Upgren and
        E. W. Weis 117

THE STROMGREN uvby SYSTEM: WHY?

        David L. Crawford 121

CCD PHOTOMETRY ON THE uvby SYSTEM

        Barbara J. Anthony-Twarog
        and Bruce A. Twarog 127

SYSTEMATIC COLOR TRANSFORMATION EFFECTS IN STROMGREN PHOTOMETRY

        J. Manfroid and
        C. Sterken 139

E(b-y) AS THE PRIMARY PARAMETER FOR POLAR ΔM_1, ΔC_1 AND Z(pc) RELATIONS

        Jens Knude 147

CCD FOUR-COLOR PHOTOMETRY OF GLOBULAR CLUSTER STARS

        A. G. Davis Philip 153

STROMGREN PHOTOMETRY OF REDDENED B-TYPE STARS

        A. J. Delgado and
        E. J. Alfaro 163

A COMPARISON OF THE SYSTEMATIC ACCURACY IN FOUR PHOTOMETRIC SYSTEMS

        J. W. Pel 165

THE EXTENSION OF THE GENEVA CATALOGUE

        Noel Cramer 173

HIGH PRECISION PHOTOMETRY IN THE STUDY OF VARIABLE STARS

        G. Burki 183

WASHINGTON CCD PHOTOMETRY: ABUNDANCE DISTRIBUTIONS IN DIFFERENT
LATITUDES TOWARD THE GALACTIC PLANE

          Neil D. Tyson   193

ULTRAVIOLET EXCESS PROPERTIES OF THE THICK DISK AND HALO
POPULATIONS

          Steven R. Majewski   209

uvby PHOTOMETRY OF A-TYPE STARS AT HIGH GALACTIC LATITUDE

          John S. Drilling and
          A. G. Davis Philip   225

THE LATE B-TYPE STARS: REFINED CLASSIFICATION, CONFRONTATION
WITH STROMGREN PHOTOMETRY AND THE EFFECTS OF ROTATION

          R. F. Garrison and
          R. O. Gray   231

PROBING THE STELLAR POPULATION OF THE OUTER GALACTIC DISK:
STAR CLUSTERS AND "WINDOWS" THROUGH THE GALACTIC PLANE

          Kenneth A. Janes   233

PRECISION PHYSICS FROM INFRARED PHOTOMETRY

          R. Michael Rich   245

STELLAR PHOTOMETRY WITH THE PC ON HST

          Patrick Seitzer   261

THE PROMISE AND CHALLENGE OF TIME-RESOLVED CCD ENSEMBLE
PHOTOMETRY

          Ronald L. Gilliland   265

THE ACTIVITIES OF THE NSSDC AND ADC IN THE AREA OF ARCHIVING
PHOTOMETRIC DATA

          Wayne H. Warren Jr.   275

SUMMARY PAPER: PRECISION PHOTOMETRY, PAST, PRESENT, FUTURE

          Kenneth A. Janes   283

Section 2 - Poster Papers

A CCD PHOTOMETER FOR THE VASSAR COLLEGE OBSERVATORY

    F. R. Chromey, Mary Ellen Hunt, Monica Edelstein and Elizabeth Bonar   293

BV PHOTOMETRY OF OB+ STARS IN THE SOUTHERN MILKY WAY

    John S. Drilling   299

PHOTOMETRIC BOXES IN THE $UVBV_1B_2V_1G$ SYSTEM

    M. Golay, B. Nicolet and N. Cramer   303

A METHOD FOR SELECTING LAMBDA BOOTIS CANDIDATES FROM STROMGREN PHOTOMETRY AND OBJECTIVE-PRISM SPECTRA OF FIELD STARS

    Richard O. Gray   309

PHOTOMETRIC ABUNDANCES FOR A AND F STARS

    B. Hauck and S. Berthet   313

TWENTY-FIVE YEARS OF $uvby$-β DATA

    B. Hauck and M. Mermilliod   315

DISTANCES AND AGES FOR A V STARS FROM $uvby$ AND $H\beta$ PHOTOMETRY

    F. Figueras, J. Torra and C. Jordi   319

V, R, I, $H\alpha$ PHOTOMETRY OF NGC 2264

    C. B. Luginbuhl, F. J. Vrba, S. E. Strom and K. M. Strom   323

COMPARISON OF TWO SEMI-EMPIRICAL CALIBRATIONS OF THE $uvby\beta$ SYSTEM

    P. North and D. Kobi   327

PHOTOMETRY OF THE YOUNG OPEN CLUSTER NGC 581

          Randy L. Phelps and
          Kenneth A. Janes   331

AUTOMATIC PHOTOMETRIC TELESCOPES AND PHOTOMETRIC ACCURACY

          C. Sterken and
          J. Manfroid   335

THE PRESENT STATUS OF INVESTIGATIONS USING THE VILNIUS
PHOTOMETRIC SYSTEM

          V. Straizys   341

CCD SCHMIDT PHOTOMETRY OF NGC 5822

          W. Weller, Barbara J.
          Anthony-Twarog and
          Bruce A. Twarog   345

     Section 3 - Banquet Talk and Photos

ITERATED FUNCTION SYSTEMS AND THE DYNAMICAL REPRESENTATION OF
NATURAL STRUCTURES

          Michael Frame   351

Photos

  Bengt Strömgren in the middle 30's   361
  Union College   363
  Fred Chromey and Mary Ellen Hunt   365
  N. Cramer and F. Figueras   366
  Harvey Richer, Arthur Upgren and David Crawford 367
  Kavan Ratnatunga and Peter Stetson   368
  Ken Janes and Steven Majewski   369
  Ata Sarajedini and Neil Tyson   370
  Chris Sterken and Bruce Twarog   371
  Kristina and Davis Philip   372

INDICES   373

  Name Index   375
  Object Index   378
  Subject Index   380

PREFACE

The first discussions concerning this meeting were held in Copenhagen in May, 1988 at the meeting held in honor of Bengt Strömgren. My idea was to hold a meeting in the United States, also in honor of Bengt Strömgren, with a main focus on Precision Photometry. CCD photometry now allows very precise measures to be made of stars of magnitudes 15 - 16, something that was impossible to do with the older single channel photometry. We planned to have papers by theorists and observers. The former would investigate the theories and see what new theoretical predictions could be made if one could increase the precision of the observations and the latter would present papers on their latest results, most of them using CCD technology. Bengt Gustafsson suggested that the subtopic of the meeting should be Astrophysics of the Galaxy, following one of Strömgren's major interests in photometry. As an indication of Strömgren's influence on this field it is interesting to note that in the name index of these proceedings Stömgren has more entries than anyone else.

The meeting was scheduled to be held in October, 1990, well before the General Assembly in Argentina. Union College, in Schenectady, New York, was selected as the host and Van Vleck Observatory, Wesleyan University was co-host. Union College and Van Vleck Observatory contributed the major part of the funds needed to organize and run the meeting. The New York Astronomical Corporation also contributed some funds. These institutions are thanked for their generous contributions. Copies of these proceedings are being sent to all the institutions on the Van Vleck Observatory exchange list.

Thirty-eight astronomers from eight countries attended the meeting. The members of the Scientific Organizing Committee were:

A. G. Davis Philip Chmn	Union College and VVO
Roger Bell	University of Maryland
David Crawford	KPNO, NOAO
Michel Grenon	Geneva Observatory
Bengt Gustafsson	Astronomical Obs., Uppsala
Bernard Hauck	Lausanne University
James Hesser	Dominion Astrophysical Obs.
Charles Perry	Louisiana State University
Robert Shobbrook	University of Sydney
Christian Sterken	Astrophys. Inst, Brussels
Vytas Straizys	Vilnius Observatory
Arthur Upgren	Van Vleck Observatory

Authors sent in their papers electronically, with confirming paper copies sent by mail. The edited manuscripts, printed on a laserjet printer, were returned to authors for proofreading. My co-editors, Arthur Upgren and Kenneth Janes, proofread the entire manuscript and they made numerous suggestions for improving the text. Bitnet provided a way to have authors quickly answer questions concerning their manuscripts and permitted a rapid means of communication among the editors.

Members of the Physics Department at Union College who helped with the meeting were Gary Reich (Chairman of the Department), Laszlo Baksay (who helped with some of the local arrangements) and Ralph Alpher (who welcomed the participants to the meeting on the opening day). Michael Frame (Mathematics Department) gave the after dinner talk at the banquet, on Iterated Function Systems. Kristina Philip worked during the meeting, did proofreading and also helped with the preparation of the indices.

A. G. Davis Philip

Schenectady
March, 1991

List of Participants

Belgium

Christian L. Sterken — Astrophysical Institute, Brussels

Canada

Robert Garrison — University of Toronto
Harvey Richer — University of British Columbia
Peter Stetson — Dominion Astrophysical Observatory

Denmark

Jens Knude — Copenhagen University Observatory

Mexico

Eugenio E. Mendoza V — National University of Mexico

Netherlands

Jan Willem Pel — Kapteyn Observatory

Spain

A. J. Delgado — Inst. of Astrophysics of Andalucia
Francesca Figueras — University of Barcelona
J. Torra — University of Barcelona

Switzerland

Gilbert Burki — Geneva Observatory
Noel Cramer — Geneva Observatory
Bernard Hauck — Lausanne University

U. S. A.

Daniel B. Caton — Appalachian State University
Frederick Chromey — Vassar College
David Crawford — Kitt Peak National Observatory
Pierre Demarque — Yale University Observatory
John S. Drilling — Louisiana State University
Ronald L. Gilliland — Space Telescope Science Institute
Richard O. Gray — Appalachian State University
Mary Ellen Hunt — Bryn Mawr College
Kenneth A. Janes — Boston University
Rebecca A. Koopmann — Yale University Observatory

Christian B. Luginbuhl	U. S. Naval Observatory
Steven Majewski	Observatories of the Carnegie Inst.
Randy L. Phelps	Boston University
A. G. Davis Philip	VVO and Union College
Kavan Ratnatunga	NASA
Mike Rich	Columbia University
Ata Sarajedini	Yale University Observatory
Allen Sweigart*	NASA
Patrick Seitzer	Space Science Telescope Institute
Bruce S. Twarog	Kansas State University
Neil Tyson	Columbia University
Arthur R. Upgren	Van Vleck Observatory
Wayne Warren	NASA
Edward W. Weis	Van Vleck Observatory
Kenneth M. Yoss	University of Illinois

*(At the last moment Sweigart was unable to attend but his paper is part of the proceedings)

Chairmen of Sessions

A. G. Davis Philip
Bruce A. Twarog
Kenneth A. Janes
Jens Knude
David L. Crawford
Bernard Hauck
Christian Sterken
Peter B. Stetson
J. W. Pel
Eugenio E. Mendoza V
John S. Drilling
Pierre Demarque
Arthur R. Upgren (Poster Session)

ON BENGT STRÖMGREN'S WORK IN PHOTOMETRY

Jens Knude

Copenhagen University Observatory

It seems perhaps a bit irrelevant to try accounting for Bengt Strömgren's efforts in the field of photometry regarding the clear and comprehensive overviews which have been published during the last 30 - 35 years, Strömgren (1956, 1963 and 1966). Bengt Strömgren's interest in photometry was determined by its application to well defined astrophysical problems. From Annual Review of Astronomy and Astrophysics 1983 Strömgren wrote:

"In developing this system of photometric photometry, I had a number of research goals in mind. For a main sequence B-star, a pinpointing of the location of the star in the HR diagram with sufficient accuracy to make possible age determinations for field stars was aimed at and achieved. Places of star birth could then be computed, at least for stars with ages less than about 200 million years. For F stars, distances and color excesses of improved accuracy could be obtained, and consequently it became possible to map the distribution of interstellar reddening matter using a relatively dense network of stars. I hoped that in this way Ambartsumian clouds, with total visual absorptions down to $0\overset{m}{.}1$, could be located and investigated, and this indeed proved to be possible. Also, for the F stars the statistical distribution of stellar metal content in well-defined volumes of space could be determined using the metal index, and in particular unbiased samples of Population II stars could be obtained, the ultimate goal being a mapping of groups of Population II stars, divided according to stellar metal content and covering as large a part of our Galaxy as possible. Finally, I hoped that the photoelectric index data obtained for stars on the ZAMS of galactic clusters would yield information concerning the relative contents of helium and heavy elements, and here also results have been obtained."

Photometry as we know it today is a powerful tool of classification (and monitoring) - the evolution of the uvbyβ system may thus be seen as an attempt to obtain an accurate and effective experimental method resulting in astrophysical parameters pertinent for individual stars as well as for large samples of stars in relation to the understanding of the Galaxy. The system's indices accordingly had to be sensitive to the variation of the astrophysical parameters as they were known from theory and previous observing and hopefully also could be useful beyond that knowledge in order to be able to make discoveries. I think it fair to state that these goals have been obtained in the case of the uvbyβ system.

From reading the sequence of papers on photometry and related topics published by Strömgren between 1933 and 1987 one gets a feeling of how Strömgren's photometric system may have evolved. The paper from 1933 in Zeitschrift für Astrophysik is a discussion of the HR diagram and an attempt to understand the location of the stars in the diagram as determined by its mass and hydrogen content. In 1932 Strömgren managed to compute a theoretical index giving a star's relative content of hydrogen. Previously stellar interiors were considered to consist almost completely of fully ionized heavier elements. And the last paper from 1987 is a thorough discussion of the relations between ages, chemical composition and kinematics for a large sample of F stars within 100 pc. One might coarsely divide Strömgren's work on stars and the Galaxy into two main parts. The first period from the late twenties to the late forties could perhaps be characterized by an emphasize on the understanding of the physics of individual stars, theoretical studies of their internal structure and atmosphere and the way nuclear processes and temporal evolution changed the position of stars in the HR diagram. In the second period from about 1950 Strömgren concentrated on the major problems of galactic structure - on the construction and applications of photometric systems.

An early problem was to understand the location of the zero - age line in the T_{eff} - M_v diagram and the width of the main sequence band for various spectral classes. The scarcity of stellar T_{eff} - M_v data in the thirties may have prompted the evolution of an accurate method to obtain data for statistically significant samples.

In his Darwin Lecture (1963b) Strömgren poses three questions;

"1) Can we distinguish observationally between stars of different ages in the main sequence band?

(2) .... can we account for the location of the so-called zero-age line as well as for the observed width of the main-sequence band?

(3) Is it possible at the present stage of development of observation and of theory to combine ages and space velocities of main sequence stars to obtain significant knowledge regarding their places of formation in the Galaxy?"

And from the 1984 paper we might add:

"(4) The reasons for the choice of a chemical composition parameter for the purpose of dividing field stars into population classes, ... in connection with the ultimate aim of providing observational tests of theoretical models of chemical and dynamical evolution of our galaxy."

The first three questions, especially the first, relate directly to the photometric accuracy. The M_v or gravity parameter of a photometric system must be of such accuracy that one can separate the upper and lower parts of the main sequence band, in other words be able to follow the effects of the thermonuclear hydrogen consumption. Rather early in the history of the uvby system Strömgren might state that the obtainable accuracy of the gravity parameter corresponds to only a tenth of the observed width of the main-sequence band from early B to late F stars. From the theory of stellar evolution ages can accordingly be estimated better than a few times 10 million years for the early B stars and better than 2 billion years for the F and early G stars. Answers to the first three questions in fact are affirmative. It was (is) a grand program: formulation of the relevant physical questions and the development of the practical means for their investigation. It takes a talent for details while still keeping the breadth of view to accomplish such an undertaking. As an example of the sense for important details the discussion of the correction for atmospheric extinction in 1937 for rather narrow-band photometry serves well.

In 1950 the first notice on photoelectric classification appears in the Transactions of the 7th IAU General Assembly, 1948 in Zurich, as a report to Commission 36 where Strömgren mentions the successful correlation of measurements of an index related to the strength of the Balmer line Hβ and spectral types in the range from B5 to M0. Quite unusually the m.e. of the observations is presented: ± 0.006 mag. This number has persisted as a typical value for four-color and Hβ photometry ever since. Apparently this was just the first result from a planned systematic effort initiated to establish a photoelectric, photometric classification scheme. In 1951 it is reported that a set of 110 well known standard stars has been observed in 25 narrow, < 100 Å, interference filters and one glass filter, located throughout the optical wave band in order to make a selection of fewer bands in an optimum way for classification purposes. The following year, 1952, a report on the capabilities of the narrow band l-c system was presented. l is an index formed from the interference filters a, b and c whereas c is an estimate of the Balmer discontinuity measured by two interference filters d and e and a glass filter f. Probable errors of one determination on an average night of l and c were ± 0.003 and ± 0.005 mag respectively.

As discussed in Strömgren (1963b) a two dimensional classification, T_{eff} and M_v, for B, A and early F stars proved feasible with the l and c indices.

TABLE 1

Filter Passbands

Filter	λ max	Half-width
a	5030	90
b	4861	35
c	4700	100
d	4520	90
e	4030	90
f	3650	350

$c = [m(f)-m(e)] - [m(e)-m(d)]$

$l = m(b) - \tfrac{1}{2} \times [m(a) + m(b)]$

A question repeatedly formulated in the early discussions of the HR diagram was whether the star's spectral properties were influenced by any parameters beyond the two basic parameters, mass and age. Strömgren discussed this problem during the Vatican study week (1957) and used a chemical composition index, m, constructed from three of the six l-c bands to show that the photometric indices might indeed be influenced by a third parameter. The lcm system was shown to provide a three dimensional classification for stars earlier than about F7: T_{eff}, M_v and relative chemical composition. However, for Strömgren the lc system seems to have served its purpose by the end of the fifties when the possibility of a three dimensional photoelectric classification had been demonstrated. The next step was the breeding of a working horse. A more effective system was required possibly because the m - index was not sensitive to the large metal variation expected for late F and early G type stars and partly because a relative gain could be obtained by introducing slightly broader bands.

During the time of the Vatican study week problems regarding the Galaxy's evolution might already have had Strömgren's interest and a photometric system able to distinguish between the various populations from a purely chemical point of view was needed. The resulting uvby system was conceived and tested during a two month stay at Lick in 1959 and a long observing period at Palomar Observatory in 1960.

Bengt Strömgren's contemplations leading to the uvby system may have formed the basis for the two thorough overviews on photometric classification in Stars and Stellar Systems (1963a) and Annual Review (1966).

The future applications of the uvby system were based on a great observing program of 1217 stars (1965) and the first versions of the basic diagrams most contemporary astronomers are so familiar with: the c_1 - (b-y) and the m_1 - (b-y) graphs can be found in the Darwin Lecture. Despite the careful selection of the bands the (b-y), m_1 and c_1 indices are affected by interstellar reddening but a classification from the reddening free indices $[m_1]$ and $[c_1]$, based on standard reddening ratios deduced from O-star observations, is presented in the Annual Review (1966) paper and has become a most important tool when reduction from standard system to intrinsic stellar properties is performed.

TABLE 2

Filter Passbands

Filter	Central λ	Half width
u	3500	300
v	4110	190
b	4670	180
y	5470	230
β_N	4861	30
β_W	4861	150

Table 2 also contains Crawford's addition of the Hβ bands. Three (four) indices were formed from the uvby(β) bands;

$$(b-y)$$
$$c_1 = (u-v) - (v-b)$$
$$m_1 = (v-b) - (b-y)$$
$$H\beta = \beta_N - \beta_W$$

To see some of the philosophy behind the choice of passbands I quote from Crawford's (1975) paper on the calibration of F stars;

(2) (b-y), a color index, reasonable free of effects of blanketing. Such blanketing begins to increase below 4500 Å.

(3) m_1, a color difference designed to measure the blanketing in the 4100-Å region with respect to a gradient defined by the b and y regions.

(4) c_1, a color difference designed to measure the strength of the Balmer discontinuity; the intensity near 3500 Å relative to a gradient defined by the v and b regions. As the blanketing at 3500 Å is roughly

twice that at 4100 Å the index is roughly free of blanketing effects. I assume that it is free of such effects unless tests indicate that the assumption is not valid.

(5) β, a narrow-band index measuring the strength of the Hβ line in a star's spectrum. It is free of effects of interstellar extinction."

The 1217 founding stars are bright V < 6.5 mag and were observed with good accuracy:

± 0.004 p.e. of (b-y)
± 0.005 p.e. of m_1
± 0.006 p.e. of c_1

The β values of the 80 standard stars define a photometric system with an accuracy of 0.003 mag or better. The (b-y) - m_1 diagram for the 1217 stars shows a varying width in m_1. The largest width is noticed for the late F and early G stars. The Darwin Lecture contains the "first" calibration of Δm_1(b-y) = m_1(standard) - m_1(observed) in terms of [Fe/H]. It may be incidental but it is interesting to note that the few stars of type $\approx$ G0 used in the first [Fe/H] - Δm_1 comparison almost divide in the three metal groups - solar, intermediate and extreme - which have been so extensively studied by means of the uvbyβ system.

After establishing the uvby system and after it was complemented with the β index, replacing the previous l-index, which allows correction for the effect of interstellar reddening - or alternatively opens the possibility of studying the local distribution of interstellar dust, several major observing programs were initiated on Strömgren's suggestion and active cooperation. To mention some:

southern earlier spectral types in the BSC
southern standard stars
northern hemisphere reddening survey
all sky F and G star programs
Galactic cap surveys

The southern bright stars are by the way part of the sample used to establish the "first" age-metallicity relation using uvbyβ data, Twarog (1980). Strömgren discussed aspects of the all sky surveys in terms of Galactic evolution in a series of three papers 1984, 1985 and 1987. The Helsinki (1984) and Cambridge (1987) papers in particular emphasize the importance of the excellent accuracy obtainable with careful uvbyβ photometry and the most recent calibrations. The effect of the error propagation on the final parameters T_{eff}, [Fe/H], M_v and age computed by Bengt Strömgren is given in Table 3.

TABLE 3

Mean Errors

Parameter	Mean error
T_{eff}	±60 K
[Fe/H]	±0.06 dex
M_v	< 0.26 mag
log A	~0.10

The mean errors include the effects of cosmic scatter and a zero-point shift in T_{eff} is less than 100 K. The last row of Table 3 indicates a typical value for Population I stars with an age of either four or eight billion years. The ages of IpII stars still present a problem. The mean error formally equals ±0.06 in log A. But changes of the abundance of C, N, O, Si and Mg relative to Fe may "cause an appreciable reduction of the theoretically calculated ages for this category of stars by log A = 0.1 - 0.2". With these provisos we present in Table 4 the beautiful variation of [Fe/H] with age, taken from Table 2, 1987

TABLE 4

Variation of [Fe/H] with Age

log<A>	N	[Fe/H]
6.0	371	+.15 -.15
7.7	282	-.16 -.29
9.5	231	-.30 -.39
10.0	204	-.40 -.49
11.9	133	-.50 -.59
13.8	49	-.60 -.69
14.7	24	-.70 -.79

The reliability of the relative chemical composition differences measured by δm_1 is really high as shown by Abt (1986).

We will now briefly return to the question of the classification of various populations within the uvbyβ system - question (4) mentioned previously. Five populations were defined or agreed upon during the

1957 Vatican meeting. As an example let us consider the category attracting the most attention recently because it is one of the testing grounds for theories on the Galaxy's formation and evolution: the intermediate population II. The original definition includes stars with -30 < W < +30 km/sec, variables with P < 250 days and spectral type earlier than M5e but does not formulate a prescription of how to classify individual field stars. Strömgren defined his IpII population as a group of stars intermediate between the disk and the halo population on chemical criteria alone, (1964, 1966, and 1969):

$$0.045 < \delta m_1(b-y) < 0.080$$

The chemical composition is measured relative to some standard line, e.g. that of Crawford (1975) valid for F stars. The lower limit was chosen because no stars among the 1217 bright stars with an absolute W velocity beyond 40 km/s had a relative metallicity $\delta m_1(b-y)$ less than 0.045 mag. The stars with $\delta m_1(b-y) > 0.045$ further showed a sharply limited distribution in the c_1 - (b-y) plane. All IpII stars had (b-y) > 0.29 implying that the population was old with log A ~9.8. The upper $\delta m_1(b-y)$ limit of 0.080 was chosen rather low so the IpII could be homogeneous. A larger limit would not have increased the number of IpII stars substantially. Furthermore no galactic cluster was known to have $\delta m_1(b-y) > 0.080$ yet extreme population II globular clusters all have $\delta m_1(b-y) > 0.080$. As is well known the chemical parameter has proven an excellent population tracer, see e.g. Table 2 in the Helsinki paper, and due to the age - [Fe/H] correlation the chemical composition may replace the age.

The practical design of photometers also had Strömgren's interest and he actively influenced the construction of the Danish four channel/two channel photometers build for simultaneous uvby and β work respectively. The simultaneous observing in the four bands is a crucial part of the procedure for obtaining accurate data and it makes atmospheric extinction corrections unnecessary if only colors are required. Strömgren also suggested the combination of the two separate instruments into one six channel photometer which may be used in almost any kind of weather for the β observations at least. The six channels are therefore very efficient instruments as required for the large survey programs. For a period filterless spectograph-photometers were also tested. They were difficult to handle but very efficient. Changing from a 50 cm telescope to 1.5 m and leaving out the uvby filters increased the counting rate by a factor of 30.

For an outline of Strömgren's ideas for the future, e.g. the calibration of the IpII stars the 1987 paper may be consulted.

Let me end this talk with a final quotation from Nissen's (1990) bibliography; "Bengt Strömgren's influence on 20th century astronomy and astrophysics goes beyond what is documented in his own publications. He initiated and directed a number of large projects in the fields of astrometry, photoelectric photometry, stellar atmospheres, stellar interiors and galactic dynamics. He very seldom wanted to be a co-author on publications resulting from these investigations and as result future generations will only be able to trace his influence through the acknowledgements in these publications."

Finally I may mention, for those interested in the history of modern science, that Bengt Strömgren's correspondence, manuscripts and early calculations have been donated by the Strömgren family to the Institute of the History of the Exact Sciences at the University of Aarhus.

REFERENCES

Abt, H. A. 1986 ApJ 309, 260
Crawford, D. L. 1975 AJ 80, 955
Nissen, P. E. and Gustafsson, B. 1990 Matematisk Fysiske Medl, Royal Danish Academy of Sciences and Letters, 12:4, 7
Strömgren, B. 1932 Z. Astrophys. 4, 118
Strömgren, B. 1933 Z. Astrophys. 7, 222
Strömgren, B. 1937 Handbuch der Experimentalphysik 26, 321
Strömgren, B. 1950 Trans. IAU VII, 404
Strömgren, B. 1951 AJ 56, 142
Strömgren, B. 1952 AJ 57, 200
Strömgren, B. 1956 Vistas Astron. 2, 1336
Strömgren, B. 1958 Scripta Varia, l'Academie Pontificale des Sciences No. 16, 385
Strömgren, B. 1963a Stars and Stellar Systems, III, University of Chicago Press, p 123
Strömgren, B. 1963b QJRAS 4, 8
Strömgren, B. 1964 Astrophysica Norvegica, 9, 333
Strömgren, B. 1966 ARAA 4, 433
Strömgren, B. 1969 Mitt. Astron. Ges. 27, 15
Strömgren, B. 1983 ARAA 21, 1
Strömgren, B. 1984 Proc. Nordic Astronomy Meeting, September 3-5, K. J. Donner, ed., University of Helsinki, Report 6/84
Strömgren, B. 1985 in IAU Symposium No. 106, The Milky Way Galaxy, van Woerden, H., Allen, R. J. and Burton, W. B., eds., Reidel, Dordrecht, p. 153
Strömgren, B. 1987 The Galaxy, Proc NATO Advanced Study Inst G. Gillmore and R. Carswell, eds., Reidel, Dordrecht, p. 229
Strömgren, B. and Perry, C. 1965 Photoelectric uvby Photometry for 1217 Stars Brighter Than V 6.5, Institute for Advanced Study, Princeton, New Jersey
Twarog, B. A. 1980 ApJ 242, 242

DISCUSSION

DEMARQUE: I might mention the excellent chapter in Chandrasekhar's Stellar Structure monograph, which is fully devoted to Strömgren's theoretical work on the hydrogen and helium abundances in stars.

JANES: I was somewhat surprised at your mention that Strömgren first calculated that the interiors of stars are primarily hydrogen. I had always understood that Cecilia Payne-Gaposhkin had proposed this about the same time. She was very vigorously attacked by Russell among others.

KNUDE: As I recall, the correspondence between Eddington and Strömgren on this issue does not mention Payne-Gaposhkin's suggestion.

PHILIP: It would be interesting if someone would bring up to date some work that Charley Perry and I did in 1976. We made a bibliography of all the four-color papers for the early 50's to 1976 and the number of papers per year was increasing linearly with time. Of course, now it would be a major task to make such a bibliography.

BURKI: I have a comment concerning the accuracy on the indices and parameters you give in your talk.

The consequence of the success of Strömgren photometry is that this system is used by many diferent observers, using various sets of filters and telescopes. Thus, the comparison of the measurements obtained by various observers generally give much larger dispersions for the measurements of the same star. (See paper by Sterken and Manfroid which appeared in Astron. and Astrophys.)

PHILIP: I have a counter example. One time Hilditch noted that for some NGP stars in common (with V mags near 14) the rms errors were $\pm$ 0.01. I was happy with this result, in that for stars this faint, using different filters and telescopes, the agreement was as good as $\pm$ 0.01 mag.

KNUDE: There are several comparisons with mean residuals in all indices less than ± 0.01 and rms ~ 0.015 mag.

HORIZONTAL-BRANCH MODELS

Allen V. Sweigart

Laboratory for Astronomy and Solar Physics,
NASA Goddard Space Flight Center

ABSTRACT: A number of observational constraints on the properties of theoretical models for globular-cluster horizontal-branch (HB) stars are reviewed with particular emphasis on the luminosity difference between the tip of the red-giant branch (RGB) and the HB, the HB luminosity function and the relative number of HB versus RGB or asymptotic-giant-branch (AGB) stars. We discuss the importance of such observational constraints in resolving the current theoretical uncertainty concerning the end of the HB phase.

1. INTRODUCTION

The horizontal branch plays a fundamental role in studies of the galactic globular clusters. This is true not only for the determination of the cluster ages and helium abundance but also for the investigation of the noncanonical effects which might arise from internal rotation, mixing during the RGB phase, hypothetical particles such as WIMPs and axions, etc. Moreover, the HB preserves a fossil record of the conditions that existed at the formation of the galaxy - a record that one would keenly like to decipher.

The sensitivity of the HB to the parameters affecting the evolution of a globular-cluster star also makes the HB an important testing ground for confronting theory with observations. This is particularly important because the HB is arguably the least understood evolutionary phase in the globular clusters. Observationally we lack satisfactory explanations for the Sandage period shift (or Oosterhoff) effect, the second parameter problem and the gaps found along the blue HB in some clusters. Theoretically we are still far from understanding the physics of convective overshooting and semiconvection within the helium-burning core.

Precision photometry using available CCD devices can contribute greatly to these problems by providing improved HB luminosities and effective temperatures and by increasing the size of the available number counts. The importance of the HB number counts both for determining the HB luminosity function and the relative number of HB versus RGB or AGB stars should be emphasized. As discussed by Renzini and Fusi Pecci (1988), the number counts depend directly on the amount of available nuclear fuel and therefore test a very fundamental aspect of the theoretical models.

In the following sections we will review a number of observational constraints on the properties of globular-cluster HB stars. These constraints will provide important checks on the reliability of the theoretical models and therefore on the globular-cluster parameters derived from them.

2. PRE-HORIZONTAL BRANCH EVOLUTION

2.1 Core Mass M_c

Many aspects of the evolution of an HB star, including its HB lifetime and track morphology, depend sensitively on the mass M_c within the helium core at the zero-age horizontal-branch (ZAHB) phase or, equivalently, on the mass within the helium core at the tip of the RGB. Evolutionary sequences for the RGB phase computed by various investigators yield essentially the same values of M_c except for the discordant results of Mazzitelli (1989), whose core masses are significantly larger by 0.02 $M_\odot$. Mazzitelli (1989) attributed much of this difference to the shell shifting algorithm commonly used in canonical RGB computations and to the explicit procedure commonly used to advance the chemical composition between models.

Numerical shifting of the hydrogen-burning shell between models is necessitated by the extreme thinness of the shell (10^{-3} $M_\odot$) during the RGB phase. Otherwise the number of models required to reach the tip of the RGB would be prohibitively large. Fig. 1 shows how M_c for a typical globular-cluster star depends on the amount of shifting of the hydrogen-burning shell per model, as expressed in terms of the shell's thickness (Sweigart 1991). Shifting which begins fairly early when the helium core contains 0.315 $M_\odot$ changes M_c by only 10^{-3} $M_\odot$ even when the shifting per model is as large as twice the shell's thickness. Shifting which begins later when the core contains 0.35 $M_\odot$, corresponding to a case considered by Mazzitelli (1989), has only a negligible effect on M_c.

In evolving a RGB star it is essential to insure that the change in the total hydrogen content of the star from one model to the next is consistent with the total energy output of the hydrogen-burning shell over the time step. It is straightforward to show that the explicit procedure for advancing the chemical composition in which the change in the hydrogen abundance depends only on the physical conditions in the previous model fulfills this requirement (Sweigart 1991). In contrast, the implicit procedure for advancing the chemical composition,

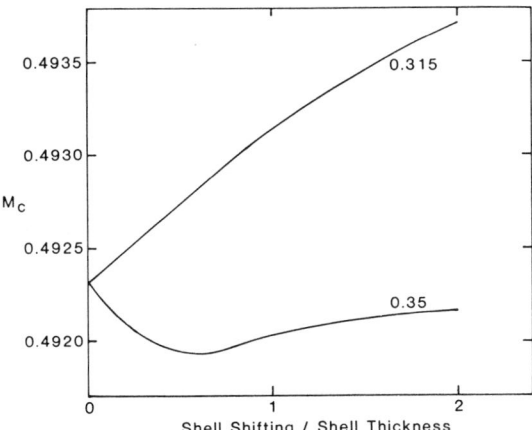

Fig. 1. Dependence of the core mass M_c at the helium flash in solar units on the amount of shifting of the hydrogen-burning shell between RGB models for a representative globular-cluster star. The two curves are labeled by the mass of the helium core in solar units at the time when the shifting begins.

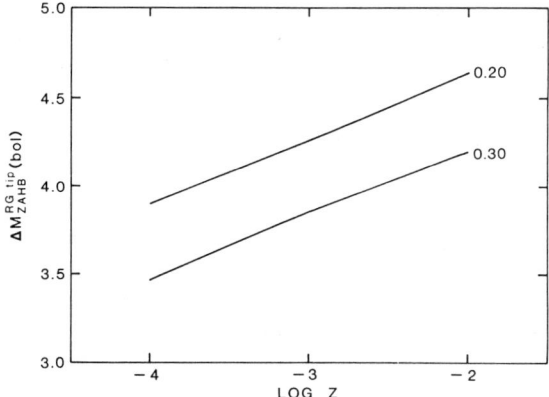

Fig. 2. Dependence of the difference $\Delta M_{ZAHB}^{RG\ Tip}(bol)$ in bolometric magnitude between the ZAHB and the tip of the RGB on the heavy-element abundance Z. Each curve is labeled by the main-sequence helium abundance Y_{MS}.

advocated by Mazzitelli (1989), systematically underestimates the amount of hydrogen depletion per model. The result is a core that grows and heats up too slowly and therefore becomes too large at the helium flash.

We conclude that both of the above effects are unimportant.

2.2 Luminosity at Tip of Red-Giant Branch

Agreement among theoreticians on the canonical values of M_c does not, however, insure agreement with nature! One still needs an observational check of the theoretical results. This is best accomplished by using the luminosity L_{tip} at the tip of the RGB because of its strong dependence on M_c. We thus need to ask how well L_{tip} can be determined observationally. One way of addressing this question is to use a relationship due to Renzini and Buzzoni (1986) between the expected number n_j of stars in a given evolutionary phase j and the duration t_j of this phase, i.e.,

$$n_j = B(t)\ L_T\ t_j, \qquad (1)$$

where $B(t)$ is the specific evolutionary flux at an age t and L_T is the total luminosity of the sampled population. The amount of time spent by a globular-cluster star within, say, 0.1 mag of the tip of the RGB is 4×10^5 yr (Sweigart and Gross 1978). Using this value for t_j in equation (1) and setting $B(t)$ equal to 2×10^{-11} stars yr^{-1} $L_\odot^{-1}$, as is appropriate for a globular-cluster age of 15 Gyr, gives 1 for the expected number of stars for each 10^5 $L_\odot$ of globular-cluster luminosity. Thus a typical globular cluster should contain a star very close to the tip of the RGB, as is, in fact, observed (Frogel, Cohen and Persson 1983).

It is convenient in checking the theory to use the difference $\Delta M_{ZAHB}^{RG\ Tip}(bol)$ in the bolometric magnitude between the ZAHB and the tip of the RGB, since this quantity is independent of the distance modulus. The values of $\Delta M_{ZAHB}^{RG\ Tip}(bol)$ obtained from the results of Sweigart and Gross (1978) and Sweigart, Renzini and Tornambe (1987) are plotted in Fig. 2 as a function of the heavy-element abundance Z for initial helium abundances Y_{MS} of 0.20 and 0.30. The ZAHB luminosity was taken at $\log T_{eff} = 3.85$. The results in Fig. 2 are well approximated by the expression

$$\Delta M_{ZAHB}^{RG\ Tip}(bol) = 6.23 - 4.26\ Y_{MS} + 0.37\ \log Z. \qquad (2)$$

The increase in $\Delta M_{ZAHB}^{RG\ Tip}(bol)$ with increasing metallicity is due about equally to a decrease in the ZAHB luminosity and an increase in the RGB tip luminosity. Recent observations of bright red giants in a number of globular clusters by Da Costa and Armandroff (1990) have shown good agreement with this predicted brightening of the RGB tip.

The dependence of $\Delta M_{ZAHB}^{RG\ Tip}(bol)$ on helium abundance also deserves comment. As Y_{MS} increases, the ZAHB becomes brighter while the RGB tip becomes fainter. Formally if one could determine $\Delta M_{ZAHB}^{RG\ Tip}(bol)$ with an accuracy of 0.1 mag at a known metallicity, one could estimate Y_{MS} to within about ± 0.02. The helium dependence of $\Delta M_{ZAHB}^{RG\ Tip}(bol)$ can also be used to investigate whether there is an anticorrelation between Y_{MS} and [Fe/H], as suggested by Sandage, Katem and Sandage (1981) to explain the period shift effect. An anticorrelation of the size

$$\Delta Y_{MS} = -0.09\ \Delta[Fe/H] \tag{3}$$

(Sandage 1990b, eq. [26]) would double the derivative of $\Delta M_{ZAHB}^{RG\ Tip}(bol)$ with respect to log Z from 0.37 to 0.75 and thus could be tested observationally (see Da Costa and Armandroff 1990).

The constraint represented by Fig. 2 tests many aspects of the RGB and HB evolution, e.g., the validity of canonical RGB physics and core masses, the metallicity dependence of the ZAHB and RGB tip luminosities and the globular-cluster helium abundance. Observational confirmation of this constraint would support the use of the RGB tip luminosity as a standard candle in the globular clusters, the potential of which has already been demonstrated by VandenBerg and Durrell (1990).

2.3 R-Method

Before leaving the RGB we should mention another constraint, namely, that provided by the ratio $R = N_{HB}/N_{RGB}$ between the observed number of HB and RGB stars. Iben (1968) first showed that the corresponding ratio of theoretical lifetimes t_{HB}/t_{RGB} is primarily a function of the helium abundance. More recently, Buzzoni et al. (1983) have derived a helium abundance of 0.23 ± 0.02 independent of metallicity from a new calibration of $R(Y_{MS})$ based on canonical models. It should be remembered, however, that the observed values of R constrain not only the helium abundance but also the theoretical ingredients that go into the $R(Y_{MS})$ calibration including the treatment of semiconvection, the values of M_c, the metallicity dependence of the HB luminosity, and the evolutionary behavior during the core-helium-exhaustion phase.

Consider, for example, the dependence of the $R(Y_{MS})$ calibration on the HB luminosity. Sandage and Cacciari (1990, eq. [6]) have shown that the period shift effect implies an increase in the bolometric magnitude $M_{bol}(RR)$ of the RR Lyrae stars of 0.45 mag per dex increase in [Fe/H]. As is well-known, such an increase in $M_{bol}(RR)$ considerably exceeds the increase of 0.19 mag/dex predicted by the theoretical models. According to Buzzoni et al. (1983, Section 2e) any difference

$\delta M_{bol}(RR)$ between the actual HB luminosity and the canonical luminosity will change the helium abundance derived from the R-method by an amount $\delta Y_{MS} = 0.12\ \delta M_{bol}(RR)$, assuming that the cause of the luminosity difference does not alter other aspects of the RGB or HB evolution. Thus the enhanced metallicity dependence of the HB luminosity implied by the period shift effect leads to a positive correlation between Y_{MS} and [Fe/H], i.e., $\delta Y_{MS} = 0.12\ (0.45 - 0.19)\ \delta[\text{Fe/H}] = 0.03\ \delta[\text{Fe/H}]$, just the opposite of the anticorrelation in equation (3). Such an increase in Y_{MS} with increasing [Fe/H] would produce a period shift that is opposite to the observed period shift, thus requiring that $M_{bol}(RR)$ change by even more than 0.45 mag/dex.

3. VERTICAL STRUCTURE OF THE HORIZONTAL BRANCH

3.1 Composition Dependence

Sandage (1987, 1990a) has recently shown that globular cluster HB's have an intrinsic luminosity width that increases with metallicity. In this section we will examine this result in light of the theoretical models and will discuss how HB luminosity functions, i.e., number of stars versus luminosity above the ZAHB, can be used to resolve some theoretical uncertainties concerning the final stages of the HB phase.

We first consider the dependence of the HB luminosity width on composition. Fig. 3 illustrates how the morphology of canonical tracks varies with helium abundance at a fixed metallicity. We readily see that the track morphology changes noticeably with increasing Y_{MS}. In particular, the blueward loops become more extensive, and the tracks deviate more and more from the ZAHB. The net effect is a substantial increase in the predicted luminosity width from 0.1 - 0.2 mag at $Y_{MS} = 0.20$ to 0.4 mag or more at $Y_{MS} = 0.30$. The luminosity width at a given effective temperature will, of course, depend on the actual mass distribution of the stars populating the HB and therefore on a cluster's overall HB morphology. This sensitivity of the HB luminosity width to Y_{MS} can be used to estimate the helium abundance. See, for example, Dorman, VandenBerg and Laskarides (1989), who derived a helium abundance of 0.24 from the observed HB luminosity width of 47 Tuc.

The strength of the hydrogen-burning shell in the models in Fig. 3 increases with increasing Y_{MS}. This leads not only to a brighter HB but also to a significantly larger growth in the mass of the helium core during the HB phase. It is primarily this growth in the mass of the core that is responsible for the longer blueward loops and larger deviations from the ZAHB in the higher Y_{MS} tracks.

The dependence of the track morphology on metallicity is next illustrated in Fig. 4. The high metallicity tracks in this figure do not reproduce the larger luminosity width noted by Sandage (1987,

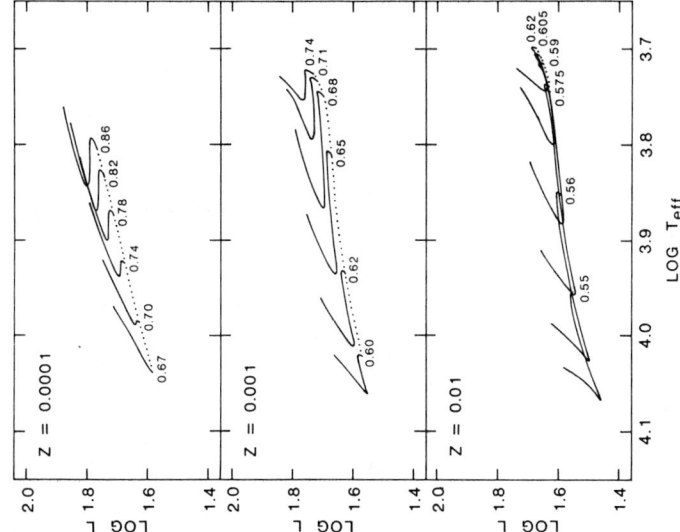

Fig. 4. Same as Fig. 3 except for various heavy-element abundances Z and a fixed helium abundance Y_{MS} of 0.25.

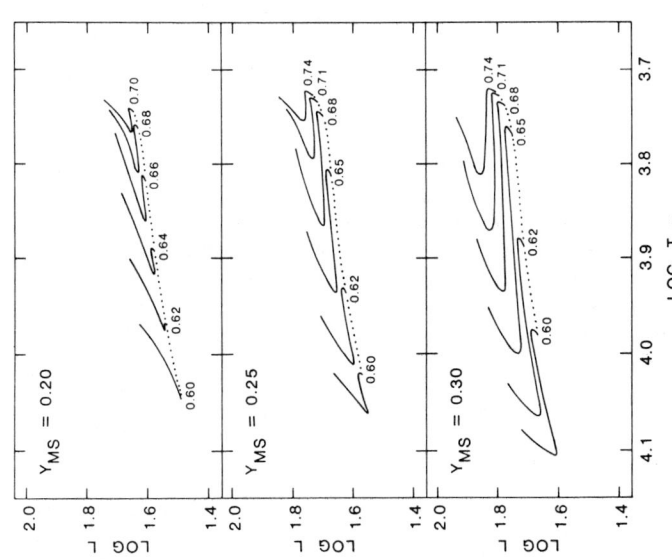

Fig. 3. Canonical HB tracks for the indicated helium abundances Y_{MS} and for a heavy-element abundance Z of 0.001 (Sweigart 1987). Each track extends from the ZAHB (dotted curves) to the point where the central helium abundance Y_c is 0.05. The ZAHB position of each track is labeled by the track mass in solar units.

1990a). In fact, the most metal-rich tracks hardly deviate from the ZAHB except near the end of their HB phase. Thus the luminosity function in a metal-rich globular cluster would be expected to show a concentration of stars at or just above the ZAHB luminosity. At lower metallicities (or at higher helium abundances, see Fig. 3) one would expect the ZAHB luminosity to be less sharply defined.

3.2 Core-Helium Exhaustion

Besides its value in constraining the globular-cluster composition, the HB luminosity function is also important for understanding the physics of convective overshooting and semiconvection within the helium-burning core. To see why this is the case, we need to review briefly how the composition distribution within the core of an HB star changes during the HB phase. According to canonical theory an HB star begins its evolution on the ZAHB with a central convective core containing 0.1 M_o and a central helium abundance Y_c of 0.95. As Y_c gradually decreases due to the helium burning, the opacity and hence the radiative gradient at the edge of the convective core increase. The resulting overadiabaticity of the radiative gradient leads to convective overshooting which causes the convective core to grow. During this overshooting phase the size of the convective core increases from 0.1 to 0.15 M_o while Y_c decreases to 0.7. Beyond this point, however, convective overshooting no longer leads to convective neutrality at the convective-core edge. As a result, an HB star then forms a semiconvective zone surrounding the convective core, as described by Castellani, Giannone and Renzini (1971). During the ensuing semiconvective phase the size of the convective core remains essentially constant while the semiconvective zone grows until it contains 0.1 M_o. All of these events are illustrated in Fig. 5 for a representative HB star.

A number of numerical techniques have been used to follow the composition changes shown in Fig. 5 (Robertson and Faulkner 1972, Castellani et al. 1985, Dorman, VandenBerg and Laskarides 1989). In general, these techniques give satisfactory results until Y_c falls below roughly 0.10. Attempts to follow the subsequent evolution during the core-helium-exhaustion phase, i.e., beyond the last composition profile in Fig. 5, frequently encounter "breathing pulses" by the convective core (Sweigart and Demarque 1973). During a breathing pulse the convective core suddenly grows so large that it engulfs the entire semiconvective zone. So much fresh helium is captured and mixed into the center that Y_c actually increases for a while.

The breathing pulses apparently arise from an instability in the size of the convective core. Suppose that convective overshooting were to cause an outward extension of the convective-core edge near the end of the HB phase. The capture of even a modest amount of helium at that time can lead to a significant relative increase in Y_c and therefore in the helium-burning luminosity. This, in turn, increases the outward flux and hence the radiative gradient throughout the core, thereby causing more overshooting and a further extension of the convective-core edge. The consequence is a convective runaway that doesn't stop until the convective core nearly doubles in size.

The breathing pulses have two important consequences. First, they bring additional helium fuel into the central regions and thus prolong the time spent near the end of the HB phase. Second, they reduce the amount of helium fuel that is left for the subsequent AGB phase and therefore decrease the predicted lifetime of this phase. The first consequence can be tested by comparing the expected number of bright HB stars with the observed HB luminosity functions. This would provide an important constraint on both the extent of the breathing pulses and the efficiency of convective overshooting. In addition, it would constrain the explanation offered by Lee, Demarque and Zinn (1990) for the period shift effect. These authors argue that the RR Lyrae stars in the metal-poor globular clusters are near the end of their HB phase and therefore lie at luminosities considerably above the ZAHB. In this way it is possible to produce significantly longer periods. However, it is not at all clear that this explanation can account for the observed number of RR Lyrae stars, especially in the metal-poor cluster M 15 (Rood and Crocker 1990). The second consequence could be tested by using the ratio $R_2 = N_{AGB}/N_{HB}$ between the observed number of AGB and HB stars (Buonanno, Corsi and Fusi Pecci 1985). This ratio is sensitive to changes in the composition profile within the core during the final part of the HB phase. Improved statistics for this ratio would be particularly valuable in placing further constraints on the extent of the breathing pulses.

It is not clear whether the breathing pulses actually occur in real HB stars or whether they are merely artifacts of the canonical algorithms for treating semiconvection and, in particular, of the canonical assumption of instantaneous overshooting. If one argues that the breathing pulses occur to the extent found in canonical models, then one needs to explain why the predicted value of R_2 (0.06) is so much less than the observed value of 0.15 (Renzini and Fusi Pecci 1988). If instead one argues that the breathing pulses do not occur (Caputo et al. 1989), then one needs to explain how convective overshooting can substantially increase the extent of the helium-depleted region during the main HB phase, as required by the observed value of R_2, while being so inefficient at the end of the HB phase.

Perhaps convective overshooting in real HB stars is neither inefficient nor instantaneous. In order to investigate this possibility, we have followed the evolution of a representative HB star through the core-helium-exhaustion phase by using a parameterization for the overshooting efficiency. By varying a free parameter F_{ov}, one can go from totally inefficient overshooting without breathing pulses to highly efficient overshooting that mimics the results obtained with

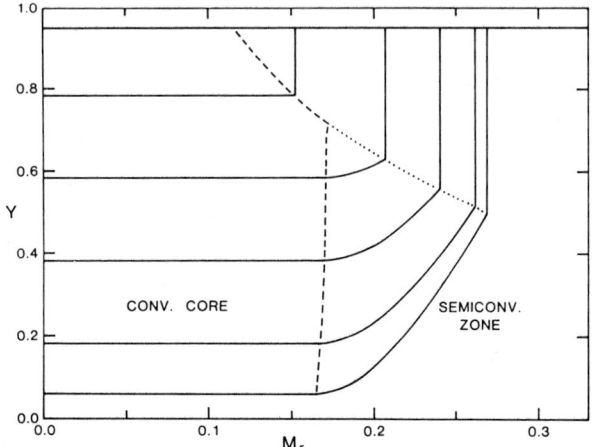

Fig. 5. Core-helium distribution at various times during the evolution of a representative HB star. The dashed and dotted curves denote the edge of the convective core and the outer edge of the semiconvective zone, respectively.

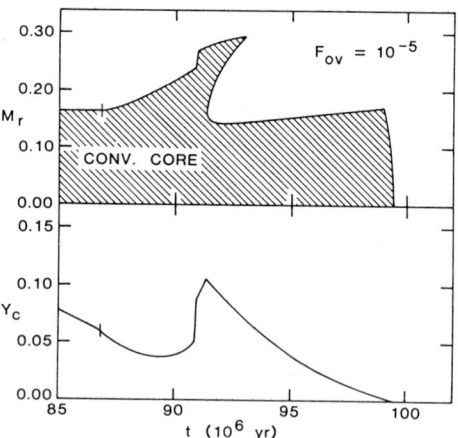

Fig. 6. Time dependence of the mass M_r in solar units within the convective core (upper panel) and of the central helium abundance Y_c (lower panel) during the core-helium-exhaustion phase of a representative HB star. The time t is measured from the ZAHB phase. The fluctuation at $t = 91 \times 10^6$ yr arose when a small fully convective shell was captured by the convective core.

the canonical algorithms for semiconvection. A typical result is shown in Fig. 6. Prior to the tick marks in this figure at time t = 87 x 10^6 yr the evolution was followed in the canonical manner, while after the tick marks the parameterization of Sweigart (1990) was used to determine the overshooting efficiency. The upper panel shows that the convective core underwent a breathing pulse during which it captured all of the material that was previously within the semiconvective zone. The resulting influx of fresh helium fuel lead to the increase in Y_c shown in the lower panel. Interestingly enough, however, this breathing pulse increased the HB lifetime by only 8%. This increase together with the expected decrease in the subsequent AGB lifetime would reduce the predicted value of R_2 by 0.02 which is within the observational uncertainty.

Whether the results in Fig. 6 or any of the other possibilities describe the evolution of real HB stars can only be determined by a careful comparison with observations. In this regard the observed HB luminosity functions and number ratios R_2 will provide key constraints in answering the question: How do HB stars really evolve?

ACKNOWLEDGEMENT

The support of NASA RTOP 188-41-51 Task 03 is gratefully acknowledged.

REFERENCES

Buonanno, R., Corsi, C. E. and Fusi Pecci, F. 1985 A&A 145, 97
Buzzoni, A., Fusi Pecci, F., Buonanno, R. and Corsi, C. E. 1983 A&A 128, 94
Caputo, F., Castellani, V., Chieffi, A., Pulone, L. and Tornambe, A. 1989 ApJ 340, 241
Castellani, V., Chieffi, A., Pulone, L. and Tornambe, A. 1985 ApJ 296, 204
Castellani, V., Giannone, P. and Renzini, A. 1971 Ap&SS 10, 355
Da Costa, G. S. and Armandroff, T. E. 1990 AJ 100, 162
Dickens, R. J., Croke, B. F., Cannon, R. D. and Bell, R. A. 1991, preprint
Dorman, B., VandenBerg, D. A. and Laskarides, P. G. 1989 ApJ 343, 750
Frogel, J. A., Cohen, J. G. and Persson, S. E. 1983 ApJ 275, 773
Fusi Pecci, F. 1987 in IAU Colloquium 95, The Second Conference on Faint Blue Stars, A. G. Davis Philip, D. S. Hayes and J. W. Liebert, eds., L. Davis Press, Schenectady, p. 107
Iben, I. Jr. 1968 Nature 220, 143
Lee, Y.-W., Demarque, P. and Zinn, R. 1990 ApJ 350, 155
Mazzitelli, I. 1989 ApJ 340, 249
Renzini, A. and Buzzoni, A. 1986 in Spectral Evolution of Galaxies, C. Chiosi and A. Renzini, eds., Reidel, Dordrecht, p 195
Renzini, A., and Fusi Pecci, F. 1988 ARAA 26, 199

Robertson, J. W. and Faulkner, D. J. 1972 ApJ 171, 309
Rood, R. T. and Crocker, D. A. 1990 in IAU Colloquium 111, The Use of Pulsating Stars in Fundamental Problems of Astronomy, E. G. Schmidt, ed., Cambridge Univ. Press, N. Y., p. 103
Sandage, A. 1987 in IAU Colloquium 95, The Second Conference on Faint Blue Stars, A. G. Davis Philip, D. S. Hayes and J. W. Liebert, eds., L. Davis Press, Schenectady, p. 41
Sandage, A. 1990a ApJ 350, 603
Sandage, A. 1990b ApJ 350, 631
Sandage, A. and Cacciari, C. 1990 ApJ 350, 645
Sandage, A., Katem, B. and Sandage, M. 1981 ApJS 46, 41
Sweigart, A. V. 1987 ApJS 65, 95
Sweigart, A. V. 1990 in Confrontation between Stellar Pulsation and Evolution, C. Cacciari and G. Clementini, eds, Astronomical Society of the Pacific, Vol 13, p 1
Sweigart, A. V. 1991, in preparation
Sweigart, A. V. and Demarque, P. 1973 in IAU Colloquium 21, Variable Stars in Globular Clusters and in Related Systems, J. D. Fernie, ed., Reidel, Dordrecht, p. 221
Sweigart, A. V. and Gross, P. G. 1978 ApJS 36, 405
Sweigart, A. V., Renzini, A. and Tornambe, A. 1987 ApJ 312, 762
VandenBerg, D. A., Bolte, M. and Stetson, P. B. 1990 AJ 100, 445
VandenBerg, D. A. and Durrell, P. R. 1990 AJ 99, 221

DISCUSSION

DEMARQUE: I was intrigued by your inclusion of the "second parameter" problem among those that lack a satisfactory explanation, since one of the main points of my paper deals with a discussion of the recent evidence which supports age as the second parameter. Do you have any comments about these recent developments?

SWEIGART: The difficulty in explaining the second parameter problem stems in part from the large sensitivity of the HB morphology at intermediate metallicities to a host of parameters that include not only age but also helium abundance, [CNO/Fe], mean rotation rate, etc., and in part from the observational difficulty in determining these parameters with the required accuracy. In fact, the second parameter may not be a single parameter at all but rather a combination of many parameters appropriately weighted for each cluster, as suggested by Buonanno, Corsi and Fusi Pecci (1985) and Fusi Pecci (1987).

There are reasons to believe that age may not be the only second parameter. Consider, for example, the problem of mass loss. A globular cluster star of intermediate metallicity will lose about 0.2 $M_\odot$ during the RGB phase. Even a small variation in this mean mass loss can dramatically alter the HB morphology. Conceivably a difference in the mean rotation rate between second parameter clusters might alter the mean mass loss by affecting the strength of the stellar winds during the RGB phase and the luminosity at the helium flash. Thus any explanation for the second parameter problem is fundamentally at the mercy of our nearly complete ignorance about what determines the mean mass loss during the RGB phase.

There is also the intriguing case of NGC 2808 which suffers from both a blue and a red second-parameter problem. How can age be the only second parameter in a cluster of supposedly coeval stars? Moreover, Fusi Pecci (1987) has pointed out that the globular clusters with an extremely blue HB tend to have very high central densities, implying that the cluster environment may also affect the HB morphology. Finally it should be remembered that the typical observational error in [Fe/H] is around 0.2, and thus second parameter clusters may not actually have identical metallicities. Little is really known about possible differences in [CNO/Fe], although [CNO/Fe] appears to be the same in the second-parameter clusters NGC 288 and NGC 362 (Dickens et al. 1991).

My point of view is that age may be a significant and perhaps in some cases the most important second parameter but probably not the only second parameter. In this regard, it should be mentioned that VandenBerg, Bolte and Stetson (1990) have found a dispersion in globular cluster ages of only 2 Gyr at [Fe/H] = -1.3, which is too small to produce the observed differences in HB morphology.

DEMARQUE: I am a little confused about your discussion of the breathing pulses at the end of your paper. How does the efficiency of core overshoot affect the quantity R_2? And how is R_2 affected by the breathing pulses? Do you always get a value of R_2 that is smaller than the observed value R_2 = 0.15?

SWEIGART: I have not yet continued the models with breathing pulses through the AGB phase and therefore cannot give a definite quantitative answer for their effect on R_2. However, it is possible to make a good estimate, as was done at the end of my paper. Essentially what happens is as follows. As the assumed overshooting efficiency increases, the breathing pulses become more extensive and bring more helium fuel into the center. This not only prolongs the HB lifetime but also decreases the amount of fuel left for the subsequent AGB phase, thereby decreasing the AGB lifetime. The net effect is a reduction in the value of R_2 below what would be found if the breathing pulses are suppressed by assuming inefficient overshooting.

PHILIP: CCD four-color photometry of BHB stars allows one to segregate stars into two groups; those evolving to the blue (on the ZAHB) and those evolving to the red, back up the asymptotic giant branch. Such measures should make it possible to calculate the width of the HB at different (b-y) colors much more accurately and thus yield a better estimate of the helium abundance. If one can measure the vertical width to ± 0.01 mag what sort of accuracy could be expected in the measure of the helium abundance for a cluster if the [Fe/H] value is already known?

SWEIGART: Since the luminosity width of the HB increases by about 0.2 mag between YMS = 0.20 and 0.30, it should be possible to measure the helium abundance to within ±0.01, assuming that one has a sufficiently large sample of HB stars. In doing this, one would want to be sure that the set of evolutionary tracks one is using is able to reproduce the cluster's overall HB morphology.

PHILIP: In your paper you discuss the R method (R = NHB/NRGB) which is another way of measuring the helium abundance. As mentioned above, one can separate stars evolving on the HB into those evolving to the blue and those evolving to the red. So a ratio of HBb/HBr could be calculated. The diagrams that you showed do not have time intervals marked, but in general I gather that evolution speeds up past the turn from blue to red. How would such a ratio be related to [Fe/H] or the helium abundance?

SWEIGART: You raise an interesting question. Using the time intervals given in Sweigart (1987), one finds that the fraction of the HB lifetime spent evolving blueward increases as either the helium abundance or [Fe/H] increases. However, the rate of evolution varies greatly along an HB track, being quite slow at the blue turnaround and becoming progressively faster during the evolution back to the AGB. Thus any interpretation of the relative numbers of blueward and redward evolving stars will require a careful modeling of the entire HB morphology to insure that one is using the proper tracks. Nevertheless it would be worthwhile to look for a signature of the composition in the HB luminosity function.

JANES: What are the properties of Population I HB stars?

SWEIGART: The answer to this question depends critically on how mass loss varies with [Fe/H]. If the mass loss at Population I compositions is the same as in the globular clusters, then a Population I HB star will sit in a clump tight against the RGB. If the mass loss is greater, then a Population I HB star can evolve along a long blueward loop as shown in Fig. 4 and become, for example, a metal-rich RR Lyrae star.

NEW LINES, NEW MODELS, NEW COLORS

Robert L. Kurucz

Center for Astrophysics

ABSTRACT: I have used my newly calculated iron group line list together with my earlier atomic and molecular line data to compute new opacities for the temperature range 2,000K to 200,000K. Calculations have been completed at the San Diego Supercomputer Center for microturbulent velocities 0, 1, 2, 4 and 8 km/s for scaled solar abundances [+1.0], [+0.5], [+0.3], [+0.2], [+0.1], [+0.0], [-0.1], [-0.2], [-0.3], [-0.5], [-1.0], [-1.5], [-2.0], [-2.5], [-3.0], [-3.5], [-4.0], [-4.5] and [-5.0] (log abundance of elements heavier than helium relative to solar). Thus far I have completed the grid of model atmospheres for [0.0], 2 km/s for the temperature range 3,500K to 50,000K, and for log g from 0.0 to 5.0. This grid will allow a consistent theoretical treatment of photometry from K stars to B stars. Preliminary results are reported for UBV and uvby photometry. A solar photospheric model computed with the new opacities matches the observed energy distribution. A grid of models will be computed for each abundance and microturbulent velocity. Models and fluxes are now being distributed on magnetic tapes. I hope to have CD-ROMs available in the near future.

1. OLD MODELS

I start by listing the shortcomings in my old models as I did at an earlier meeting (Kurucz 1987) but this time I can report that I have corrected most of them.

My old models (Kurucz 1979a,b) were produced as long as 18 years ago on computers that are primitive by today's standards. The number of optical depth layers was limited by small memories and slow processors. Now I can compute with many more layers and go to shallower optical depths. This greatly improves the numerical accuracy of the calculated radiation field at wavelengths that have very high or rapidly varying opacity. Fortunately, such wavelengths do not very much affect the structure of a model. There was also a limit to the number of frequencies that I could afford to compute. Now I can use up to 1221 wavelength intervals from 9 to 160,000 nm. I am now able to use 1 nm resolution in the ultraviolet for better comparison to satellite observations.

My old models for F and G stars are systematically in error and predict color indices that are off by as much as 0.05 mag. I assumed that the error was caused by problems in the mixing length treatment of convection and by the omission of molecular line opacity in the coolest models. My theoretical model for the solar photosphere has several per cent error in the flux in the red and, of course, cannot reproduce the molecular features in the ultraviolet. Improvements in my treatment of convection, even going so far as having hot and cold streams, have reduced the error somewhat for the hotter convective models. I have also added approximate overshooting and have tried various increases in the mixing-length to scale-height ratio. I now use 1.25. However, I am now convinced that most of the error comes from missing line opacity, including missing atomic line opacity, which turns out to be significant at all effective temperatures. I hope that Nordlund and others will be able to produce models with realistic convection cells to take care of the convection physics, so now I am concentrating on the opacity. I discuss molecular and atomic opacity in the next section.

We do not know much about microturbulent velocity. It has been decreasing as a function of time as the models have improved. In the Sun it is depth-dependent varying from about 0.5 to 1.8 km/s. The models assume a constant value. Twenty five years ago, I arbitrarily chose 2 km/s as a nice round number. In some stars it can be much larger or much smaller. Opacity, radiative acceleration, and model structure vary considerably with the microturbulent velocity. It may be that 1 or 1.5 km/s is a good choice for high gravity models. George Michaud tells me that the existence of some types of diffusion implies microturbulent velocities less than 100 m/s. Microturbulent velocity may vary strongly with phase and with depth in pulsating stars, thereby strongly affecting the atmospheric structure and colors. I plan to investigate this through spectrum synthesis. My new grids of models have microturbulent velocity as a parameter, so it must be specified when choosing a model.

The helium abundance is another arbitrary number. I chose 10% by number. Others use a 10% He/H ratio. I have switched to a smaller value because I think it more probable. Small errors in the helium abundance produce errors in the density, electron number, and opacity and consequently produce systematic errors in the derived stellar parameters. The "solar" metal abundances have also changed with time and are not yet final. I now use Anders and Grevesse (1989) abundances.

My old low gravity models have systematic errors because of non-LTE and sphericity effects. So do my new models.

2. NEW LINES

I reported on my line and opacity calculations at a NATO workshop in Trieste in September (Kurucz 1991). The details of my line lists and the opacities can be found in that paper. Here I will give only a brief outline.

My earlier model calculations used the distribution-function line opacity computed by Kurucz (1979ab) from the line data of Kurucz and Peytremann (1975). We had computed gf values for 1.7 million atomic lines for sequences up through nickel using scaled-Thomas-Fermi-Dirac wave functions and eigenvectors determined from least squares Slater parameter fits to the observed energy levels. That line list has provided the basic data and has since been combined with a list of additional lines, corrections, and deletions with the help of Barbara Bell and Terry Varner at the Center for Astrophysics. The line data are being continually, but slowly, improved. We collect all published data on gf values and include them in the line list whenever they appear to be more reliable than the current data. I have also completely recomputed Fe II (Kurucz 1981).

After the Kurucz-Peytremann calculations were published, I started work on line lists for diatomic molecules beginning with H_2, CO (Kurucz 1977), and SiO (Kurucz 1980). Next, Lucio Rossi of the Istituto Astrofisica Spaziale in Frascati, John Dragon of Los Alamos and I computed line lists for electronic transitions of CH, NH, OH, MgH, SiH, CN, C_2 and TiO. In addition to lines between known levels, these lists include lines whose wavelengths are predicted and are not good enough for detailed spectrum comparisons but are quite adequate for statistical opacities. Work is continuing on other molecules and molecular ions, and on the vibration-rotation spectra. I also have data for terrestrial atmospheric molecules.

In 1983 I recomputed the opacities using the additional atomic and molecular data described above which totaled 17,000,000 lines. These new opacities were used to produce improved empirical solar models (Avrett, Kurucz and Loeser 1984), but were found to still not have enough lines. For example, there were several regions between 200 and 350 nm where the predicted solar intensities are several times higher than observed, say, 85% blocking instead of the 95% observed. The integrated flux error of these regions is several per cent of the total. In a flux constant theoretical model this error is balanced by a flux error in the red. The model thus predicts the wrong colors. In detailed ultraviolet spectrum calculations, half the intermediate strength and weak lines are missing. After many experiments, I determined that this discrepancy is caused by missing iron group atomic lines that go to excited configurations that have not been observed in the laboratory. Most laboratory work has been done with emission sources that cannot strongly populate these configurations. Stars, however, show these lines in absorption without difficulty. Including these additional lines produces a dramatic increase in opacity, both in the Sun and in hotter stars. A stars have the same lines as the Sun but more flux in the ultraviolet to block. In B stars and in O stars there are large effects from third and higher iron group ions. Envelope opacities that are used in interior and pulsation models are also strongly affected.

I was granted a large amount of computer time at the San Diego Supercomputer Center by NSF to carry out new calculations. To compute the iron group line lists I determined eigenvectors by combining least squares fits for levels that have been observed with computed Hartree-Fock integrals (scaled) for higher configurations including as many configurations as I can fit into a Cray. All configuration interactions are included. My computer programs have evolved from Cowan's (1968) programs. Transition integrals are computed with scaled-Thomas-Fermi-Dirac wave functions and the whole transition array is produced for each ion. The forbidden transitions can be computed as well. Radiative, Stark and van der Waals damping constants and Lande g values are automatically produced for each line. The first nine ions of Ca through Ni produced 42,000,000 lines. I will recompute the energy levels and line lists when new analyses become available and I will make the predictions available to laboratory spectroscopists. I plan to do the heavier and lighter elements as a background project.

The models I can compute now are not valid for M stars. I eventually need line lists for the triatomic molecules, but I hope that other people will do the work before I have to learn the physics. I am working on the low temperature bands now, however, for atmospheric transmission.

In late 1988 I used the line data described above to compute new solar abundance opacity tables for use in my modeling. The calculations involved 58,000,000 lines, 3,500,000 wavelength points, 56 temperatures from 2,000K to 200,000K, 21 log pressures from -2 to +8 and 5 microturbulent velocities 0, 1, 2, 4, 8 km/s, and took a large amount of computer time. The opacity is tabulated both as 12-step distribution functions for intervals on the order of 1 to 10 nm, and as opacity sampling where, simply, every hundredth wavelength point in the calculation was saved. There are actually two sets of distribution functions, a higher resolution version with 1212 "little" intervals, and a lower resolution version with 328 "big" intervals. The "little" wavelength intervals are nominally 1 nm in the ultraviolet and 2 nm in the visible. The opacities were tested by computing a solar model as described below.

Since the beginning of 1990 I have been able to take tremendous advantage of the new Cray YMP at the San Diego Supercomputer Center. In a few months I finished more than I had expected to do in two years. I computed opacities ranging from 0.00001 solar to 10 times solar, enough to compute model atmospheres ranging from the oldest Population II stars to high abundance Am and Ap stars. The hardest part was transmitting the results (200 tapes) back to Cambridge over Internet, but even that usually worked quite well. The exact abundances are [+1.0], [+0.5], [+0.3], [+0.2], [+0.1], [+0.0], [-0.1], [-0.2], [-0.3], [-0.5], [-1.0], [-1.5], [-2.0], [-2.5], [-3.0], [-3.5], [-4.0], [-4.5], [-5.0] and [+0.0, no He]. The final files for each abundance require

NEW LINES, NEW MODELS, NEW COLORS   31

three 6250 bpi VAX backup tapes. I have begun to distribute copies of the tapes. I plan to produce 600 megabyte CD-ROMs of these opacities that can be read on any workstation with a CD reader.

I am open to suggestions for computing opacities for other abundance mixes. At the present time I am planning to compute C/O variations, Population II opacities with enhanced light even elements, and some Am and Ap mixes.

3. NEW MODELS

I have rewritten my model atmosphere program to use the new line opacities, additional continuous opacities, and an approximate treatment of convective overshooting. The opacity calculation was checked by computing a small grid of solar models with various microturbulent velocities and mixing-length-to-scale-height ratios. I adopted a solar model shown in Fig. 1 that matches the observed irradiance (Neckel and Labs 1984, Labs et al. 1987) with V_{turb} = 1.5 km/s and l/H = 1.25. I am confident that I have solved the missing opacity problem. Thus far I have computed (on Vaxstations) a grid of 400 solar abundance, 2 km/s models covering the effective temperature range from K stars to O stars. The models are listed in Table 1. Models cooler than 9,000K are convective with l/H = 1.25. The range of this grid should allow photometric calibrations consistent for both cool and hot stars. Figs. 2 to 5 show sample energy distributions from the grid. I have begun to distribute tapes of the models and the flux predicted from each model at 1221 wavelengths in the range .01 to 160 micrometers. The range is enough to treat ionization in H II regions and to calibrate the infrared.

I have begun to compute grids of models on the Cray using the 2 km/s grid for starting models. The solar abundance grid is probably completed for five microturbulent velocities 0, 1, 2, 4 and 8 km/s, but just checking the output is a tremendous amount of work for me since there are 2000 models. I plan to do all the abundances for which I have opacities. I will compute the fluxes, predicted photometry, Balmer line profiles, and limb darkening for each model. Most users will be able to find what they need by simple interpolation. I expect to be able to compute a complete, full-resolution spectrum for any of the models which can be compared directly to high resolution observations, or degraded to low resolution, say, 1 Å.

4. NEW COLORS

I have computed preliminary UBV and uvby colors for the solar abundance grid described above following Buser and Kurucz (1978) and Relyea and Kurucz (1978) including the typographic correction from Lester et al. (1986). The colors were normalized by finding the model in the grid that best interpolates the spectrophotometry of Vega (Hayes and Latham 1975, Tug, White and Lockwood 1977) and that best matches

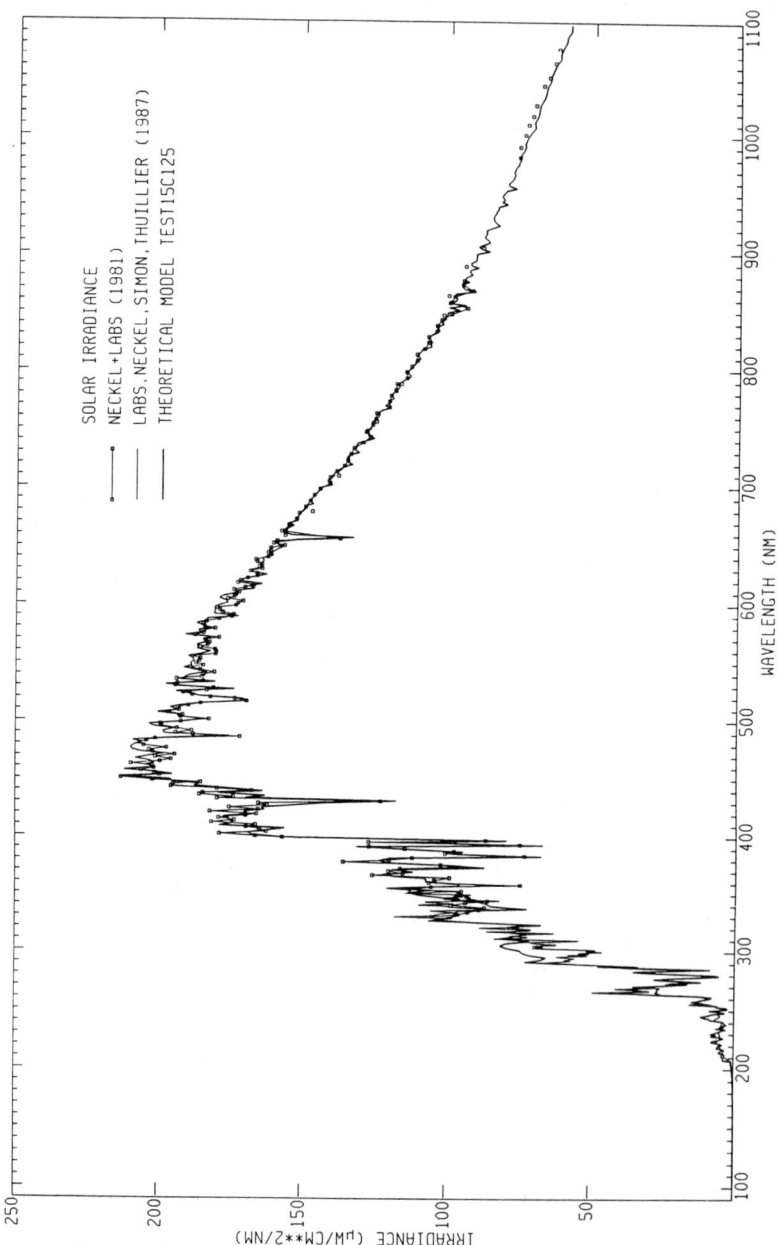

Fig. 1. Predicted solar irradiance compared to observed.

NEW LINES, NEW MODELS, NEW COLORS

TABLE 1

Predicted bolometric corrections, UBV and uvby colors for the new solar abundance, 2 km/s grid. (The solar model is 1.5 km/s.)

Teff	log g	BC	U-B	B-V	u-v	b-y	m1	c1
3500	0.00	-2.028	2.406	1.863	4.653	1.215	0.818	0.585
3500	0.50	-2.044	2.175	1.749	4.407	1.146	0.689	0.735
3500	1.00	-2.074	1.938	1.640	4.134	1.090	0.541	0.869
3500	1.50	-2.096	1.715	1.537	3.854	1.045	0.397	0.969
3500	2.00	-2.140	1.508	1.442	3.569	1.009	0.252	1.046
3500	2.50	-2.181	1.334	1.361	3.304	0.982	0.128	1.082
3500	3.00	-2.178	1.209	1.320	3.093	0.970	0.053	1.045
3500	3.50	-2.007	1.156	1.362	3.009	0.985	0.084	0.869
3500	4.00	-1.834	1.121	1.401	2.941	0.994	0.127	0.697
3500	4.50	-1.756	1.097	1.433	2.871	1.000	0.162	0.544
3500	5.00	-1.727	1.091	1.469	2.804	1.007	0.203	0.381
3750	0.00	-1.480	2.494	1.797	4.637	1.131	1.028	0.317
3750	0.50	-1.469	2.276	1.717	4.437	1.071	0.947	0.399
3750	1.00	-1.469	2.063	1.644	4.225	1.023	0.851	0.476
3750	1.50	-1.477	1.863	1.576	4.009	0.983	0.752	0.537
3750	2.00	-1.479	1.682	1.511	3.791	0.947	0.661	0.572
3750	2.50	-1.498	1.523	1.455	3.577	0.918	0.575	0.589
3750	3.00	-1.518	1.387	1.407	3.375	0.894	0.502	0.579
3750	3.50	-1.522	1.276	1.383	3.197	0.887	0.444	0.532
3750	4.00	-1.460	1.184	1.394	3.055	0.910	0.387	0.458
3750	4.50	-1.384	1.111	1.401	2.931	0.927	0.345	0.386
3750	5.00	-1.347	1.058	1.402	2.824	0.931	0.322	0.316
4000	0.00	-1.121	2.259	1.662	4.176	1.037	0.969	0.164
4000	0.50	-1.110	2.065	1.591	4.008	0.980	0.919	0.207
4000	1.00	-1.106	1.883	1.529	3.847	0.936	0.860	0.253
4000	1.50	-1.101	1.719	1.474	3.693	0.899	0.801	0.292
4000	2.00	-1.108	1.576	1.429	3.551	0.871	0.743	0.321
4000	2.50	-1.118	1.454	1.391	3.414	0.848	0.691	0.334
4000	3.00	-1.130	1.350	1.360	3.281	0.828	0.648	0.327
4000	3.50	-1.146	1.260	1.339	3.151	0.814	0.612	0.297
4000	4.00	-1.154	1.176	1.330	3.023	0.812	0.573	0.251
4000	4.50	-1.132	1.091	1.334	2.894	0.834	0.503	0.218
4000	5.00	-1.096	1.014	1.330	2.772	0.851	0.432	0.203
4250	0.00	-0.853	1.915	1.528	3.687	0.954	0.825	0.125
4250	0.50	-0.842	1.754	1.462	3.536	0.899	0.799	0.139
4250	1.00	-0.838	1.602	1.405	3.392	0.856	0.761	0.156
4250	1.50	-0.833	1.462	1.353	3.260	0.819	0.721	0.176
4250	2.00	-0.839	1.344	1.310	3.146	0.792	0.684	0.192
4250	2.50	-0.848	1.248	1.275	3.050	0.769	0.656	0.199
4250	3.00	-0.859	1.172	1.249	2.968	0.750	0.636	0.194
4250	3.50	-0.874	1.113	1.233	2.896	0.736	0.625	0.172
4250	4.00	-0.890	1.064	1.226	2.830	0.727	0.621	0.132
4250	4.50	-0.902	1.013	1.228	2.755	0.728	0.605	0.086
4250	5.00	-0.896	0.950	1.232	2.660	0.746	0.544	0.078
4500	0.00	-0.653	1.569	1.392	3.295	0.870	0.677	0.199
4500	0.50	-0.643	1.446	1.337	3.163	0.824	0.663	0.188
4500	1.00	-0.639	1.328	1.288	3.038	0.786	0.641	0.181
4500	1.50	-0.633	1.217	1.242	2.921	0.752	0.616	0.181
4500	2.00	-0.637	1.121	1.203	2.820	0.726	0.592	0.182
4500	2.50	-0.644	1.040	1.168	2.731	0.702	0.572	0.180
4500	3.00	-0.653	0.975	1.140	2.657	0.682	0.561	0.169
4500	3.50	-0.665	0.928	1.120	2.600	0.666	0.560	0.146
4500	4.00	-0.680	0.894	1.110	2.557	0.653	0.570	0.110
4500	4.50	-0.696	0.867	1.108	2.521	0.644	0.586	0.058
4500	5.00	-0.708	0.840	1.115	2.480	0.644	0.590	0.009
4750	0.00	-0.498	1.260	1.252	2.991	0.777	0.556	0.323
4750	0.50	-0.491	1.164	1.207	2.868	0.742	0.544	0.293
4750	1.00	-0.485	1.068	1.165	2.749	0.712	0.527	0.269
4750	1.50	-0.484	0.985	1.130	2.647	0.687	0.511	0.249
4750	2.00	-0.487	0.912	1.099	2.558	0.665	0.497	0.232
4750	2.50	-0.491	0.849	1.071	2.481	0.645	0.485	0.217
4750	3.00	-0.497	0.798	1.046	2.416	0.627	0.480	0.200
4750	3.50	-0.506	0.760	1.026	2.362	0.610	0.482	0.176
4750	4.00	-0.517	0.733	1.011	2.322	0.595	0.494	0.143
4750	4.50	-0.531	0.716	1.004	2.294	0.582	0.516	0.097
4750	5.00	-0.545	0.702	1.004	2.271	0.573	0.543	0.038
5000	0.00	-0.372	1.001	1.108	2.752	0.680	0.465	0.461
5000	0.50	-0.370	0.917	1.072	2.628	0.655	0.450	0.417
5000	1.00	-0.369	0.839	1.041	2.515	0.635	0.434	0.376
5000	1.50	-0.371	0.770	1.014	2.415	0.618	0.419	0.338
5000	2.00	-0.374	0.711	0.991	2.329	0.604	0.407	0.306
5000	2.50	-0.377	0.664	0.971	2.261	0.590	0.401	0.276
5000	3.00	-0.382	0.627	0.953	2.203	0.576	0.401	0.248
5000	3.50	-0.388	0.599	0.938	2.157	0.561	0.408	0.218
5000	4.00	-0.395	0.582	0.926	2.123	0.547	0.422	0.183
5000	4.50	-0.405	0.572	0.918	2.099	0.534	0.444	0.143
5000	5.00	-0.416	0.567	0.915	2.082	0.521	0.473	0.092
5250	0.00	-0.273	0.787	0.970	2.562	0.592	0.392	0.592
5250	0.50	-0.276	0.718	0.944	2.445	0.575	0.379	0.535
5250	1.00	-0.280	0.649	0.920	2.330	0.561	0.362	0.480
5250	1.50	-0.284	0.587	0.900	2.228	0.550	0.348	0.430
5250	2.00	-0.288	0.535	0.882	2.141	0.541	0.336	0.384
5250	2.50	-0.292	0.495	0.869	2.070	0.533	0.330	0.341
5250	3.00	-0.297	0.463	0.858	2.012	0.525	0.330	0.301
5250	3.50	-0.301	0.442	0.849	1.969	0.515	0.338	0.262
5250	4.00	-0.306	0.432	0.843	1.939	0.504	0.353	0.221
5250	4.50	-0.312	0.429	0.839	1.919	0.493	0.376	0.179
5250	5.00	-0.319	0.431	0.837	1.907	0.481	0.406	0.132

Teff	log g	BC	U-B	B-V	u-v	b-y	m1	c1
5500	0.00	-0.184	0.672	0.812	2.539	0.485	0.355	0.857
5500	0.50	-0.202	0.554	0.817	2.301	0.501	0.319	0.658
5500	1.00	-0.209	0.497	0.802	2.191	0.494	0.306	0.590
5500	1.50	-0.215	0.442	0.790	2.086	0.489	0.292	0.523
5500	2.00	-0.222	0.393	0.779	1.994	0.484	0.281	0.462
5500	2.50	-0.228	0.354	0.772	1.918	0.480	0.276	0.405
5500	3.00	-0.233	0.323	0.765	1.854	0.475	0.274	0.353
5500	3.50	-0.238	0.302	0.761	1.805	0.469	0.280	0.304
5500	4.00	-0.242	0.292	0.759	1.772	0.462	0.294	0.258
5500	4.50	-0.246	0.289	0.759	1.750	0.454	0.314	0.212
5500	5.00	-0.250	0.295	0.761	1.740	0.445	0.342	0.163
5750	0.00	-0.119	0.546	0.678	2.467	0.405	0.309	1.037
5750	0.50	-0.131	0.498	0.673	2.342	0.404	0.297	0.937
5750	1.00	-0.153	0.386	0.692	2.095	0.432	0.261	0.706
5750	1.50	-0.163	0.335	0.685	1.989	0.430	0.249	0.628
5750	2.00	-0.171	0.289	0.681	1.891	0.430	0.240	0.549
5750	2.50	-0.178	0.247	0.678	1.804	0.430	0.234	0.476
5750	3.00	-0.185	0.214	0.677	1.732	0.428	0.232	0.409
5750	3.50	-0.191	0.189	0.676	1.675	0.425	0.236	0.350
5750	4.00	-0.196	0.173	0.677	1.632	0.421	0.246	0.295
5750	4.50	-0.199	0.166	0.680	1.602	0.417	0.262	0.242
5750	5.00	-0.202	0.169	0.685	1.587	0.410	0.287	0.191
5777	4.4377	-0.197	0.133	0.664	1.558	0.410	0.248	0.241
6000	0.00	-0.073	0.449	0.548	2.425	0.329	0.266	1.234
6000	0.50	-0.084	0.419	0.557	2.310	0.337	0.262	1.109
6000	1.00	-0.095	0.375	0.559	2.183	0.342	0.251	0.995
6000	1.50	-0.121	0.262	0.589	1.924	0.378	0.216	0.734
6000	2.00	-0.131	0.217	0.589	1.823	0.381	0.207	0.645
6000	2.50	-0.141	0.174	0.590	1.728	0.383	0.201	0.557
6000	3.00	-0.149	0.137	0.593	1.646	0.385	0.200	0.475
6000	3.50	-0.156	0.106	0.597	1.577	0.385	0.202	0.401
6000	4.00	-0.162	0.082	0.601	1.523	0.384	0.209	0.334
6000	4.50	-0.167	0.068	0.606	1.483	0.381	0.222	0.274
6000	5.00	-0.170	0.063	0.613	1.458	0.378	0.242	0.217
6250	0.50	-0.050	0.360	0.444	2.302	0.271	0.227	1.303
6250	1.00	-0.059	0.330	0.460	2.179	0.284	0.225	1.158
6250	1.50	-0.071	0.287	0.468	2.048	0.294	0.216	1.026
6250	2.00	-0.099	0.175	0.504	1.788	0.333	0.184	0.751
6250	2.50	-0.111	0.130	0.510	1.684	0.339	0.178	0.647
6250	3.00	-0.121	0.089	0.517	1.591	0.344	0.176	0.551
6250	3.50	-0.131	0.051	0.523	1.510	0.347	0.176	0.461
6250	4.00	-0.139	0.019	0.530	1.442	0.348	0.181	0.381
6250	4.50	-0.145	-0.003	0.537	1.391	0.348	0.191	0.310
6250	5.00	-0.149	-0.018	0.546	1.353	0.347	0.207	0.244
6500	0.50	-0.026	0.309	0.337	2.301	0.209	0.199	1.484
6500	1.00	-0.034	0.298	0.361	2.192	0.226	0.199	1.340
6500	1.50	-0.044	0.266	0.384	2.062	0.244	0.198	1.177
6500	2.00	-0.057	0.223	0.398	1.927	0.256	0.191	1.029
6500	2.50	-0.086	0.110	0.437	1.669	0.296	0.164	0.747
6500	3.00	-0.099	0.064	0.448	1.564	0.304	0.160	0.633
6500	3.50	-0.111	0.021	0.457	1.471	0.310	0.160	0.529
6500	4.00	-0.122	-0.018	0.466	1.391	0.315	0.163	0.435
6500	4.50	-0.130	-0.050	0.475	1.325	0.317	0.170	0.350
6500	5.00	-0.137	-0.075	0.485	1.273	0.318	0.181	0.274
6750	0.50	-0.010	0.257	0.252	2.287	0.162	0.175	1.612
6750	1.00	-0.018	0.267	0.267	2.205	0.171	0.176	1.508
6750	1.50	-0.025	0.251	0.297	2.085	0.192	0.179	1.343
6750	2.00	-0.035	0.217	0.325	1.951	0.211	0.180	1.166
6750	2.50	-0.050	0.173	0.343	1.815	0.226	0.176	1.009
6750	3.00	-0.081	0.057	0.385	1.563	0.265	0.153	0.725
6750	3.50	-0.096	0.008	0.398	1.455	0.275	0.151	0.603
6750	4.00	-0.110	-0.037	0.409	1.362	0.282	0.151	0.495
6750	4.50	-0.122	-0.079	0.419	1.281	0.287	0.154	0.397
6750	5.00	-0.131	-0.115	0.429	1.214	0.290	0.161	0.310
7000	0.50	-0.002	0.200	0.183	2.251	0.126	0.152	1.692
7000	1.00	-0.008	0.229	0.194	2.201	0.130	0.158	1.622
7000	1.50	-0.015	0.235	0.215	2.106	0.142	0.163	1.494
7000	2.00	-0.023	0.217	0.248	1.981	0.163	0.169	1.315
7000	2.50	-0.034	0.177	0.281	1.843	0.185	0.172	1.127
7000	3.00	-0.051	0.130	0.303	1.705	0.202	0.170	0.960
7000	3.50	-0.083	0.012	0.344	1.464	0.239	0.150	0.685
7000	4.00	-0.099	-0.039	0.359	1.357	0.249	0.149	0.559
7000	4.50	-0.115	-0.088	0.371	1.263	0.257	0.150	0.447
7000	5.00	-0.128	-0.132	0.383	1.182	0.263	0.154	0.347
7250	0.50	0.000	0.138	0.137	2.176	0.107	0.132	1.697
7250	1.00	-0.005	0.185	0.133	2.173	0.098	0.140	1.695
7250	1.50	-0.010	0.209	0.151	2.106	0.105	0.150	1.593
7250	2.00	-0.017	0.210	0.176	2.004	0.118	0.158	1.448
7250	2.50	-0.026	0.185	0.214	1.874	0.142	0.166	1.256
7250	3.00	-0.040	0.141	0.250	1.735	0.165	0.170	1.062
7250	3.50	-0.059	0.089	0.275	1.596	0.182	0.170	0.889
7250	4.00	-0.092	-0.031	0.312	1.367	0.217	0.152	0.628
7250	4.50	-0.110	-0.085	0.328	1.261	0.228	0.152	0.500
7250	5.00	-0.126	-0.137	0.341	1.169	0.236	0.153	0.388

Teff	log g	BC	U-B	B-V	u-v	b-y	m1	c1
7500	0.50	-0.008	0.056	0.132	2.016	0.117	0.108	1.563
7500	1.00	-0.011	0.128	0.086	2.110	0.077	0.121	1.711
7500	1.50	-0.014	0.169	0.098	2.076	0.078	0.134	1.650
7500	2.00	-0.018	0.188	0.121	1.998	0.087	0.147	1.528
7500	2.50	-0.027	0.188	0.148	1.897	0.099	0.161	1.376
7500	3.00	-0.037	0.153	0.193	1.763	0.127	0.169	1.170
7500	3.50	-0.053	0.103	0.228	1.624	0.150	0.173	0.977
7500	4.00	-0.075	0.045	0.254	1.486	0.167	0.174	0.801
7500	4.50	-0.106	-0.075	0.289	1.271	0.199	0.157	0.556
7500	5.00	-0.126	-0.132	0.305	1.168	0.210	0.157	0.431
7750	1.00	-0.026	0.063	0.059	2.005	0.069	0.104	1.657
7750	1.50	-0.026	0.120	0.055	2.018	0.058	0.118	1.664
7750	2.00	-0.027	0.155	0.075	1.967	0.063	0.135	1.570
7750	2.50	-0.033	0.168	0.103	1.884	0.073	0.151	1.433
7750	3.00	-0.043	0.160	0.133	1.781	0.086	0.167	1.272
7750	3.50	-0.058	0.133	0.164	1.664	0.101	0.182	1.097
7750	4.00	-0.073	0.063	0.215	1.512	0.138	0.180	0.874
7750	4.50	-0.106	-0.058	0.252	1.290	0.170	0.165	0.617
7750	5.00	-0.127	-0.121	0.272	1.177	0.184	0.165	0.476
8000	1.00	-0.055	-0.017	0.052	1.847	0.075	0.086	1.523
8000	1.50	-0.047	0.062	0.023	1.931	0.046	0.103	1.633
8000	2.00	-0.046	0.110	0.035	1.913	0.044	0.120	1.582
8000	2.50	-0.047	0.138	0.061	1.852	0.051	0.140	1.468
8000	3.00	-0.054	0.145	0.092	1.767	0.063	0.159	1.320
8000	3.50	-0.064	0.126	0.130	1.660	0.081	0.177	1.143
8000	4.00	-0.078	0.078	0.174	1.533	0.107	0.186	0.943
8000	4.50	-0.100	0.015	0.207	1.398	0.129	0.189	0.759
8000	5.00	-0.131	-0.104	0.241	1.195	0.159	0.176	0.524
8250	1.00	-0.104	-0.107	0.058	1.641	0.090	0.066	1.326
8250	1.50	-0.079	-0.002	0.004	1.820	0.041	0.089	1.557
8250	2.00	-0.073	0.058	0.004	1.837	0.031	0.107	1.559
8250	2.50	-0.071	0.099	0.025	1.801	0.034	0.128	1.475
8250	3.00	-0.074	0.120	0.055	1.736	0.043	0.150	1.347
8250	3.50	-0.082	0.118	0.090	1.650	0.056	0.173	1.192
8250	4.00	-0.093	0.087	0.130	1.543	0.076	0.190	1.009
8250	4.50	-0.108	0.031	0.173	1.417	0.103	0.198	0.814
8250	5.00	-0.132	-0.039	0.205	1.284	0.123	0.200	0.635
8500	1.00	-0.171	-0.207	0.061	1.407	0.100	0.051	1.102
8500	1.50	-0.120	-0.066	-0.005	1.693	0.042	0.078	1.451
8500	2.00	-0.109	0.005	-0.018	1.748	0.023	0.095	1.509
8500	2.50	-0.104	0.053	-0.004	1.735	0.022	0.115	1.459
8500	3.00	-0.104	0.084	0.023	1.688	0.028	0.139	1.354
8500	3.50	-0.107	0.095	0.056	1.620	0.038	0.164	1.215
8500	4.00	-0.116	0.083	0.093	1.534	0.051	0.188	1.052
8500	4.50	-0.127	0.041	0.137	1.426	0.074	0.205	0.864
8500	5.00	-0.144	-0.024	0.176	1.300	0.100	0.210	0.677
8750	1.50	-0.167	-0.130	-0.010	1.557	0.044	0.069	1.329
8750	2.00	-0.151	-0.050	-0.033	1.652	0.019	0.086	1.439
8750	2.50	-0.144	0.005	-0.026	1.661	0.014	0.104	1.423
8750	3.00	-0.141	0.044	-0.004	1.631	0.016	0.127	1.342
8750	3.50	-0.141	0.066	0.027	1.579	0.023	0.154	1.223
8750	4.00	-0.145	0.066	0.063	1.509	0.035	0.182	1.074
8750	4.50	-0.155	0.041	0.103	1.420	0.050	0.208	0.902
8750	5.00	-0.165	-0.014	0.147	1.309	0.076	0.220	0.716
9000	1.50	-0.218	-0.191	-0.014	1.419	0.045	0.063	1.200
9000	2.00	-0.196	-0.102	-0.043	1.551	0.017	0.079	1.358
9000	2.50	-0.188	-0.042	-0.043	1.583	0.007	0.096	1.375
9000	3.00	-0.184	0.003	-0.026	1.570	0.007	0.118	1.318
9000	3.50	-0.182	0.032	0.002	1.531	0.012	0.144	1.215
9000	4.00	-0.183	0.042	0.037	1.473	0.021	0.173	1.082
9000	4.50	-0.188	0.029	0.075	1.401	0.034	0.203	0.925
9000	5.00	-0.199	-0.007	0.114	1.313	0.048	0.231	0.752
9250	2.00	-0.244	-0.151	-0.051	1.449	0.015	0.074	1.269
9250	2.50	-0.235	-0.087	-0.056	1.502	0.003	0.089	1.315
9250	3.00	-0.230	-0.038	-0.043	1.505	0.000	0.110	1.282
9250	3.50	-0.228	-0.004	-0.019	1.478	0.004	0.135	1.200
9250	4.00	-0.227	0.014	0.013	1.433	0.011	0.164	1.081
9250	4.50	-0.228	0.011	0.051	1.371	0.021	0.196	0.936
9250	5.00	-0.235	-0.016	0.091	1.295	0.034	0.227	0.770
9400	3.95	-0.255	-0.005	-0.003	1.411	0.004	0.156	1.089
9500	2.00	-0.293	-0.199	-0.057	1.346	0.014	0.070	1.177
9500	2.50	-0.283	-0.131	-0.066	1.419	-0.001	0.084	1.249
9500	3.00	-0.278	-0.080	-0.058	1.436	-0.005	0.103	1.238
9500	3.50	-0.276	-0.041	-0.037	1.422	-0.004	0.127	1.175
9500	4.00	-0.274	-0.016	-0.007	1.388	0.002	0.155	1.072
9500	4.50	-0.274	-0.011	0.029	1.338	0.011	0.188	0.938
9500	5.00	-0.276	-0.030	0.070	1.272	0.023	0.223	0.779
9750	2.00	-0.343	-0.244	-0.062	1.248	0.012	0.067	1.087
9750	2.50	-0.331	-0.174	-0.074	1.334	-0.003	0.081	1.177
9750	3.00	-0.327	-0.120	-0.069	1.365	-0.009	0.098	1.186
9750	3.50	-0.325	-0.078	-0.052	1.364	-0.010	0.120	1.141
9750	4.00	-0.323	-0.049	-0.025	1.340	-0.006	0.148	1.054
9750	4.50	-0.321	-0.037	0.010	1.300	0.002	0.181	0.933
9750	5.00	-0.322	-0.047	0.049	1.244	0.012	0.216	0.785

Teff	log g	BC	U-B	B-V	u-v	b-y	m1	c1
10000	2.00	-0.392	-0.287	-0.067	1.155	0.011	0.065	1.002
10000	2.50	-0.380	-0.214	-0.080	1.252	-0.006	0.078	1.106
10000	3.00	-0.376	-0.159	-0.078	1.294	-0.013	0.094	1.130
10000	3.50	-0.374	-0.115	-0.064	1.303	-0.014	0.114	1.101
10000	4.00	-0.373	-0.082	-0.040	1.289	-0.012	0.141	1.030
10000	4.50	-0.371	-0.064	-0.008	1.259	-0.006	0.173	0.923
10000	5.00	-0.371	-0.068	0.031	1.212	0.003	0.209	0.786
10500	2.00	-0.491	-0.365	-0.076	0.989	0.007	0.062	0.850
10500	2.50	-0.478	-0.290	-0.091	1.097	-0.010	0.074	0.968
10500	3.00	-0.475	-0.233	-0.092	1.155	-0.019	0.088	1.013
10500	3.50	-0.474	-0.187	-0.083	1.180	-0.022	0.106	1.009
10500	4.00	-0.474	-0.151	-0.065	1.183	-0.021	0.129	0.965
10500	4.50	-0.473	-0.125	-0.038	1.168	-0.018	0.158	0.886
10500	5.00	-0.471	-0.116	-0.002	1.139	-0.011	0.194	0.772
11000	2.50	-0.576	-0.357	-0.101	0.961	-0.014	0.071	0.845
11000	3.00	-0.573	-0.300	-0.104	1.024	-0.023	0.085	0.900
11000	3.50	-0.573	-0.254	-0.097	1.060	-0.028	0.101	0.912
11000	4.00	-0.574	-0.216	-0.083	1.076	-0.029	0.121	0.888
11000	4.50	-0.574	-0.187	-0.060	1.074	-0.027	0.147	0.831
11000	5.00	-0.573	-0.171	-0.030	1.059	-0.022	0.180	0.741
11500	2.50	-0.674	-0.417	-0.110	0.844	-0.019	0.069	0.741
11500	3.00	-0.670	-0.362	-0.113	0.908	-0.028	0.082	0.799
11500	3.50	-0.670	-0.316	-0.108	0.948	-0.032	0.097	0.817
11500	4.00	-0.672	-0.278	-0.096	0.971	-0.034	0.116	0.806
11500	4.50	-0.674	-0.248	-0.078	0.980	-0.034	0.139	0.767
11500	5.00	-0.674	-0.228	-0.052	0.976	-0.031	0.169	0.697
12000	2.50	-0.771	-0.469	-0.118	0.746	-0.023	0.068	0.655
12000	3.00	-0.766	-0.416	-0.122	0.808	-0.032	0.080	0.711
12000	3.50	-0.767	-0.372	-0.118	0.848	-0.037	0.094	0.732
12000	4.00	-0.769	-0.335	-0.107	0.875	-0.039	0.111	0.728
12000	4.50	-0.772	-0.305	-0.091	0.889	-0.039	0.133	0.699
12000	5.00	-0.773	-0.283	-0.069	0.893	-0.038	0.160	0.645
12500	2.50	-0.867	-0.514	-0.126	0.662	-0.027	0.066	0.582
12500	3.00	-0.862	-0.463	-0.131	0.722	-0.036	0.078	0.637
12500	3.50	-0.862	-0.422	-0.127	0.761	-0.041	0.091	0.658
12500	4.00	-0.865	-0.387	-0.117	0.787	-0.043	0.108	0.656
12500	4.50	-0.868	-0.357	-0.103	0.805	-0.044	0.129	0.634
12500	5.00	-0.870	-0.335	-0.083	0.813	-0.043	0.154	0.590
13000	2.50	-0.960	-0.555	-0.134	0.589	-0.031	0.065	0.520
13000	3.00	-0.956	-0.505	-0.139	0.648	-0.041	0.076	0.574
13000	3.50	-0.957	-0.466	-0.135	0.685	-0.045	0.089	0.594
13000	4.00	-0.959	-0.433	-0.126	0.710	-0.047	0.105	0.593
13000	4.50	-0.962	-0.405	-0.113	0.728	-0.048	0.125	0.573
13000	5.00	-0.964	-0.383	-0.094	0.740	-0.048	0.149	0.536
14000	2.00	-1.156	-0.708	-0.125	0.344	-0.018	0.053	0.273
14000	2.50	-1.140	-0.626	-0.148	0.465	-0.039	0.064	0.414
14000	3.00	-1.138	-0.576	-0.153	0.524	-0.048	0.074	0.471
14000	3.50	-1.140	-0.540	-0.150	0.559	-0.053	0.086	0.491
14000	4.00	-1.142	-0.511	-0.142	0.582	-0.055	0.101	0.489
14000	4.50	-1.144	-0.487	-0.130	0.598	-0.056	0.118	0.471
14000	5.00	-1.147	-0.467	-0.114	0.610	-0.056	0.140	0.439
15000	2.50	-1.308	-0.688	-0.160	0.359	-0.046	0.062	0.325
15000	3.00	-1.310	-0.637	-0.167	0.423	-0.056	0.072	0.389
15000	3.50	-1.314	-0.602	-0.164	0.458	-0.060	0.083	0.410
15000	4.00	-1.318	-0.575	-0.157	0.479	-0.062	0.097	0.408
15000	4.50	-1.320	-0.554	-0.145	0.493	-0.063	0.113	0.391
15000	5.00	-1.322	-0.538	-0.130	0.503	-0.063	0.133	0.360
16000	2.50	-1.464	-0.745	-0.171	0.264	-0.052	0.060	0.246
16000	3.00	-1.471	-0.690	-0.179	0.335	-0.062	0.070	0.319
16000	3.50	-1.479	-0.655	-0.177	0.373	-0.067	0.081	0.343
16000	4.00	-1.484	-0.629	-0.170	0.394	-0.069	0.093	0.343
16000	4.50	-1.487	-0.610	-0.159	0.407	-0.070	0.109	0.327
16000	5.00	-1.489	-0.596	-0.145	0.416	-0.070	0.128	0.298
17000	2.50	-1.608	-0.800	-0.180	0.177	-0.056	0.058	0.171
17000	3.00	-1.621	-0.739	-0.190	0.257	-0.068	0.068	0.256
17000	3.50	-1.633	-0.702	-0.189	0.297	-0.073	0.078	0.285
17000	4.00	-1.641	-0.676	-0.182	0.320	-0.075	0.090	0.288
17000	4.50	-1.646	-0.658	-0.172	0.334	-0.076	0.105	0.274
17000	5.00	-1.649	-0.646	-0.159	0.342	-0.076	0.123	0.246
18000	2.50	-1.741	-0.852	-0.186	0.094	-0.058	0.056	0.097
18000	3.00	-1.760	-0.784	-0.200	0.185	-0.073	0.066	0.198
18000	3.50	-1.777	-0.745	-0.200	0.230	-0.078	0.076	0.234
18000	4.00	-1.789	-0.719	-0.194	0.255	-0.081	0.087	0.240
18000	4.50	-1.796	-0.701	-0.184	0.269	-0.081	0.101	0.228
18000	5.00	-1.800	-0.690	-0.171	0.278	-0.081	0.118	0.202
19000	2.50	-1.865	-0.904	-0.190	0.013	-0.059	0.052	0.025
19000	3.00	-1.889	-0.827	-0.208	0.118	-0.077	0.064	0.143
19000	3.50	-1.911	-0.785	-0.210	0.168	-0.083	0.073	0.186
19000	4.00	-1.927	-0.759	-0.204	0.195	-0.086	0.084	0.196
19000	4.50	-1.937	-0.740	-0.195	0.211	-0.087	0.098	0.187
19000	5.00	-1.942	-0.729	-0.183	0.220	-0.087	0.114	0.164

Teff	log g	BC	U-B	B-V	u-v	b-y	m1	c1
20000	3.00	-2.008	-0.868	-0.215	0.055	-0.080	0.062	0.090
20000	3.50	-2.036	-0.823	-0.218	0.111	-0.088	0.071	0.142
20000	4.00	-2.056	-0.795	-0.214	0.141	-0.090	0.082	0.156
20000	4.50	-2.069	-0.777	-0.205	0.158	-0.091	0.094	0.150
20000	5.00	-2.077	-0.765	-0.194	0.168	-0.091	0.110	0.130
21000	3.00	-2.119	-0.907	-0.220	-0.004	-0.082	0.059	0.039
21000	3.50	-2.152	-0.859	-0.226	0.057	-0.091	0.069	0.099
21000	4.00	-2.177	-0.829	-0.222	0.091	-0.094	0.079	0.119
21000	4.50	-2.193	-0.810	-0.215	0.109	-0.096	0.091	0.116
21000	5.00	-2.203	-0.798	-0.204	0.120	-0.096	0.106	0.099
22000	3.00	-2.225	-0.943	-0.226	-0.058	-0.084	0.056	-0.005
22000	3.50	-2.261	-0.893	-0.232	0.007	-0.094	0.067	0.058
22000	4.00	-2.290	-0.861	-0.230	0.043	-0.098	0.077	0.083
22000	4.50	-2.310	-0.841	-0.223	0.064	-0.099	0.088	0.085
22000	5.00	-2.322	-0.829	-0.213	0.076	-0.100	0.102	0.070
23000	3.00	-2.327	-0.974	-0.231	-0.105	-0.086	0.053	-0.042
23000	3.50	-2.364	-0.924	-0.238	-0.040	-0.096	0.064	0.021
23000	4.00	-2.396	-0.891	-0.237	-0.001	-0.101	0.074	0.049
23000	4.50	-2.419	-0.870	-0.231	0.022	-0.103	0.085	0.054
23000	5.00	-2.434	-0.857	-0.222	0.035	-0.104	0.099	0.043
24000	3.00	-2.424	-1.002	-0.237	-0.146	-0.088	0.050	-0.072
24000	3.50	-2.463	-0.952	-0.244	-0.082	-0.098	0.062	-0.011
24000	4.00	-2.496	-0.920	-0.243	-0.043	-0.103	0.072	0.017
24000	4.50	-2.522	-0.898	-0.238	-0.018	-0.106	0.083	0.026
24000	5.00	-2.539	-0.885	-0.230	-0.003	-0.107	0.096	0.018
25000	3.00	-2.512	-1.028	-0.242	-0.186	-0.090	0.047	-0.102
25000	3.50	-2.557	-0.976	-0.250	-0.118	-0.101	0.059	-0.036
25000	4.00	-2.592	-0.946	-0.249	-0.081	-0.106	0.069	-0.010
25000	4.50	-2.619	-0.924	-0.244	-0.056	-0.108	0.080	-0.001
25000	5.00	-2.638	-0.910	-0.237	-0.040	-0.110	0.092	-0.007
26000	3.00	-2.594	-1.055	-0.244	-0.226	-0.089	0.043	-0.136
26000	3.50	-2.646	-0.998	-0.257	-0.150	-0.104	0.057	-0.057
26000	4.00	-2.683	-0.969	-0.255	-0.114	-0.108	0.067	-0.033
26000	4.50	-2.711	-0.949	-0.251	-0.090	-0.111	0.077	-0.025
26000	5.00	-2.731	-0.935	-0.244	-0.074	-0.112	0.089	-0.030
27000	3.50	-2.728	-1.019	-0.263	-0.182	-0.107	0.054	-0.077
27000	4.00	-2.770	-0.989	-0.262	-0.144	-0.111	0.064	-0.052
27000	4.50	-2.799	-0.970	-0.257	-0.122	-0.113	0.074	-0.046
27000	5.00	-2.819	-0.957	-0.250	-0.106	-0.115	0.086	-0.051
28000	3.50	-2.802	-1.040	-0.269	-0.213	-0.109	0.050	-0.098
28000	4.00	-2.851	-1.008	-0.269	-0.172	-0.114	0.062	-0.069
28000	4.50	-2.882	-0.990	-0.264	-0.150	-0.116	0.072	-0.063
28000	5.00	-2.903	-0.978	-0.257	-0.136	-0.118	0.083	-0.069
29000	3.50	-2.874	-1.060	-0.273	-0.243	-0.110	0.047	-0.119
29000	4.00	-2.925	-1.026	-0.275	-0.200	-0.117	0.059	-0.085
29000	4.50	-2.960	-1.008	-0.270	-0.177	-0.119	0.069	-0.079
29000	5.00	-2.983	-0.997	-0.263	-0.163	-0.120	0.080	-0.085
30000	3.50	-2.946	-1.078	-0.275	-0.269	-0.110	0.044	-0.140
30000	4.00	-2.995	-1.044	-0.281	-0.227	-0.119	0.055	-0.101
30000	4.50	-3.033	-1.025	-0.277	-0.202	-0.122	0.066	-0.093
30000	5.00	-3.058	-1.015	-0.270	-0.188	-0.123	0.077	-0.099
31000	3.50	-3.023	-1.094	-0.276	-0.291	-0.110	0.042	-0.157
31000	4.00	-3.063	-1.061	-0.285	-0.251	-0.121	0.052	-0.117
31000	4.50	-3.101	-1.042	-0.282	-0.227	-0.124	0.063	-0.107
31000	5.00	-3.128	-1.032	-0.276	-0.213	-0.126	0.074	-0.112
32000	4.00	-3.130	-1.077	-0.289	-0.274	-0.121	0.049	-0.131
32000	4.50	-3.168	-1.058	-0.288	-0.250	-0.126	0.060	-0.119
32000	5.00	-3.195	-1.047	-0.282	-0.235	-0.128	0.071	-0.124
33000	4.00	-3.201	-1.090	-0.291	-0.293	-0.122	0.046	-0.144
33000	4.50	-3.233	-1.072	-0.292	-0.271	-0.127	0.057	-0.132
33000	5.00	-3.260	-1.062	-0.287	-0.257	-0.129	0.067	-0.134
34000	4.00	-3.275	-1.102	-0.293	-0.308	-0.122	0.044	-0.155
34000	4.50	-3.300	-1.085	-0.295	-0.289	-0.128	0.053	-0.142
34000	5.00	-3.325	-1.076	-0.292	-0.276	-0.131	0.064	-0.144
35000	4.00	-3.353	-1.111	-0.295	-0.321	-0.123	0.043	-0.163
35000	4.50	-3.369	-1.097	-0.298	-0.305	-0.128	0.051	-0.152
35000	5.00	-3.390	-1.088	-0.296	-0.293	-0.132	0.061	-0.153
37500	4.50	-3.553	-1.118	-0.304	-0.334	-0.130	0.046	-0.168
37500	5.00	-3.562	-1.112	-0.303	-0.327	-0.133	0.053	-0.170
40000	4.50	-3.748	-1.130	-0.309	-0.351	-0.133	0.044	-0.176
40000	5.00	-3.747	-1.127	-0.309	-0.348	-0.135	0.049	-0.179
42500	5.00	-3.935	-1.137	-0.314	-0.362	-0.138	0.048	-0.184
45000	5.00	-4.116	-1.145	-0.319	-0.373	-0.141	0.047	-0.189
47500	5.00	-4.288	-1.152	-0.323	-0.382	-0.143	0.048	-0.193
50000	5.00	-4.451	-1.158	-0.326	-0.390	-0.145	0.048	-0.197

TABLE 2

Predicted bolometric corrections, UBV and uvby colors for the new solar abundance, 0, 1, 2, 4 and 8 km/s grids for sample temperatures 6,000K and 9,500K.

Teff	log g	Vt	BC	U-B	B-V	u-v	b-y	m1	c1
6000	0.00	0.00	-0.081	0.370	0.503	2.326	0.317	0.202	1.287
6000	0.00	1.00	-0.077	0.399	0.520	2.363	0.321	0.227	1.265
6000	0.00	2.00	-0.073	0.449	0.548	2.425	0.329	0.266	1.234
6000	0.00	4.00	-0.064	0.543	0.594	2.545	0.337	0.335	1.198
6000	0.00	8.00	-0.054	0.687	0.656	2.736	0.345	0.429	1.187
6000	1.00	0.00	-0.105	0.293	0.517	2.067	0.330	0.195	1.013
6000	1.00	1.00	-0.119	0.377	0.568	2.181	0.348	0.253	0.977
6000	1.00	2.00	-0.095	0.375	0.559	2.183	0.342	0.251	0.995
6000	1.00	4.00	-0.085	0.473	0.604	2.323	0.352	0.313	0.991
6000	1.00	8.00	-0.074	0.629	0.667	2.547	0.363	0.402	1.014
6000	2.00	0.00	-0.141	0.142	0.558	1.709	0.373	0.165	0.631
6000	2.00	1.00	-0.137	0.169	0.570	1.749	0.376	0.181	0.633
6000	2.00	2.00	-0.131	0.217	0.589	1.823	0.381	0.207	0.645
6000	2.00	4.00	-0.120	0.311	0.624	1.965	0.388	0.256	0.676
6000	2.00	8.00	-0.107	0.461	0.674	2.193	0.394	0.331	0.740
6000	3.00	0.00	-0.160	0.066	0.567	1.537	0.377	0.167	0.447
6000	3.00	1.00	-0.156	0.091	0.577	1.575	0.380	0.178	0.455
6000	3.00	2.00	-0.149	0.137	0.593	1.646	0.385	0.200	0.475
6000	3.00	4.00	-0.137	0.228	0.624	1.789	0.390	0.244	0.518
6000	3.00	8.00	-0.122	0.380	0.669	2.019	0.395	0.316	0.595
6000	4.50	0.00	-0.178	0.013	0.586	1.398	0.373	0.203	0.244
6000	4.50	1.00	-0.175	0.031	0.593	1.425	0.377	0.208	0.254
6000	4.50	2.00	-0.167	0.068	0.606	1.483	0.381	0.222	0.274
6000	4.50	4.00	-0.156	0.149	0.632	1.608	0.387	0.258	0.315
6000	4.50	8.00	-0.140	0.291	0.670	1.820	0.389	0.326	0.387
9500	2.00	0.00	-0.305	-0.199	-0.056	1.346	0.014	0.070	1.175
9500	2.00	1.00	-0.301	-0.199	-0.056	1.346	0.014	0.070	1.175
9500	2.00	2.00	-0.293	-0.199	-0.057	1.346	0.014	0.070	1.177
9500	2.00	4.00	-0.275	-0.200	-0.058	1.346	0.014	0.067	1.181
9500	2.00	8.00	-0.245	-0.202	-0.059	1.343	0.015	0.063	1.186
9500	4.00	0.00	-0.282	-0.016	-0.006	1.387	0.003	0.155	1.070
9500	4.00	1.00	-0.280	-0.016	-0.007	1.387	0.002	0.155	1.070
9500	4.00	2.00	-0.274	-0.016	-0.007	1.388	0.002	0.155	1.072
9500	4.00	4.00	-0.261	-0.016	-0.008	1.389	0.001	0.155	1.075
9500	4.00	8.00	-0.239	-0.015	-0.011	1.391	0.001	0.151	1.085

the Balmer line profiles (Peterson 1969) and then by forcing the computed colors for that model to match the observed colors. That model has T_{eff} = 9,400 K and log g = 3.95 just as in the earlier grid (Kurucz 1979a). However, remember that the helium abundance, metal abundances, line opacity and resolution are different. Models with lower microturbulent velocity are not observationally distinguishable in the visible. This model is not a physical model for Vega because Vega does not have solar abundances (for example, see Adelman and Gulliver 1990), but even if the physical parameters are somewhat off, the temperature-pressure structure must be correct in the continuum and Balmer-line-wing-forming layers. A correct model will have a similar structure but perhaps for a somewhat different effective temperature and gravity.

The bolometric corrections were normalized so that the smallest correction (at 7,250, 0.50) would be zero. The Vega model and the solar model are also included in Table 1. The Vega model is shown in Fig. 6. The bolometric correction for the Sun is -0.197 and the (B-V) index is 0.664. Table 2 lists samples of microturbulent velocity variations in all the colors. Fig. 7 is a sample (c_1) vs (b-y) diagram for the old grid, the new grid at 2 km/s, plus microturbulent velocity variations at 6,000K. There are large changes in the color calibration relative to the old; the (b-y) index has changed by 0.05 for G stars and c_1 has changed by 0.05 for F stars. The color calibration for the early A stars has barely changed. Microturbulent velocity strongly affects the colors and cannot be ignored.

ACKNOWLEDGEMENTS

This work is supported in part by NASA grants NSG-7054, NAG5-824, and NAGW-1486, and has been supported in part by NSF grant AST85-18900. The most important contribution to this work is a large grant of Cray computer time at the San Diego Supercomputer Center.

REFERENCES

Adelman, S. J. and Gulliver, A.F. 1990 ApJ 348, 712
Anders, E. and Grevesse, N. 1989 Geochimica et Cosmochimica Acta 53 197
Avrett, E. H., Kurucz, R. L.,and Loeser, R. 1984 BAAS 16, 450
Buser, R. and Kurucz, R. L. 1978 A&A 70, 555
Cowan, R. D. 1968 J. Opt. Soc. Am. 58, 808
Kurucz, R. L. 1977 SAO Special Report No 374
Kurucz, R. L. 1979a ApJS 40, 1
Kurucz, R. L. 1979b in Problems of Calibration of Multicolor Photometric Systems, A. G. D. Philip, ed., Dudley Obs. Report No. 14, p. 363
Kurucz, R. L. 1980 BAAS 11, 710
Kurucz, R. L. 1981 SAO Special Report No. 390,

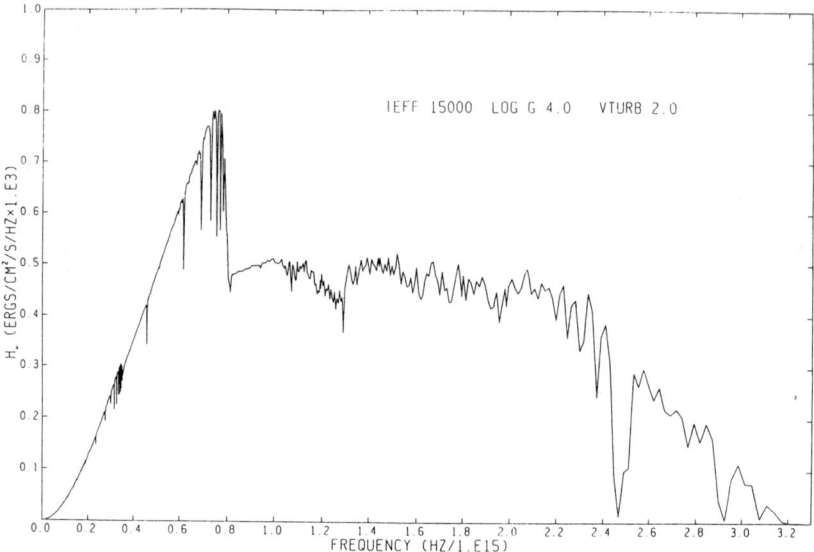

Fig. 2. Flux distribution for new model 15,000 K, log g 4, 2 km/s, a typical B star. The small dip at the Balmer discontinuity may be an artifact.

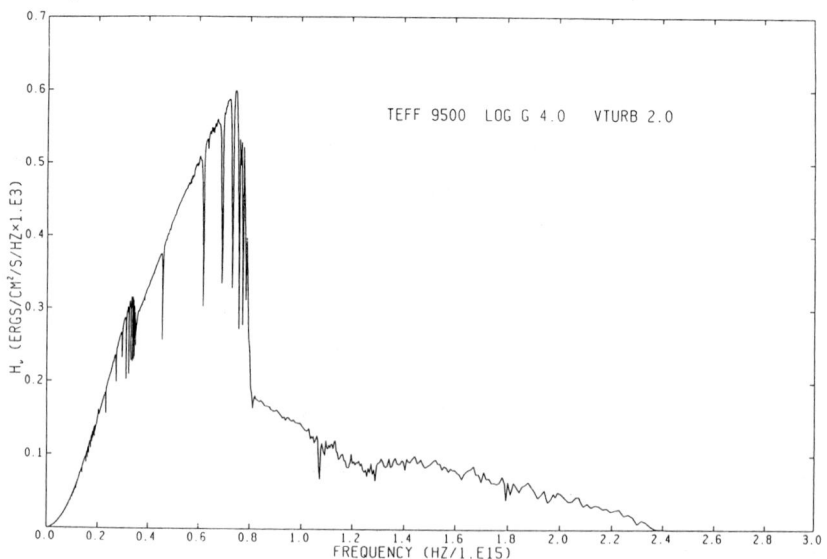

Fig. 3. Flux distribution for new model 9,000K, log g 4, 2 km/s, an early A star. The small dip at the Balmer discontinuity may be an artifact.

NEW LINES, NEW MODELS, NEW COLORS

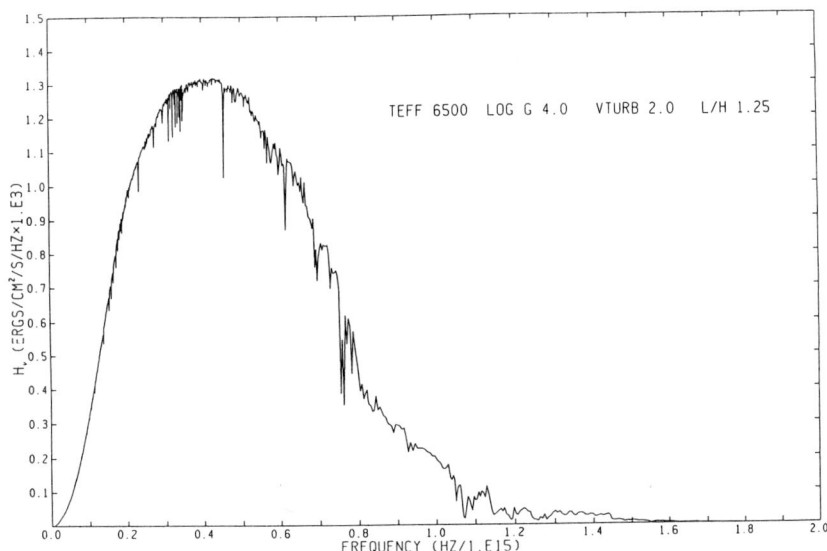

Fig. 4. Flux distribution for new model 6,500K, log g 4, 2 km/s, l/H 1.25, approximately Procyon.

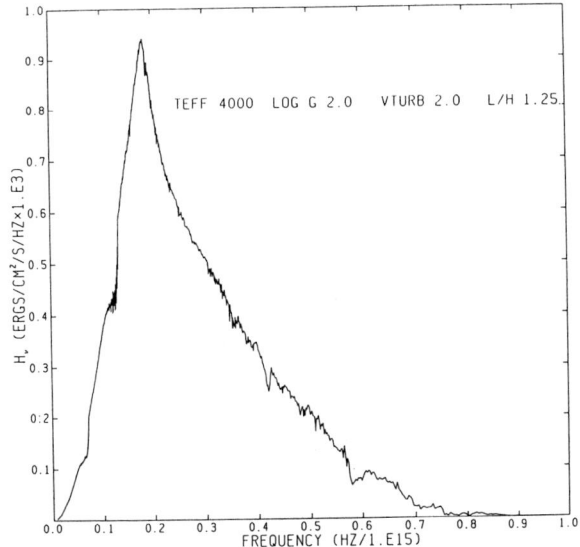

Fig. 5. Flux distribution for new model 4,000K, log g 1.5, 2 km/s, l/H 1.25, a K giant. Note the CO bands in the infrared.

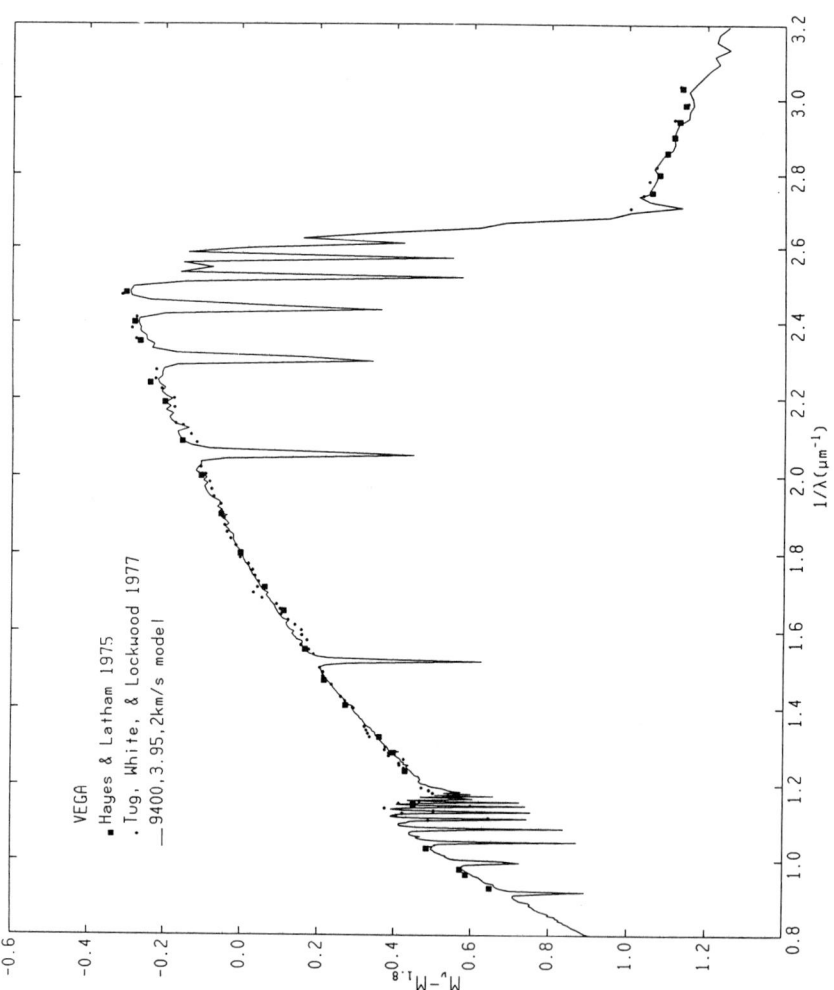

Fig. 6. Observed and predicted spectrophotometry relative to 1.8 inverse micrometers (555.6 nm) for Vega. The computed slopes of the Balmer, Paschen, and Brackett continua and the Balmer discontinuity agree with observations.

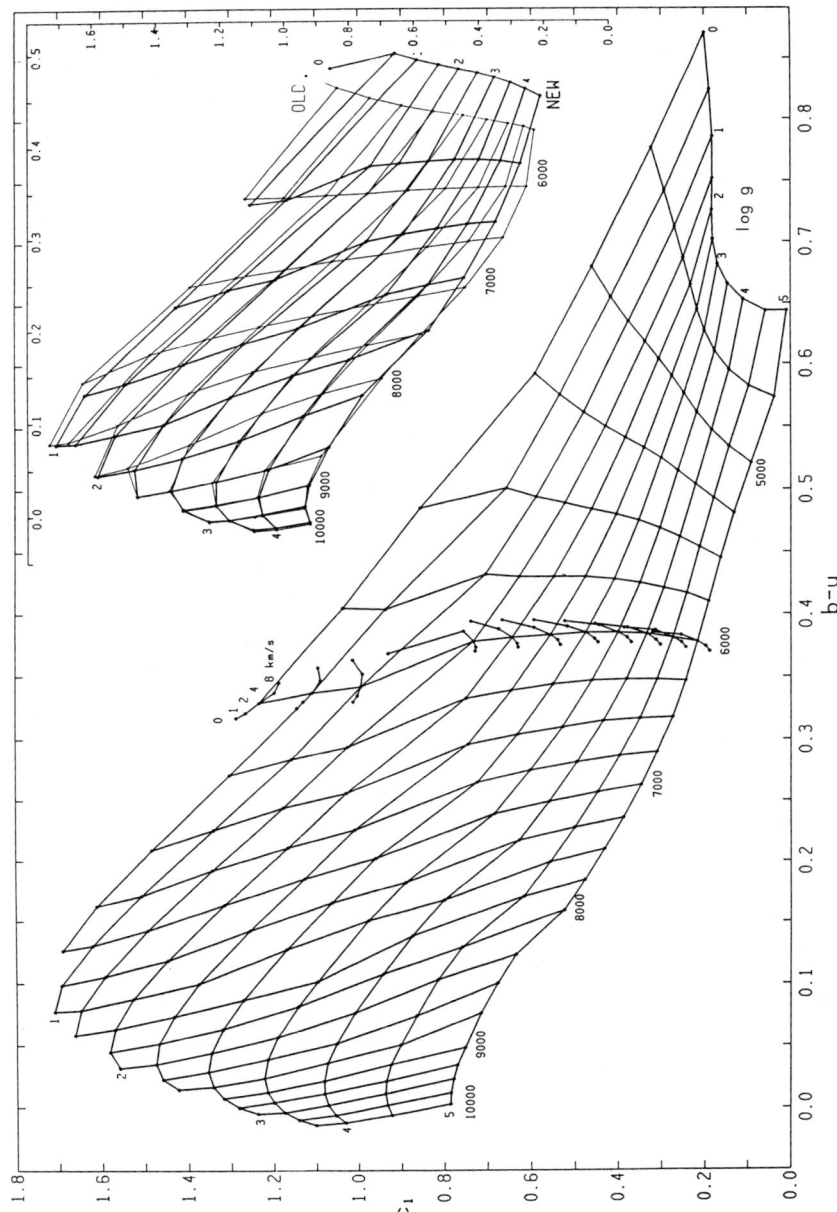

Fig. 7. (c_1) vs (b-y) diagrams for the 2 km/s, solar abundance grid and for the old grid (Relyea and Kurucz 1978). The effect of changing the microturbulent velocity to 0, 2, 4 or 8 km/s is also indicated for 6,000K.

Kurucz, R. L. 1987 In IAU Colloquium No. 95, The Second Conference on Faint Blue Stars, A. G. Davis Philip, D. S. Hayes and J. W. Liebert, eds., L. Davis Press, Schenectady, p. 129
Kurucz, R. L. 1991 in Stellar Atmospheres: Beyond Classical Models L. Crivellari, I. Hubeny and D.G. Hummer, eds., NATO ASI Series, Kluwer, Dordrecht, 1991, in press.
Kurucz, R. L. and Peytremann, E 1975 SAO Special Report No. 362
Labs, D., Neckel, H., Simon, P. C. and Thuillier, G. 1987 Solar Physics 107, 203
Lester, J. B., Gray, R. O. and Kurucz, R. L. 1986 ApJS 61, 509
Neckel, H. and Labs, D. 1984 Solar Physics 90, 205
Peterson, D. M. 1969 SAO Special Report No. 293
Relyea, L. J. and Kurucz, R. L. 1978 ApJS 37, 45
Tug, H., White, N. M. and Lockwood, G. W. 1977 A&A 61, 679

DISCUSSION

HAUCK: The comparison of both sets of models in the c_1 vs $(b-y)_0$ plane is very interesting, but what is the situation in the m_1 vs $(b-y)_0$ diagram?

KURUCZ: I have not yet had a chance to plot it. I calculated the colors only yesterday and plotted them overnight. I probably will be able to show you a plot before the end of the meeting.

RICH: In the galactic bulge, the giants have IR colors too blue for their abundances. Have you looked into this with the new models?

KURUCZ: Not yet. I have computed only solar abundance models thus far but I will produce the appropriate models soon.

DEMARQUE: 1. How important are Non-LTE effects? 2. Why did you calculate Rosseland mean opacities for the optically thin layers?

KURUCZ: 1. Increasing the line opacity tends to thermalize the radiation field and reduce the non-LTE effects. Every time Avrett has added more line opacity from my calculations, his solar models have come closer to LTE. Of course, when there are very strong radiation fields, as in hot stars, or when the density becomes high in hot stars, or when the density becomes low as in supergiants, non-LTE effects must be considered. Up to now, I have thought the opacity was a bigger problem. 2. In atmospheres I use them for scaling starting models. But I mainly computed them for interiors opacities. They should be much better than the Los Alamos opacities.

THE AGE SPREAD AMONG GALACTIC GLOBULAR CLUSTERS

Pierre Demarque

Center for Solar and Space Research
Yale University

ABSTRACT: Recent progress in estimating the relative ages of Galactic globular clusters is described. The evidence for an age spread of several Gyr among halo clusters is discussed. This result has important implications for our understanding of the formation and evolution of the Galaxy. Finally, current problems and work in progress on absolute ages are reviewed, primarily within the context of the age spread.

1. INTRODUCTION

The globular cluster system provides the most ancient record of our Galaxy's history. This fossil record can be used to establish the chronology of the Galactic halo and to test theories of galaxy formation which predict very different timescales for its formation. In the classical picture developed by Eggen et al. (1962, hereafter ELS), the globular cluster system was formed rapidly, in a time of the order of a 100 Myr. At the other extreme, models which involve a more chaotic process of halo formation, possibly involving the agglomeration of separate fragments over an extended period of time, could accommodate a halo chronology of several Gyr (see e.g. Searle and Zinn 1978). In the ELS picture, all globular clusters in the halo are coeval at the accuracy with which we measure ages. On the other hand, an age spread of several Gyr is expected in less organized models of galaxy formation. Thanks to the advent of CCD detectors and the spectacular recent advances in precision photometry, some data are now sufficiently accurate to allow us to draw conclusions concerning the relative ages of globular clusters in our Galaxy.

2. THE EXISTENCE OF AN AGE SPREAD

2.1 Age as the "second parameter": a test of the theory of stellar evolution

One of the exciting developments of the last few years in globular cluster research has been the confirmation of the long-conjectured (and often contested) interpretation of the second parameter in HB morphology as age (Searle and Zinn 1978). This idea had also been discussed by Rood and Iben (1968) on different grounds. But the confirmation required the use of CCD detectors, which made possible the simultaneous high precision photometry of both HB and T-O stars in globular clusters. There are now several pieces of evidence which support this conclusion. Stetson et al. (1989) have pointed out that the peculiarity of the CM-Diagram for Pal 12, which exhibits a redder HB than suggested by its metallicity, and at the same time a brighter main sequence T-O than M 5, could be due to a younger age for that cluster. Perhaps the most convincing evidence for interpreting the second parameter age as is given by NGC 288 and NGC 362, two clusters with nearly the same metallicity, and strikingly different HB morphology (Bolte 1989). High precision photometry of the T-O luminosities and morphology of the two clusters by Green and Norris (1990), yields an age difference between the two clusters of the order of 3 Gyr, a result which is consistent with recent synthetic HB models (Lee et al. 1990, hereafter LDZ). Figs. 1 and 2 illustrate the comparison. More recently, further confirmation has been provided by Buonanno et al. (1990) who have measured a CCD CM-Diagram for Ruprecht 106, a metal-poor cluster with a red HB, and a main sequence T-O suggesting an age 4 - 5 Gyr younger than the other metal-poor clusters.

It might be argued that some other parameter, or a combination of several parameters other than age could be responsible for the observed differences between two second parameter CM-Diagrams. It has indeed long been known that in principle, a number of parameters could affect HB morphology: differences in helium abundances, differences in CNO abundances or even variations in the HB helium core mass (and/or mass loss) due to differences in internal rotation rates on the giant branch (Mengel and Gross 1976, Renzini 1977, Demarque 1980). It is possible that some or all or these factors enter at some level in determining HB morphology. However, a detailed analysis of HB morphology based on synthetic HB models reveals that age is the only second parameter candidate that is compatible with both standard stellar evolution theory and pulsation theory (LDZ). Differences in helium and CNO abundances can be ruled out. Synthetic HB models show that the observed differences in RR Lyrae period distributions between Type I and Type II Oosterhoff-Sawyer groups are incompatible with a difference in helium abundance between the two groups (LDZ). And for a given age, it is not possible to reconcile the differences in HB morphology and turnoff luminosities between NGC 288 and NGC 362 in terms of a difference in CNO content. This is because variations in CNO abundances affect the T-O and HB in the opposite way of what is needed to explain the observations (Demarque et al. 1988). Finally, an explanation of the HB morphologies based on differences in internal rotation rates seem to be ruled out by recent work on the evolution of low-metallicity rotating stars (Deliyannis et al. 1989).

2.2 Is there a mean age-metallicity relation?

The possible existence of an age metallicity-relation among

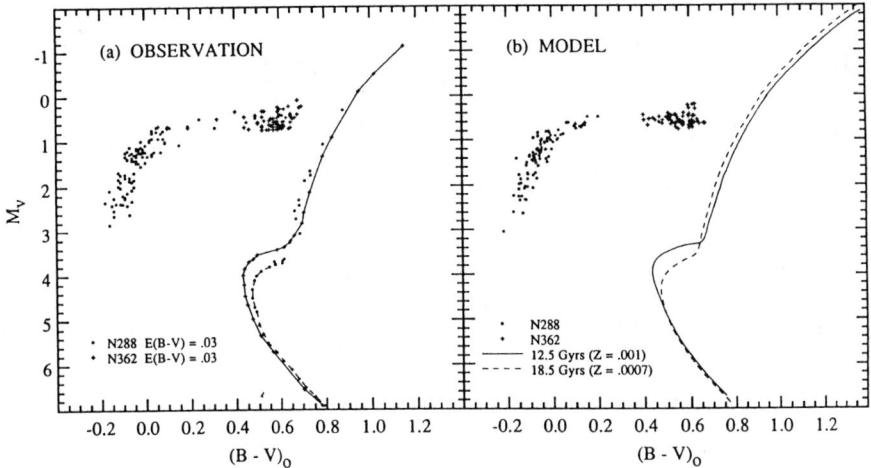

Fig. 1. Sketch of the CM-Diagrams of NGC 288 and NGC 362, after the work of LDZ. Note that NGC 362 exhibits both a redder HB and a more luminous T-O than NGC 288.

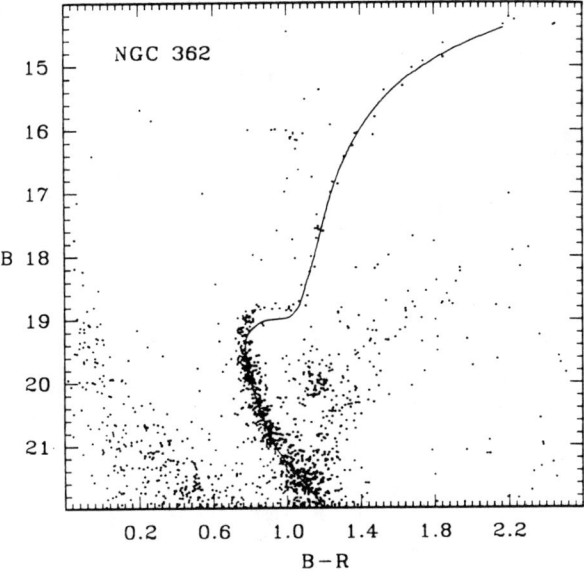

Fig. 2. Precision comparison of the CM-Diagrams of NGC 288 and NGC 362 (Green and Norris 1990). Note the difference in turnoff positions and shapes. The ridge line for NGC 288 has been shifted in reddening and distance modulus to match the main sequence and giant branch of NGC 362.

globular clusters has also often been discussed and reviewed during the last few years (see e.g. Demarque 1980, Zinn 1986, Sarajedini and King 1989). In the previous section, we have identified age as the second parameter, i.e. there is a spread in globular cluster ages for a given metallicity (where metallicity is the first parameter). It follows that any relation between metallicity and age within the Galactic halo could only hold as a mean relation, with an intrinsic scatter corresponding to the second parameter. Three issues must be considered in discussing the existence of an age-metallicity relation: (1) the dependence on the metallicity [Fe/H] of the absolute magnitude M_v of the RR Lyrae variables, (2) the relation between [O/Fe] and [Fe/H] for globular cluster stars (which may itself exhibit an intrinsic scatter), and (3) the theoretical calibration of the age scale. Sandage (1981) has pointed out the importance of taking into account the metallicity dependence of the RR Lyrae absolute magnitudes. He derived a steep slope (0.36) for the M_v vs. [Fe/H] relation. Using the ΔV technique, this slope yields no significant age spread for the halo clusters (ignoring possible [O/Fe] variations). More recently LDZ, on the basis of synthetic HB models, have derived a slope in the range 0.17 - 0.20 (for standard HB models, with a constant [O/Fe] ratio). A slope near 0.20 has also been derived from applications of the Baade-Wesselink method to field RR Lyrae stars (Jones et al. 1988, Cacciari et al. 1989, Liu and Janes 1990). Using the LDZ slope, Sarajedini and King (1989) have derived an age-metallicity relation. But because the correlation has a shallow slope, their result must be viewed as still tentative. This is particularly so because the theoretical calibrations depend sensitively on [O/Fe], and may also be affected by the model uncertainties discussed in Section 3 below.

2.3 The magnitude of the age spread

We have seen that in the context of understanding the process of formation of the Galactic halo, the first question is whether the age spread is of the order of 100 Myr (ELS), or much longer (see e.g. Rood and Iben 1968, Searle and Zinn 1978). Evidence for an age spread significantly in excess of 1 Gyr therefore favors a qualitatively different picture of the halo development than ELS.

There is however some uncertainty concerning the magnitude of the age difference between NGC 288 and NGC 362. While the observed difference in T-O luminosity between NGC 288 and NGC 362 is consistent, within the uncertainties, with an age spread of 3 Gyr, the LDZ synthetic HB models suggest an even larger spread, perhaps as large as 5 Gyr. This discrepancy has been interpreted by VandenBerg et al. (1990) as evidence that age is not likely to be the second parameter. In the absence of a better alternative, it seems more probable that this age discrepancy is a reflection of the extreme sensitivity of HB morphology to details in the parameters of the models. Uncertainties in globular cluster intrinsic parameters or in the models, which may be of minor importance in other contexts, are amplified on the HB.

Understanding the precise chronology of the halo will require more refined observations and improvements in the standard theory. This is currently being done. For example, Lee (1990) has recently discussed HB population models based on improved stellar models from Dorman et al. (1991), in which the HB age difference is reduced to about 3 Gyr between NGC 288 and NGC 362, in satisfactory agreement with the T-O age difference found by Green and Norris (1990).

There are still a number of obvious improvements in the physics that need to be made in order to understand better the HB morphology of globular clusters. At this point, we know little about mass loss mechanisms on the giant branch. The effects of internal rotation on HB morphology need to be explored with more realistic models than originally done by Renzini (1977), who used the models of Mengel and Gross (1976). The reader is referred to the recent work of Deliyannis et al. (1989) and Pinsonneault et al. (1991) on the effects of rotation in halo stars. Thus it is hoped that the inclusion in the models of physical processes so far neglected in evolutionary calculations will help us understand the remaining mysteries of HB morphology. But it must be emphasized that at the current level of understanding of stellar evolution, the standard theory already gives a straightforward explanation of the NGC 288 - NGC 362 observations. Further, as shown in LDZ, age provides the only interpretation of the second parameter that is consistent with both HB and main sequence T-O theory.

There are other uncertainties which may affect the relative ages of globular clusters. They are discussed in Section 3.

3. CURRENT ISSUES IN THE AGE CALIBRATION OF GLOBULAR CLUSTERS

It has sometimes been claimed that the progress in precision photometry with CCD detectors and in the calculation of isochrone turnoff shapes make possible fits with theoretical isochrones with accuracies better than 1 Gyr. These optimistic statements are only correct under the assumption that the theoretical isochrones used are free of errors, either in their input parameters, such as chemical abundances, or in the physical modeling itself. Unfortunately, there are still uncertainties in both the input composition parameters and the physical modeling of theoretical isochrones, which could change the cluster chronology by much more than 1 Gyr.

3.1 Observational problems

The two major observational uncertainties in determining absolute ages for globular clusters are well-known. They are: (1) chemical composition and (2) distance modulus (and the associated problem of reddening corrections).

3.1.1 Chemical composition

Chemical composition uncertainties are important because stellar models are sensitive to chemical abundances which affect the opacities, the mean molecular weight, and the nuclear processes that determine the rate of stellar evolution. Two parameters are normally used to quantify globular cluster abundances: [Fe/H] and [O/Fe] (see e.g. Simoda and Iben 1968, Renzini 1977, Demarque 1980). Much progress has been made in evaluating [Fe/H] (Zinn and West 1984). At this writing, significant uncertainties in the [O/Fe] ratio still plague us. There is disagreement on the form of the [O/Fe] vs [Fe/H] (Abia and Rebolo 1989; Gratton 1990). There may in addition be a real scatter in this relation (see the discussion of Sarajedini 1990). We have seen already that variations in [O/Fe] do not affect specifically the NGC 288 - NGC 362 comparison, and therefore the existence of an age spread. However, the form of the [O/Fe] vs. [Fe/H] relation will have an impact on the calibration of absolute ages. If there is an intrinsic spread in [O/Fe] among clusters of the same metallicity, it could also affect the details of halo chronology.

3.1.2 Distance modulus

In principle, distance moduli can be estimated from main sequence fitting, and from the RR Lyrae absolute magnitude calibration. Because of the current large uncertainties in subdwarf parallaxes (van Altena et al. 1988), and of the difficulties of globular cluster turnoff photometry, much interest has focused on the RR Lyrae variables as distance indicators (Sandage 1981). Relative distances can be evaluated using the RR Lyrae variables, but absolute distances are still uncertain. There is still a discrepancy of about 0.2 mag. between the LDZ relation zero-point and that from the Baade-Wesselink method (Liu and Janes 1990). Interstellar reddening poses additional problems. For all these reasons, dating methods have been devised that are independent of the cluster's distance modulus and reddening (Sarajedini 1991). But each of these methods needs to be calibrated with theoretical isochrones.

3.2 Theoretical problems

Finally, let me mention briefly the current theoretical activity in this field. The efforts focus on understanding the effects of two important physical processes not included in standard stellar models on the shapes and lifetimes of evolutionary tracks near the turnoff. These processes are: (1) internal rotation and (2) the diffusion (more specifically the gravitational settling) of helium. Their consequences have so far been only partially explored. One also needs to understand the interaction between these two mechanisms and with [O/Fe] variations, because there is evidence that non-linear feedbacks can take place. The existence of the age spread is not likely to be affected by these uncertainties. However, the absolute chronology of clusters will be modified, and possibly the magnitude of the age spread. There may also be distortions (contractions and expansions) within the current age scale. In particular, metallicity- and age-dependent changes can be expected, since in the study of both

THE AGE SPREAD AMONG GALACTIC GLOBULAR CLUSTERS

rotation and diffusion so much depends on the depth of the surface convection zone. The existence of an age-metallicity relation among Galactic globular clusters depends therefore also in part on the resolution of these problems.

3.2.1 Rotation

Because rotation causes internal motions, it has long been believed that it could retard main sequence turnoff and thus affect estimates of evolutionary lifetimes near the main sequence, and that the importance of the effect is sensitive to the distribution of angular momentum within the star (Mengel and Gross 1976, Law 1980). Evolutionary calculations including the effects of angular momentum loss and redistribution during evolution have recently been performed for globular cluster stars (Deliyannis et al. 1989). It is found that although a substantial amount of angular momentum remains trapped in the interior at the turnoff, it is insufficient to modify the lifetimes and shapes of evolutionary tracks for halo stars near the turnoff in an appreciable way. However, this angular momentum plays an essential role in causing internal mixing: it explains in a natural way the observed CNO anomalies among globular cluster giants.

Preliminary considerations suggest that internal rotation should not affect the helium core mass at the onset of the helium flash, and as a result plays little role in determining HB luminosity and morphology (Pinsonneault et al. 1991). Even using the Mengel-Gross models as a guide, which do not include angular momentum transfer and losses during evolution, and therefore overestimate the core rotation rate at the helium flash, the amount of internal angular momentum still in the models beyond turnoff is too small to affect the core mass at the helium flash. However, detailed calculations need to be performed including the rotational history along the giant branch and the HB as well. The possibility still exists that although the structural effects of rotation are negligible, internal mixing could modify advanced evolution. Rotational mixing on the giant branch could effectively decrease the helium core mass at the flash and thus modify HB morphology and luminosity (Lee and Demarque 1990). Further rotational mixing on the HB could also modify its morphology in subtle ways. At the same time, some constraints on the efficiency of rotational mixing in the outer layers can be placed in BHB stars where there is evidence for surface helium depletion (Sargent and Searle 1967), best explained by atmospheric diffusion (Michaud et al. 1983).

3.2.2 Diffusion

Near the turnoff, helium diffusion acts in two different ways: (1) in the core, where downward diffusion accelerates the exhaustion of hydrogen near the center and shortens the lifetime for turnoff (first discussed by Stringfellow et al. 1983), and (2) in the envelope, where helium is progressively drained from the surface convection zone into the radiative layers below by gravitational settling. The helium depletion in the convection zone results in an increased radius for the models, and therefore changes the shapes of the theoretical isochrones. Recent calculations of the first effect, with improved diffusion coefficients and stellar models, show a decrease of the order of 6% in turnoff age at a given luminosity, for a given chemical composition (Proffitt et al. 1990, Chaboyer et al. 1990). Internal rotation could modify this estimate downward, but the presence of a mean molecular weight gradient is believed to reduce circulation to a minimum near the center. The second effect is primarily in the radii of the models, and results in a redder isochrones for a given set of assumptions (Deliyannis et al. 1990, Proffitt et al. 1990). The decrease in age estimates (1 - 3 Gyr) depends on the dating technique, and is quite uncertain at this point. Constraints on the efficiency of helium diffusion, which may be inhibited by rotationally induced turbulence, can also be put from observations of lithium abundances, since lithium diffuses at nearly the same rate as helium and can be observed directly in these stars. Since the efficiency of diffusion is dependent on the depth of the convection zone, it requires an accurate knowledge of opacities in the layers near the base of the convection zone, where the temperature reaches a few million degrees. This is precisely the region where O contributes most to the opacity. The increase in opacity due to enhanced O then deepens the convection zone. The efficiency of helium diffusion thus decreases when [O/Fe] increases. This feedback limits the decrease in age due to diffusion in the envelope when [O/Fe] is large (Chaboyer et al. 1990).

4. FUTURE NEEDS FOR PRECISION PHOTOMETRY

Further progress in analyzing the chronology of globular clusters will require a major concentration on photometry of faint stars in globular clusters. The globular clusters in the Galactic bulge will be particularly interesting in the context of the age spread. Observing the T-O of these clusters with precision will require the full spatial resolution originally specified for the HST. Absolute ages will be determined more reliably with the help of high precision luminosity functions as well as CM-Diagrams. Giant branch and HB observations on the same system can provide consistency tests (Sarajedini 1990). In addition, precision spectroscopy will be needed to measure both [Fe/H] and [O/Fe] parameters in dwarf stars within globular clusters. The importance of helium diffusion on the calibration can best be estimated from lithium abundance determinations covering the full temperature range of the Spite plateau on the main sequence (Deliyannis et al. 1990, Chaboyer et al. 1990).

5. ACKNOWLEDGEMENTS

Support for this research from NASA grants NAGW-777 and NAGW-778 is gratefully acknowledged.

REFERENCES

Abia, C. and Rebolo, R. 1989 ApJ 347, 186
Bolte, M. 1989 AJ 97, 1688
Buonanno, R., Buscema, G., Fusi Pecci, F., Richer, H. B. and Fahlman, G. G. 1990 in The Formation and Evolution of Star Clusters, Astron. Soc. Pacific Conf. Series, No. 13, K. Janes, ed., Astron. Soc. Pacific, Chelsea, Mich., p. 240
Cacciari, C., Clementini, G. and Buser, R. 1989 A&A 216, 80
Chaboyer, B., Deliyannis, C. P., Demarque, P., Pinsonneault, M. H. and Sarajedini, A. 1990 BAAS 22, 1205
Deliyannis, C. P., Demarque, P. and Pinsonneault, M. H. 1989 ApJL 347, L 73
Deliyannis, C. P., Demarque, P. and Kawaler, S. D. 1990 ApJS 73,21
Demarque, P. 1980 in IAU Symposium No. 85, Star Clusters, J. E. Hesser, ed., D. Reidel, Dordrecht, p. 281
Demarque, P., Lee, Y. -W., Zinn, R.and Green, E. M. 1988 in The Abundance Spread Within Globular Clusters: Spectroscopy of Individual Stars, G. Cayrel de Strobel, M. Spite and T. L. Evans, eds., Paris Obs, p. 97
Dorman, B., Lee, Y. -W. and VandenBerg, D. A. 1991 ApJ, in press.
Eggen, O. J., Lynden-Bell, D. and Sandage, A. 1962 ApJ 136, 748 (ELS)
Gratton, L. 1990 in Evolution of Stars: The Photospheric Connection, IAU Symposium No 145, G. Michaud and A. Tutukov, eds, in press
Green, E. M. and Norris, J. E. 1990 ApJL 353, L 17
Jones, R. V., Carney, B. W. and Latham, D. W. 1988 ApJ 332, 206
King, C. R., Da Costa, G. S. and Demarque, P. 1985 ApJ 299, 674
Law, W. -Y. 1980 Ph.D. dissertation, Yale University
Lee, Y. -W. 1990 in The Formation and Evolution of Star Clusters, Astron. Soc. Pacific Conf. Series, No. 13, K. Janes, ed., Astron. Soc. Pacific, Chelsea, Mich., p. 205
Lee, Y. -W. and Demarque, P. 1990 ApJS 73, 709
Lee, Y. -W., Demarque, P. and Zinn, R. 1990 ApJ 350, 155 (LDZ)
Liu, T. and Janes, K. A. 1990 ApJ 354, 273
Mengel, J. G. and Gross, P. G. 1976 ApSpaceSci 41, 407 Michaud, G., Vauclair, G. and Vauclair, S. 1983 ApJ 267, 256 Pinsonneault, M. H., Deliyannis, C. P. and Demarque, P. 1991 ApJ 367,239
Proffitt, C. R., Michaud, G. and Richer, J. 1990 in Cool Stars and Stellar Systems, G.Wallerstein, ed, Astron. Soc. Pacific Conf. Series 9, p. 351
Renzini, A. 1977 in Advanced Stages of Stellar Evolution, 7th Course, SAAS-FEE Lectures,.P. Bouvier and A. Maeder, eds., p. 149
Rood, R. and Iben, I. Jr. 1968 ApJ 154, 215
Sandage, A. R. 1981 ApJ 248, 161
Sarajedini, A. 1991 in Precision Photometry: Astrophysics of the Galaxy, A. G. D. Philip, A. R. Upgren and K. A. Janes, eds., L. Davis Press, Schenectady, p. 55
Sarajedini, A. and King, C. R. 1989 AJ 98, 1624
Sargent, W. L. W. and Searle, L. 1967 ApJL 150, L 33
Searle, L. and Zinn, R. 1978 ApJ 225, 358

Simoda, M. and Iben, I. Jr. 1968 ApJ 152, 509
Stetson, P. B., VandenBerg, D. A., Bolte, M. J., Hesser, J. E. and Smith, G. H. 1989 AJ 97, 1360
Stringfellow, G. S., Bodenheimer, P., Noerdlinger, P. D. and Arigo, R. J. 1983 ApJ 264, 228
van Altena, W. F., Lee, J. T., Hanson, R. B. and Lutz, T. E. 1988 in Calibration of Stellar Ages, A. G. D. Philip, ed., L. Davis Press, Schenectady, p. 175
VandenBerg, D. A., Bolte, M. J. and Stetson, P. B. 1990 AJ 100, 445
Zinn, R. J. 1986 in Stellar Populations, STScI Symp. Ser., C. A. Norman, A. Renzini and M. Tosi, ed.s, Cambridge Univ Press, p. 73
Zinn, R. J. and West, M. J. 1984 ApJS 55,45

GLOBULAR CLUSTER PHOTOMETRY NEAR THE TURN-OFF: RELATIVE AGES AND THE HORIZONTAL BRANCH

Ata Sarajedini

Yale University Observatory

ABSTRACT: With the widespread use of CCD detectors for precise stellar photometry, the morphology of the main sequence turn-off region in Galactic globular clusters has become better defined. Due to the extreme sensitivity of the turn-off region to variations in cluster age, this has allowed unprecedented precision in the determination of relative cluster ages. One method recently devised to study relative ages is based on the color difference between the turn-off and the point at which the subgiant branch turns upward, henceforth referred to as the "turn-up." Although it is possible to define this color difference in any combination of filters, work done so far has utilized the (B-V) color difference and is thus denoted by $\Delta(B-V)$. The advantages of $\Delta(B-V)$ are that it is independent of reddening, distance modulus, and photometric zeropoint. However, perhaps the most important advantage is that $\Delta(B-V)$ is highly insensitive to cluster metallicity over a wide range of metallicities.

We have compiled $\Delta(B-V)$ measurements for approximately 20 clusters. The distribution of these values and how they vary with abundance gives information on the relative ages of the globular clusters and thus the formation history of the Galactic halo. In addition, we can use these $\Delta(B-V)$ values along with other observational data to study the luminosity behavior of the horizontal branch with metallicity.

1. INTRODUCTION

When the CCD was first introduced into astronomical photometry, it was heralded as the savior of *globular cluster* photometric studies. This is a device that allowed astronomers to probe the fainter extensions of a globular cluster's main sequence and thus derive valuable information about its turn-off, and hence its age (Hesser 1988). Previous photographic work was only able to provide main sequence data for the nearest clusters and even then it was difficult to determine whether the scatter seen in the data was due to variations in the intrinsic cluster parameters or merely to errors in the photometry.

Several years passed during which various authors applied this new technology along with equally new reduction techniques (West and Kruszewski 1981, Buonanno et al. 1983, Lupton and Gunn 1986, Penny and Dickens 1986, Stetson 1987 and references therein) to obtaining "deep" photometry for Galactic globular clusters in the standard UBV system. Many authors then went on to draw conclusions regarding cluster parameters (such as age, metallicity, reddening and distance modulus) from their CCD photometry. The main aim of these efforts was to date these clusters and thereby determine whether the halo of our Galaxy formed over a period of < 1 Gyr (Eggen, Lynden-Bell and Sandage 1962) or significantly > 1 Gyr (Searle and Zinn 1978). However, as more and more cluster photometry was gathered, and the same clusters were studied by different investigators, it was soon clear that there was something amiss. Color-magnitude diagram (CMD) principal sequences derived for the same clusters have turned out *not* to be coincident, even within the expected errors (Straniero and Chieffi 1990, VandenBerg, Bolte and Stetson 1990). As a result, values of the cluster parameters mentioned above depend on the photometry one uses, even though all of the data have been gathered with CCDs.

The reasons for these discrepancies are not completely clear. However, there are at least three possible sources of error. 1) The computational procedure one uses to derive instrumental magnitudes from stellar images on a crowded CCD frame. 2) The treatment given to the derivation of aperture corrections for a CCD image, which are tricky to derive in crowded regions or sparse regions which provide too few stars for such a derivation. 3) The technique used to calibrate the instrumental photometry and bring it onto a given standard system. Considering all of these problems, the question becomes: how do we overcome them and utilize the CCD to its fullest potential?

Each of these problems has recently been given an improved treatment. The relevant references are as follows: Stetson (1990a) for problem #1, Stetson (1990b) for problem #2, and Massey et al. (1989), Stetson and Harris (1988), Stetson et al. (1989), and Sarajedini and Da Costa (1990) for problem #3; If these improvements are followed, we will no doubt have more accurate photometry for the globular clusters.

On the other hand, when the question of the relative ages of these stellar systems is considered, it has recently become clear that perfectly calibrated photometry is not needed. In fact, the method described herein can be applied to any CCD photometry, calibrated or not, that is devoid of internal distortion effects.

2. METHOD

The relative age diagnostic we will discuss was independently arrived at by Sarajedini and Demarque (1990, SD) and VandenBerg, Bolte and Stetson (1990, VBS); however, we will limit our discussion to the treatment of SD. The quantity $\Delta(B-V)$, as illustrated in Fig. 1 is

GLOBULAR CLUSTER PHOTOMETRY NEAR THE TURN-OFF

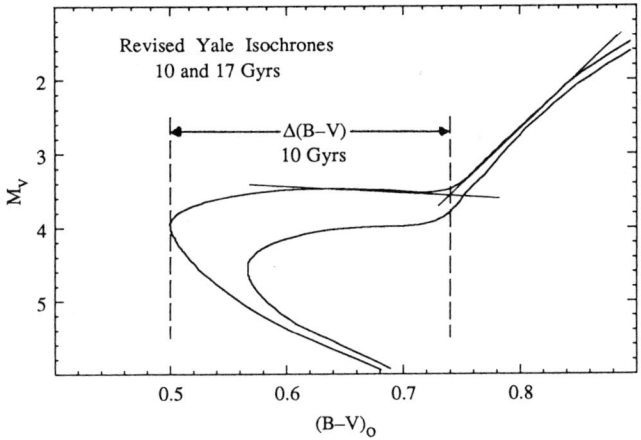

Fig. 1. The age sensitive parameter $\Delta(B-V)$ is defined as the difference in color between the turn-off and the subgiant branch (SGB) turn-up. The turn-off is the bluest apparent color of the main sequence. The SGB turn-up is determined as shown and described in the text.

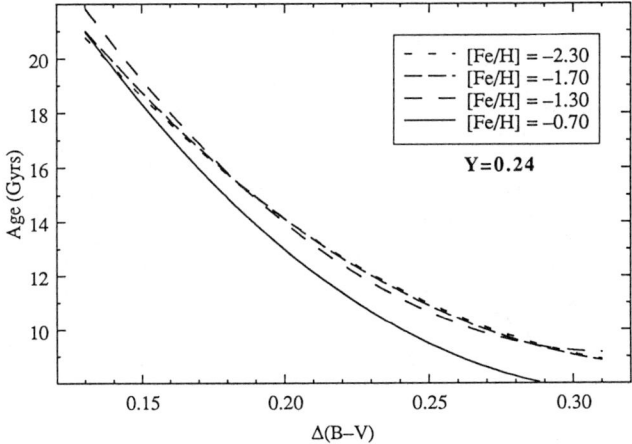

Fig. 2. The variation of $\Delta(B-V)$ with age from the Revised Yale Isochrones of Green et al. (1987). Note the insensitivity of $\Delta(B-V)$ to metallicity for $-1.30 \geq [Fe/H] \geq -2.30$.

defined to be the difference in apparent color between the main sequence turn-off (TO) and the color at which the subgiant branch (SGB) turns up, henceforth referred to as the "turn-up." In other words,

$$\Delta(B - V) \equiv (B - V)_{SGB} - (B - V)_{TO},$$

where $(B-V)_{TO}$ is determined as the bluest apparent color of the main sequence, i.e. at the turn-off point. The quantity $(B-V)_{SGB}$ is derived by constructing two straight lines, one coincident with the lower giant branch and the other coincident with the subgiant branch, as shown in Fig. 1; the color at which these two lines intersect is then the value of $(B-V)_{SGB}$. Using straight lines to describe features which are in fact curved may at first seem imprecise; however, if the lines are constructed in close enough proximity to the turn-up, very little uncertainty is introduced. In fact, when we compare the $\Delta(B-V)$ values for the 12 clusters in common between this study and that of Racine (1990), the maximum difference between the two sets of data is 0.01 mag. with an average difference of 0.00 ±0.01 mag. Therefore, we have chosen to be conservative and adopt a value of 0.01 mag. as the error in $\Delta(B-V)$. However, it should be noted that we believe this to be a maximum possible error.

The quantity $\Delta(B-V)$ is clearly dependent on cluster age, with $\Delta(B-V)$ becoming smaller for older clusters. In addition, it is completely independent of other quantities which are inherently difficult to determine, these include the cluster reddening, distance modulus, and the zeropoint of the photometric data used to determine $\Delta(B-V)$. In other words, problems two and three above are immediately overcome when $\Delta(B-V)$ is used. Furthermore, the theoretical isochrones used by SD, VBS, and Straniero and Chieffi (1990, SC) all confirm that *$\Delta(B-V)$ is highly insensitive to cluster metal abundance*, from [Fe/H] ~ -2.3 to [Fe/H] ~ -1.3. This point is well illustrated in Fig. 2, where the three dashed curves represent the behavior of $\Delta(B-V)$ with variations in age for $-2.30 \leq$ [Fe/H] ≤ -1.30 from the Revised Yale Isochrones (RYI, Green et al. 1987) for Y = 0.24. The solid curve in Fig. 2 is the relationship for [Fe/H] = -0.70. Since the metallicity dependence of $\Delta(B-V)$ at [Fe/H] = -0.70 is not severe and since there are only two clusters in our sample (47 Tuc and Pal 12) with metallicities greater than -1.30 (-0.71 and -1.14, respectively, Zinn and West 1984), we will henceforth consider $\Delta(B-V)$ to be indicative of age and not be significantly influenced by differences in metal abundance. It will become clear later that this is a reasonable assumption partially because if we take into account the effects of metallicity using the RYI, the age spread actually increases. Finally, it is worth noting that $\Delta(B-V)$ is also insensitive to helium abundance (see Fig. 3 of VBS) so we therefore assume Y = 0.24 for the Galactic globular clusters.

There are at least two potential problems with the use of $\Delta(B-V)$. 1) Enhancements in the oxygen abundances (as measured by [O/Fe]) of globular cluster turn-off stars will make the turn-offs redder at a given age (Simoda and Iben 1968, Renzini 1977), essentially mimicking the effects of age. That is to say, an increase in age makes Δ(B-V) smaller and so too does an increase in [O/Fe]. 2) Inclusion of helium diffusion in the models of Pop II stars shows that such diffusion makes the turn-offs redder (Stringfellow et al. 1983, Deliyannis et al. 1990, Profitt and Michaud 1990) also mimicking the effects of age. These two uncertainties are potential problems only if their effects vary with metal abundance. In other words, the absolute ages may be affected, but if there is no general trend with metallicity, then the relative ages are unaffected. The question then becomes: to what degree are these two effects a function of metallicity in the Galactic globular clusters? We return to this question in the next section.

In addition, when interpreting the observed values of Δ(B-V) in terms of theoretical isochrones, we must keep in mind several inherent uncertainties. Fortunately, the question of stellar rotation and its effects on the turn-off has already been answered by Deliyannis et al. (1989). They find that rotation has no significant effect on the determination of ages using the turn-off region. However, there remains the question of which opacities are appropriate to use in the construction of theoretical isochrones and which value of the mixing length parameter is appropriate for Pop II stars, and whether mixing length itself depends on other parameters (e.g. age, Y, [Fe/H]). At this point, we are left to make the assumption that these uncertainties have little or no effect on our result.

3. RELATIVE AGES

Fig. 3 shows the observed behavior of Δ(B-V) with cluster [Fe/H]. The values of Δ(B-V) are measured from the best-defined CMDs available in the literature and include values taken from Racine (1990). The cluster metal abundances are from Zinn and West (1984). There is a clear trend of increasing Δ(B-V) (i.e. decreasing age) with increasing [Fe/H] with values of Δ(B-V) ranging from 0.17 to 0.27. This implies the existence of a significant age range among the Galactic globular clusters and the presence of a significant age-metallicity relation.

Four important clusters are indicated in Fig. 3. First, the most metal rich cluster in the sample is 47 Tucanae with a Δ(B-V) = 0.25. Next, the youngest cluster is Pal 12, which has been found to be 25% - 30% younger than the average of 47 Tuc and Pal 5 (Stetson et al. 1989). We find Pal 12 has a Δ(B-V) = 0.27 and is therefore about 15% younger than the average of 47 Tuc and Pal 5. Finally, NGC 288 has a completely blue horizontal branch (HB) whereas NGC 362, which has a similar metallicity, possesses a completely red HB. Therefore, these two clusters represent a "second parameter" pair in which another factor in addition to metal abundance is controlling the HB morphology. In the case of NGC 288/362, there exists a difference in age of ~ 3 Gyrs as indicated in Fig. 3 and corroborated by Bolte (1989), Green and Norris (1990), VBS, SD and Lee (1990). Therefore, the reason for the

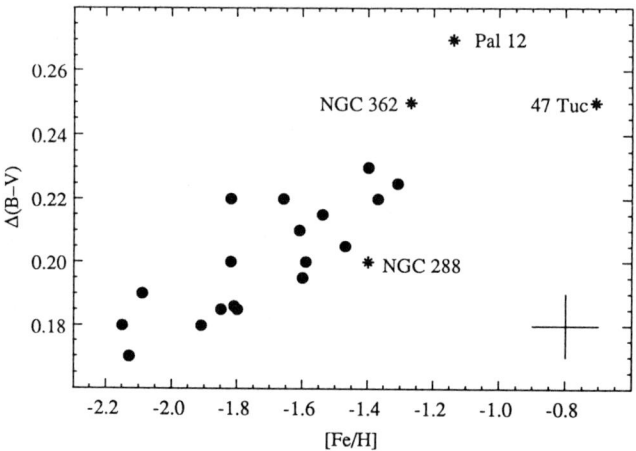

Fig. 3. The variation of observed Δ(B-V) with cluster [Fe/H]. Since Δ(B-V) is a sensitive age indicator, our data argues in favor of an age-metallicity relation. An average error bar is indicated. The clusters plotted as asterisks are discussed in the text.

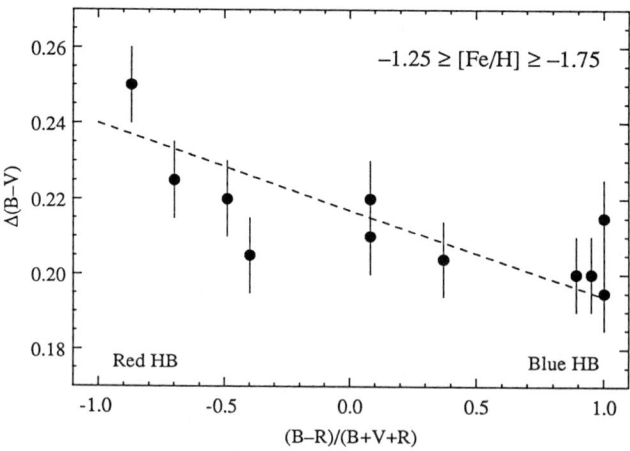

Fig. 4. Age as the second parameter illustrated by plotting Δ(B-V) (age indicator) versus the HB index of Lee et al. (1988). In a narrow metallicity range, clusters with red HBs are in general younger than those with blue HBs.

discrepant HB morphologies of NGC 288/362 is the fact that NGC 288 is older than 362 by ~ 3 Gyrs.

To investigate the second parameter problem further, we have selected clusters from our sample with [Fe/H] between -1.25 and -1.75 so as to minimize the effects of [Fe/H] variations on HB morphology. We then plot the Δ(B-V) values of these clusters as a function of their HB morphologies as measured by the index of Lee et al. (1988). Fig. 4 shows such a comparison along with 2 σ error bars in Δ(B-V) of 0.02 mag. Note that the metallicity sensitivity of Δ(B-V) in this narrow range of [Fe/H] is essentially nonexistent. In addition, the uncertainties introduced due to the variation of oxygen enhancement and helium diffusion with metallicity are minimized, again because of the narrow range in abundance. The dotted line represents the weighted least squares fit to the data. The relation between Δ(B-V) and HB morphology is significant at the > 95% level. Thus, the data for clusters in our sample is consistent with the idea that "age" is the second parameter, as first suggested by Searle and Zinn (1978). On the other hand, if the youngest cluster (NGC 362) is removed from Fig. 4, the significance of the relation drops below the 95% level.

As was mentioned in section 2, the effect of variations in the degree of oxygen enhancement and/or the degree of helium diffusion can essentially mimic the behavior of age variations as measured by Δ(B-V). We now consider these two effects in more detail.

3.1 Oxygen Abundances

It is now fairly well established that the field halo stars and most likely the globular cluster stars are enhanced in oxygen relative to the solar value (Pilachowski, Sneden and Wallerstein 1983, Gratton 1987, Bond and Luck 1988, Pilachowski 1988, Gratton and Ortolani 1989, Abia and Rebolo 1989 and Gratton 1990). However, there is a point of controversy revolving around whether this enhancement increases with decreasing metal abundance. For a fuller discussion of these points, see Sarajedini and King (1989) and Sarajedini and Demarque (1990), but for now, we note that, of the studies just mentioned, only the one by Abia and Rebolo (AR, 1989) finds a trend of increasing [O/Fe] with decreasing [Fe/H]. The remaining studies conclude that the [O/Fe] of the globular clusters is enhanced above the solar value by ~ 0.5 dex with no metallicity trend.

For the sake of argument, let us assume that the conclusions of AR are at least qualitatively correct such that [O/Fe] = 1.0 at [Fe/H] = -2.25 and [O/Fe] = 0.5 at [Fe/H] = -0.65 as assumed by VandenBerg (1985). From Fig. 1 of VandenBerg (1985), it is clear that, at a given age, the change in the color of the turn-off is *identical* at [O/Fe] = 1.0 and [O/Fe] = 0.5. Therefore, if we correct the values of Δ(B-V) for enhancements in oxygen, the relative ages will not be affected.

Finally, Chieffi, Straniero and Salaris (1990, CSS) go so far as to question the need for isochrones which include only enhancements in oxygen. They argue that elements like O, Ne, Mg, Si, S and Ca are all equally enhanced above the solar value. Therefore, any isochrones that scale [O/Fe] above the solar value must also scale [Ne/Fe], [Mg/Fe], [Si/Fe], etc. As a result, an oxygen-enhanced isochrone will simply be shifted to the red relative to a scaled-solar abundance isochrone, as shown in Fig. 1 of CSS. The point that CSS make therefore is that oxygen enhancement has little effect on relative ages derived using $\Delta(B-V)$.

3.2 Helium Diffusion

The study of surface helium abundances in relatively old halo stars is made difficult by the paucity of helium lines in the spectra of these stars. However, it has recently become clear (Deliyannis et al. 1990, Proffitt and Michaud 1990) that the abundance of lithium can be used as a tracer of the helium abundance. Therefore, it is possible to gauge the degree of helium diffusion by studying the surface abundance of lithium. The lithium observations compiled by Deliyannis et al. (1990) show that the observed lithium abundances of field stars reach a plateau for stars with effective temperature $\geq$ 5900 K regardless of their abundance up to [Fe/H] ~ -1.3 (e.g. see Fig. 4 in Deliyannis et al. 1990). This becomes even more striking when stars with [Fe/H] = -3.3 are included (Deliyannis and Demarque 1990a). However, lithium diffusion becomes more pronounced toward the hot end of the plateau (Deliyannis et al. 1990); thus the flatness of the lithium plateau suggests that the degree and therefore the effects of helium diffusion on $\Delta(B-V)$ are relatively small (see SD and Deliyannis and Demarque 1990b). Furthermore, the effects of helium diffusion do not appear to vary significantly with metal abundance as illustrated in Fig. 1 of Deliyannis and Demarque (1990a).

Chaboyer et al. (1990) have recently computed a set of theoretical isochrones which include the effects of helium diffusion. They find that, as expected, helium diffusion does indeed decrease the value of $\Delta(B-V)$; however, the variation with metal abundance is very small. That is to say, helium diffusion is not likely to change the relative ages by more that ~ 0.5 Gyr over a range in [Fe/H] from -2.30 to -1.30.

4. HORIZONTAL BRANCH

In Sarajedini and Lederman (1990), we developed a technique by which the slope of the HB luminosity - metallicity relation was determined. If we denote that slope by δ, then

$$\delta = \frac{\Delta(\Delta V_{Bump}^{HB}) - 0.22\Delta([Fe/H]^2) - 1.50\Delta[Fe/H] - 0.029\Delta t_9}{[Fe/H]_{47\,Tuc} - [Fe/H]}. \quad (1)$$

In this expression for δ, the quantities designated by "Δ" represent values relative to 47 Tuc. For example, $\Delta[Fe/H] \equiv [Fe/H] - [Fe/H]_{47 Tuc}$, and so on. The quantity ΔV(Bump-HB) refers to the difference in V magnitude between the bump in the red giant branch luminosity function and the HB (King, Da Costa and Demarque 1985; Fusi Pecci et al. 1990). Finally, t_9 represents the age of a cluster in units of gigayears as derived from Δ(B-V). We can now use the observational data shown in Table 1 of Sarajedini and Lederman (1990) to compute δ. The mean value in the last column of that table is $<\delta> = 0.15 \pm 0.05$, where 0.05 is the standard deviation of d values around the mean. Any and all errors made in the determination of the quantities that go into equation 1 are reflected in this standard deviation. Our result for the slope of the HB luminosity - metallicity relation is in accord with the results of Lee, Demarque and Zinn (1990, and references therein), Cacciari et al. (1989), Liu and Janes (1990) and Fusi Pecci et al. (1990).

In view of the results of Sandage (1990, and references therein) and Buonnano et al. (1989), we now ask the question: what does our method predict for the relative ages if $\delta = 0.35 \pm 0.05$? In other words, what combination of relative ages will yield a slope of 0.35 with a dispersion of 0.05? Table 1 shows the appropriate relations that must be used and the results for each cluster. We have essentially reversed equation 1 and solved for Δt. The last column of Table 1 shows that the clusters in our sample must on average be ~ 2 Gyrs younger than 47 Tuc if δ is to be 0.35.

The past work on this subject has linked the slope of the HB luminosity - metallicity relation to the existence of an age - metallicity relation among the Galactic globular clusters (Zinn 1986). If the slope is zero, i.e. M_v(RR) = constant, then there exists a strong relation between age and metal abundance; if the slope is ~ 0.20, the relation is still significant but its magnitude has decreased; finally, if the slope is ~ 0.35, the age - metallicity relation no longer exists (see Sarajedini and King 1989). This expected behavior allows a test of the method used herein to derive δ. Fig. 5 illustrates the relation between age and metallicity for those clusters listed in Table 1. The metallicity is plotted on the abscissa and the age relative to 47 Tuc is plotted on the ordinate. The importance of Fig. 5 is that the variation of the age - metallicity relation as described by our technique is the same as what one expects from previous investigations. Therefore, when δ is small, the age - metallicity relation is strong and decreases as δ increases becoming nonexistent when $\delta = 0.27 \pm 0.05$. The verification of this qualitative behavior shows that equation 1 is in fact a robust method by which to determine the slope of the HB luminosity - metallicity relation.

5. SUMMARY AND CONCLUSIONS

We have used the relative age diagnostic of Sarajedini and Demarque (1990) (Δ(B-V)) to study the age spectrum of the Galactic globular clusters and its implications for the luminosity of the

TABLE 1

What if $\delta = 0.35 \pm 0.05$?

Cluster NGC	Name	δ	Equation for Δt if $\delta = 0.35 \pm 0.05$	Δt
104	47 Tuc			
1261		0.38	6.34 - 20.7 δ	-1.5
2808		0.35	6.49 - 22.8 δ	-1.5
5272	M 3	0.35	7.93 - 32.8 δ	-3.6
5904	M 5	0.39	7.07 - 23.8 δ	-2.2
6341	M 92	0.28	10.1 - 49.8 δ	-3.8
6397		0.28	8.88 - 41.3 δ	-2.7
6752		0.42	9.80 - 28.7 δ	-2.3

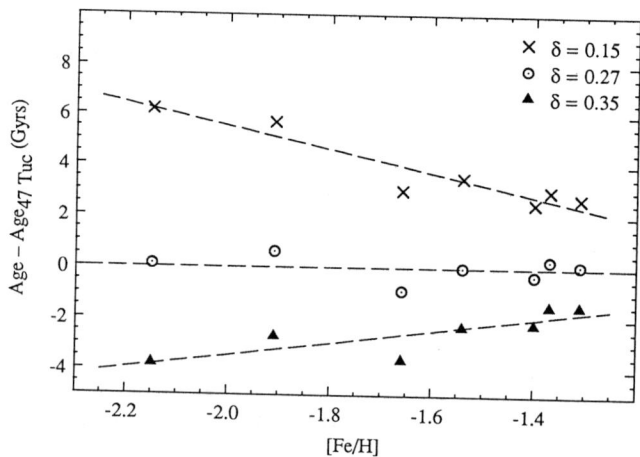

Fig. 5. The relation between age and metallicity for the clusters in Table 1. The age of each cluster is measured relative to that of 47 Tuc. Note the behavior of the age-metallicity relation as the slope of the HB luminosity - metallicity relation (δ) varies from 0.15 to 0.27 to 0.35.

horizontal branch (HB). We find that there appears to be a significant age-metallicity relation among the 22 clusters in our sample, with metal-poor clusters being older than metal-rich ones. The effects of oxygen enhancements and helium diffusion on the relative ages seem to be negligible. In addition, there are strong indications that the age of a cluster is the second parameter that controls the behavior of the HB morphology. Thus, metal abundance is the primary parameter and age is the secondary one.

With regard to the luminosity behavior of the HB, we have used the magnitude difference between the red giant branch clump and the HB along with values of $\Delta(B-V)$ and [Fe/H] to derive the slope of the HB luminosity-metallicity relation. For 8 clusters with such data, we find the slope to be 0.15 ± 0.05.

6. ACKNOWLEDGEMENTS

The author is indebted to C. Deliyannis, P. Demarque, B. Chaboyer and A. Layden for constructive input regarding the contents of this paper. Appreciation is also extended to A. G. D. Philip for inviting me to present this research.

REFERENCES

Abia, C. and Rebolo, R. 1989 ApJ 347, 186 (AR)
Bolte, M. J. 1989 AJ 97, 1688
Bond, H. E. and Luck, R. E. 1988 in IAU Symposium No 132, The Impact of Very High Signal to Noise Spectroscopy on Stellar Physics, G. Cayrel de Strobel and M. Spite, eds., Reidel, Dordrecht, p. 477
Buonanno, R., Buscema, G., Corsi, C. E., Iannicola, G. and Fusi Pecci, F. 1983 A&AS 51, 83
Buonanno, R., Corsi, C. E. and Fusi Pecci, F. 1989 A&A 216, 80
Cacciari, C., Clementini, G. and Buser, R. 1989 A&A 209, 154
Chaboyer, B., Deliyannis, C. P., Demarque, P., Pinsonneault, M. H. and Sarajedini, A. 1990, in preparation
Chieffi, A., Straniero, O. and Salaris, M. 1990 in The Formation and Evolution of Star Clusters, Astron. Soc. Pacific Conf. Series, No. 13, K. Janes, ed., Astron. Soc. Pacific, Chelsea, Mich., p. 219
Deliyannis, C. P. and Demarque, P. 1990a, preprint
Deliyannis, C. P. and Demarque, P. 1990b, in preparation
Deliyannis, C. P., Demarque, P. and Pinsonneault, M. H. 1989 ApJL 347, L 73
Deliyannis, C. P., Demarque, P. and Kawaler, S. D. 1990 ApJS 73, 21
Eggen, O. J., Lynden-Bell, D. and Sandage, A. 1962 ApJ 136, 748
Fusi-Pecci, F., Ferraro, F. R., Crocker, D. A., Rood, R. T. and Buonanno, R. 1990 A&A, in press
Gratton, R. G. 1987 A&A 177, 177
Gratton, R. G. 1990 in IAU Symposium No 145, Evolution of Stars: The Photospheric Abundance Connection, G. Michaud and A. Tutukov, eds., Kluwer, Dordrecht
Gratton, R. G. and Ortolani, S. 1989 A&A 211, 41

Green, E. M., Demarque, P. and King, C. R. 1987 The Revised Yale Isochrones and Luminosity Functions, Yale University Observatory, New Haven
Green, E. M. and Norris, J. E. 1990 ApJL 353, L 17
Hesser, J. E. 1988 in IAU Symposium No. 126, Globular Cluster Systems in Galaxies, J. E. Grindlay and A. G. D. Philip, eds., Kluwer, Dordrecht, p. 61
King, C. R., Da Costa, G. S. and Demarque, P. 1985 ApJ 299, 674
Lee, Y. -W. 1990 in The Formation and Evolution of Star Clusters, Astron. Soc. Pacific Conf. Series, No. 13, K. Janes, ed., Astron. Soc. Pacific, Chelsea, Mich., p. 205
Lee, Y. -W., Demarque, P. and Zinn, R. J. 1988 in Calibration of Stellar Ages, A. G. D. Philip, ed., L. Davis Press, Schenectady p. 149
Lee, Y. -W., Demarque, P. and Zinn, R. J. 1990 ApJ 350, 155
Liu, T. and Janes, K. A. 1990 ApJ 354, 273
Lupton, R. H. and Gunn, J. E. 1986 AJ 91, 317
Massey, P., Garmany, C. D., Silkey, M. and DeGioia-Eastwood, K. 1989 AJ 97, 107
Penny, A. J. and Dickens, R. J. 1986 MNRAS 220, 845
Pilachowski, C. A. 1988 ApJL 326, L 57
Pilachowski, C. A., Sneden, C. and Wallerstein, G. 1983 ApJS 52, 241
Proffitt, C. R. and Michaud, G. 1990 ApJ, in press
Racine, R. 1990, private communication
Renzini, A. 1977 in Advanced Stages of Stellar Evolution, 7th Course of the Swiss Society of Astronomy and Astrophysics, Saas-Fee, P. Bouvier and A. Maeder, eds., Geneva Observatory Publication, p. 149
Sandage, A. 1990 ApJ 350, 631
Sarajedini, A. and King, C. R. 1989 AJ 98, 1624
Sarajedini, A. and Demarque, P. 1990 ApJ, Dec. 10th issue (SD)
Sarajedini, A. and Lederman, A. 1990 in The Formation and Evolution of Star Clusters, Astron. Soc. Pacific Conf. Series, No. 13, K. Janes, ed., Astron. Soc. Pacific, Chelsea, Mich., p. 293
Sarajedini, A. and Da Costa, G. S. 1990, in preparation
Searle, L. and Zinn, R. J. 1978 ApJ 225, 357
Simoda, M. and Iben, I. 1968 ApJ 152, 509
Stetson, P. B. 1987 PASP 99, 191
Stetson, P. B. 1990a The Techniques of Least Squares and Stellar Photometry with CCDs, Lectures presented at V Escola Avancada de Astrofisica, preprint
Stetson, P. B. 1990b, preprint
Stetson, P. B. and Harris, W. E. 1988 AJ 96, 909
Stetson, P. B., VandenBerg, D. A., Bolte, M. J., Hesser, J. E. and Smith, G. H. 1989 AJ 97, 1360
Straniero, O. and Chieffi, A. 1990 ApJS, in press (SC)
Stringfellow, G. S., Bodenheimer, P., Noerdlinger, P. D. and Arigo, R. J. 1983 ApJ 264, 228
VandenBerg, D. A. 1985 in Production and Distribution of C, N, O Elements, I. J. Danziger, F. Matteucci and K. Kjar. eds., Garching bei Munchen, p. 73

VandenBerg, D. A., Bolte, M. J. and Stetson, P. B. 1990 AJ 100, 445
    VBS
West, R. and Kruszewski, A. 1981 Irish AJ 15, 25
Zinn, R. J. 1986 in Stellar Populations, Space Telescope Science
    Institute Symposium Series, C. A. Norman, A. Renzini and M. Tosi,
    eds., Cambridge University, Cambridge, p. 73
Zinn, R. J. and West, M. J. 1984 ApJS 55, 45

DISCUSSION

RICH: What do you get if you plot $\Delta(B - V)$ vs $\Delta V(Bump - TO)$?

SARAJEDINI: Since $\Delta(B - V)$ shows a dependence on observed metallicity (i.e. and age-metallicity relation) and $\Delta V(Bump-TO)$ is probably related to metallicity also. I would imagine that $\Delta(B - V)$ and $\Delta V(Bump - TO)$ should show a relation.

RICH: What about looking at the difference between the luminosity of the TO and the luminosity of the RGB bump?

SARAJEDINI: As you mentioned, this difference $L(TO) - L(Bump)$ is an interesting value to look at.

RATNATUNGA: How big is the current sample of clusters being used for this anaysis and the prospects of expanding it to a larger sample?

SARAJEDINI: We have 22 clusters with measured values of $\Delta(B - V)$. We hope to increase this to about 30 using observations. I have obtained other data that will be published in the interim.

ON THE COLOR AND POPULATION GRADIENTS IN THE CENTRAL-CUSP GLOBULAR CLUSTER M 15

Peter B. Stetson

Dominion Astrophysical Observatory
Herzberg Institute of Astrophysics
National Research Council of Canada

ABSTRACT: I present preliminary color-magnitude diagrams and integrated color profiles for the innermost two arcminutes of the central-cusp globular cluster M 15. These data will be used to investigate the nature and the cause of the color gradients reported for this and similar clusters.

1. INTRODUCTION

Here is a review of some of the discoveries that have been made in the last two years concerning the central-cusp[1] globular clusters. This is not meant to be an exhaustive list of every paper that has been written about these objects, or of everyone who has studied them; I intend only to remind you of important discoveries which will be relevant to my discussion later, while acknowledging some of the most active contributors to the field. The items are not in chronological order, but rather in a sequence which seems logical to me.

• M 15 is bluer in the center in the (U-B) color index[2]. More specifically, the average color of the unresolved faint cluster stars is bluer within $6\rlap{.}''6$ of the cluster center than at a radius of 15". (Bailyn et al. 1989)

• M 15 is also bluer in the center in (B-V), and furthermore it is also *fainter* in Hα (as compared to R) at the center. (Cederbloom et al. 1990)

[1]This phrase is assumed by many dynamicists to be equivalent to "post-core-collapse." My personal preference is to avoid this nomenclature, in the belief that classification schemes should be based upon observed, not inferred, properties.

[2]When I speak of *bluer* in reference to the (U-B) index, I mean "brighter at shorter wavelengths", not "brighter in B".

A. G. Davis Philip, A. R.
Upgren and K. A. Janes (eds.)
PRECISION PHOTOMETRY
69 - 88 © 1991
L. Davis Press

Peter B. Stetson
Dominion Astrophysical Obs.
5071 W. Saanich Rd.
Victoria, BC
V8X 4M6 Canada

- Similar blueward inward trends in (B-V), (V-R), (B-I), etc. are seen in other central-cusp globular clusters: M 30, NGC 4147 (a possible, not definite, member of the class), NGC 6397, NGC 6624. (Djorgovski et al. 1988)

- Giants are scarce in the center of M 30, as normalized either to the blue HB stars or to the total light. (Piotto et al. 1988)

- There is an excess of bright blue stragglers at the very center of NGC 6397. (Auriére et al. 1990)

When they first discovered the (U-B) gradient, Bailyn et al. postulated that it was produced by a centrally concentrated population of cataclysmic variables. Observations in a narrow Hα filter were then undertaken to test this hypothesis by searching for emission-line objects. Contrary to expectation, these data showed that the center of M 15 was *deficient* in Hα flux compared to the outskirts, which strongly implied that the centrally concentrated population must be dominated by stars of intermediate spectral type - somewhere between late B and early F. This conclusion would be consistent with the gradients in (B-V), (V-R), etc., seen in M 15 and other central-cusp clusters. Auriére et al.'s CM-Diagram for the inner regions of the cusp cluster NGC 6397 actually displayed an excess of bright blue stragglers right at the cluster center. Djorgovski et al. (1990a, b) were able to generalize these conclusions to a larger set of clusters: blueward inward color gradients are found in all nine of the central-cusp clusters that have been searched to date (including the marginal member of the class, NGC 4147), but not in any of the non-cusp, "King-model" clusters. Likewise, wherever the data allow such an analysis, the cusp clusters show a central deficiency of bright giants - and sometimes of faint giants as well - compared to the underlying cluster light. The jury is still out on whether the HB stars are likewise depleted toward the center, but Djorgovski et al. (1990a, b) do find blue stragglers near the center of NGC 4147 and suspect them near the center of M 30.

So maybe, everything is starting to come together. Certainly, it is not yet known for sure whether blue stragglers are single stars that have prolonged their main-sequence lifetimes by internal mixing, low-mass stars that have recently been promoted to higher masses by binary mass-exchange, coalesced stars, or something else. In fact, several mechanisms might be operating in parallel to produce blue stragglers in the cores of globular clusters. Nevertheless, whatever the truth eventually turns out to be, all the theoretical expectations and observational indications are that blue stragglers have masses consistent with their positions on the main sequence. Thus, they are more massive than the other visible cluster dwarfs, giants and HB stars. Central-cusp clusters have evidently undergone a great deal of dynamical evolution, so it's not surprising that the most massive visible stars, the blue stragglers, are concentrated to the center. With a special population of bright main sequence stars only in the center of the cluster, the ratio of giant stars to total cluster light might well decline toward the center. The only fly in this ointment is the observation of Piotto et al. (1988), that in M 30 the *HB stars* - which in this cluster are predominantly blue - are more centrally concentrated than the giants. Similar results are reported for NGC 6293, 6397 and maybe 4147 by Djorgovski et al. (1990b). Since everyone believes that HB stars are in a *post*-GB phase of evolution and are, if anything, less massive than the giants, these observations are hard to reconcile with mass segregation. However, they are consistent with the spectroscopic work of Rose et al. (1987), who found that the integrated spectrum of the center of M 30 was much richer in the light of A-type stars than the center of M 15; Rose et al. concluded that the effect could be produced *either* (1) by an excess of blue stragglers *or* (2) by a deficiency of red giants *plus* a shift of the HB from the blue toward the red in the center of M 30, as compared to both the outskirts of M 30 and the center of M 15. Rose et al. found a similar, though weaker, effect in the marginal central-cusp cluster, NGC 4147, again consistent with later imaging data. I think that until recently it was unclear whether the giant-deficient cores of M 30 and NGC 4147 are due to some unanticipated physical process or to a bizarre statistical fluke (cf. Section VII of Rose et al., and the abstract of Piotto et al. (1988) as well as their Section V). Recent discovery of the same effect in other - in some cases larger - clusters has made it seem more real. In an attempt to further clarify the population gradients in central-cusp clusters, I undertook to obtain the deepest possible CM-Diagram for the center of the class prototype, M 15.

2. COLOR-MAGNITUDE DIAGRAMS

The observational material which has been reduced so far for this project is detailed in Table 1. All these CCD frames were obtained on Mauna Kea. The first data set was obtained in 1986 with the IfA/Galileo 500^2 CCD on the University of Hawaii 2.2-m telescope. The second set was obtained in 1986 with the standard prime-focus camera ("FOCAM") on the Canada-France-Hawaii Telescope ("CFHT") and a 640 x 1024 double-density RCA CCD. The last data set was obtained in 1989 with the new DAO/CFHT high-resolution camera ("HRCam") and a different double-density RCA chip. I have also listed co-investigators for the various observing runs; these people are all carrying out their own analyses of various parts of the data, and in most cases they were involved in the actual observations.

All data for each bandpass were combined into a single instrumental system by derivation of frame-to-frame zero-point differences - differential color terms were ignored, which should be adequate at the level of precision required here. The three bandpasses were then intercompared to produce a {v, (b-r)} instrumental color-magnitude diagram. The (b-r) color index has two advantages. First and most important, my (b-r) index is roughly 1.4 times more sensitive to temperature than (B-V); thus, for distinguishing the color difference between two stars, it is as effective as doubling the number of observations in b and v. Second, the use of (b-r) makes the abscissa of the color-magnitude diagram statistically independent of the ordinate: in the standard (V, (B-V)) CM-Diagram random errors in the V magnitude produce an inclined error ellipse which can sometimes fool the eye near the magnitude limit.

TABLE 1

Observational Material

Telescope & Detector	Frames & Filter	Seeing (")	With ...
UH 2.2-m TI 500^2	8 b 8 v	0.81 - 1.24 0.83 - 1.02	J. Rose
CFHT FOCAM RCA2	10 b 8 v 14 r	0.65 - 0.97 0.58 - 1.00 0.54 - 1.32	H. Cohn J. Grindlay J. Hesser P. Luggar
CFHT HRCam RCA4	2 b 2 v 2 r	0.42 - 0.53 0.42 - 0.52 0.39 - 0.44	The above + C. Bailyn R. McClure

Fig. 1 shows my derived instrumental {v, (b-r)} CM-Diagram for the central 121" (radius) of M 15. Note that 121" represents the distance from cluster center of the last star in the farthest corner of the most off-center CCD frame - the sample is not by any means complete out to 121". In fact, I have complete areal coverage out only to a radius of some 80", and the magnitude limit depends somewhat upon radius even within this distance. However the magnitude limit is well below the HB at all radii.

Globular-cluster cognoscenti may be somewhat surprised by the CM-Diagram presented in Fig. 1. M 15 is a classic very-metal-poor cluster, and in all previous work its HB stars have seemed strongly concentrated toward the blue end. For instance, Fig. 2 shows a plot to the same scale of the excellent photographic data of Buonanno et al. (1983), for an annular region with inner and outer radii of 1!9 and 5!0 - hence, almost completely outside my own region. I have fleshed out the instability strip by plotting, as triangles, mean colors and apparent magnitudes for RR Lyraes in the same annular region, taken from the paper by Bingham et al. (1984). As you can see, the gross appearance of the HB is quite different inside and outside of 2', with Buonanno et al. and Bingham et al.'s outer-region data showing a much weaker red HB component; what red stars are present look more like the base of the AGB than like the red end of the HB. The absence of a red

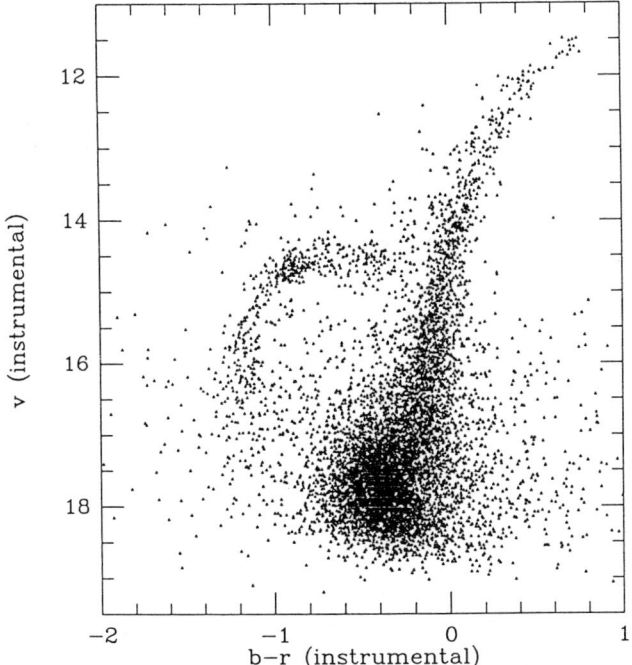

Fig. 1. Instrumental (v,(b-r)) color-magnitude diagram for the central two arcminutes of M 15, derived from the CCD images listed in Table 1.

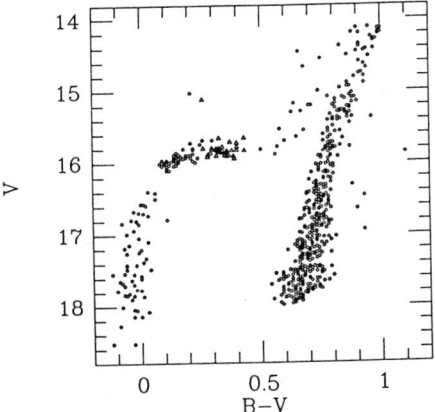

Fig. 2. Color-magnitude diagram for the horizontal-branch/giant-branch region of M 15, for an annulus limited by radii of 1'.9 and 5'.0, from the data of Buonanno et al. 1983 (open circles) and the RR Lyrae data of Bingham et al. 1984 (small triangles).

HB in M 15 is seen not only in the Buonanno et al. data: the early international-system CM-Diagram for this cluster by Arp 1955 closely resembles the Buonanno et al. diagram. To be sure, the HB sample of Sandage (1969) does appear to contain a rather higher proportion of stars on the red end than the other studies do, but Sandage chose his stars to define the limits of the instability strip; hence, they may not be a statistically representative sample. In any case, Sandage himself believed at the time that these stars represented the base of the asymptotic GB and *not* a red HB, as Fig. 11 of Sandage (1970) attests.

Is this apparent gradient in the red:blue ratio on the HB statistically significant? If it is, does the ratio continue to grow *inside* 2'? Is it accompanied by a gradient in the GB:HB ratio, as Piotto et al. found for M 30? To address these questions I examined the radial distribution of stars in the magnitude range $15.40 \leq V \leq 16.20$ (in the Buonanno et al., Bingham et al. data; this corresponds to $14.16 \leq v \leq 14.96$ in my instrumental system). The faint limit was chosen to pass through the gap which separates M 15's classical BHB from its diffuse, extended BHB; this magnitude level should be well above my completeness limit at all radii. I then counted stars in each of three color boxes, delimited by (B-V) = 0.05, 0.35, 0.65, 0.95 ((b-r) = -1.06, -0.64, -0.22, +0.21), which contain the conventional BHB plus part of the instability strip, the red HB plus the rest of the instability strip, and the GB at the level of the HB. The counts for the Buonanno et al. plus Bingham et al. data and for mine are shown in the first two lines of Table 2. The ratio of red:blue HB stars is roughly 1:4 in the outer part of the cluster, and roughly 1:2 in the inner part; these results differ at about a 2.7 σ significance level. On the other hand, there is no evidence at all for a change in the ratio of giant stars to (blue + red) HB stars.

Does the ratio continue to grow inside 2'? If I consider only stars with instrumental v magnitudes brighter than 15.0, a sample which should be essentially complete at all radii, I find that the median clusterocentric distance of my stars is 30". If I then divide my sample into halves at this radius, I come up with the diagrams shown in Fig. 3a and b and the star counts in lines 3 and 4 of Table 2. Although the CM-Diagrams suggest that in the center of the cluster the RHB may be shifting back away from the GB and toward the instability strip, when the stars are counted the net trend is still in the direction of a larger fraction of red HB stars in the central zone. However, in this case the difference is not statistically significant. These data do show a marginal (1.25 σ) difference in the ratio of GB to HB stars, in the sense that giants avoid the center in comparison with the HB stars - a trend that is in the same sense as that found in M 30 by Piotto et al., and in other central-cusp clusters by Djorgovski et al. (1990a, b). But since the difference seen here is so weak and does not continue outward beyond 2', it does not deserve touting.

TABLE 2

Star Counts on the Horizontal Branch and Giant Branch of M 15.

Radius	BHB	RHB	GB	RHB:BHB	GB:HB
1.″9 - 5.′0*	54 + 18	5 + 15	62 + 0	0.28 ± 0.07	0.67 ± 0.11
0″ - 121″	195	103	217	0.53 ± 0.06	0.73 ± 0.06
30″ - 121″	98	49	119	0.50 ± 0.09	0.81 ± 0.10
0″ - 30″	97	54	98	0.56 ± 0.09	0.65 ± 0.08
13″ - 121″	154	73	173	0.47 ± 0.07	0.76 ± 0.08
0″ - 13″	41	30	44	0.73 ± 0.18	0.62 ± 0.08

*Data taken from Buonanno et al. 1983 and Bingham et al. 1984.

I will soon present a datum which will suggest that M 15's population inside a radius of 13″ may differ from that farther out. When the sample is divided at *this* radius, I obtain the CM-Diagrams shown in Fig. 3c and d and the star counts given in lines 5 and 6 of Table 2. Again, there is a strong trend for the ratio of red:blue HB stars to increase, and a weak trend for the ratio of GB to (blue + red) HB stars to decrease toward the cluster center. However, examination of the scatter in Fig. 3d suggests that at v ~ 15 in the central 13″ of M 15, the photometric scatter is starting to become significant. Perhaps the apparent population of RHB stars is becoming spuriously augmented by stars scattered in from the more densely populated GB and BHB. I myself think that this is not the case, because there is still a clear gap between the GB and the RHB, and the scatter does not seem sufficient to nearly equalize the number of blue and red HB stars. I accept that this point may not yet be proven. So, in summary I would say that there is convincing evidence that the HB morphology inside 2′ is different from that farther out, with relatively more red HB stars in the inner zone. There is suggestive evidence that the relative frequency of red HB stars grows still further in the very inner part of the cluster. There is no convincing evidence for a central depletion of the giant stars at the magnitude level of the HB.

Fig. 3c and d also suggest that there may be a trend for the cluster center to contain a deficiency of stars on the extended blue HB, and an excess of blue and yellow stragglers - defined for the moment as any star above the turnoff, to the blue of the subGB, and off the HB. However, the data for stars this faint are subject to photometric errors and incompleteness that will be a strong function of radius and color. Likewise, the very center of the cluster appears to have an excess of supra-HB stars. However, it's conceivable that these stars appear anomalously bright because they are actually unresolved

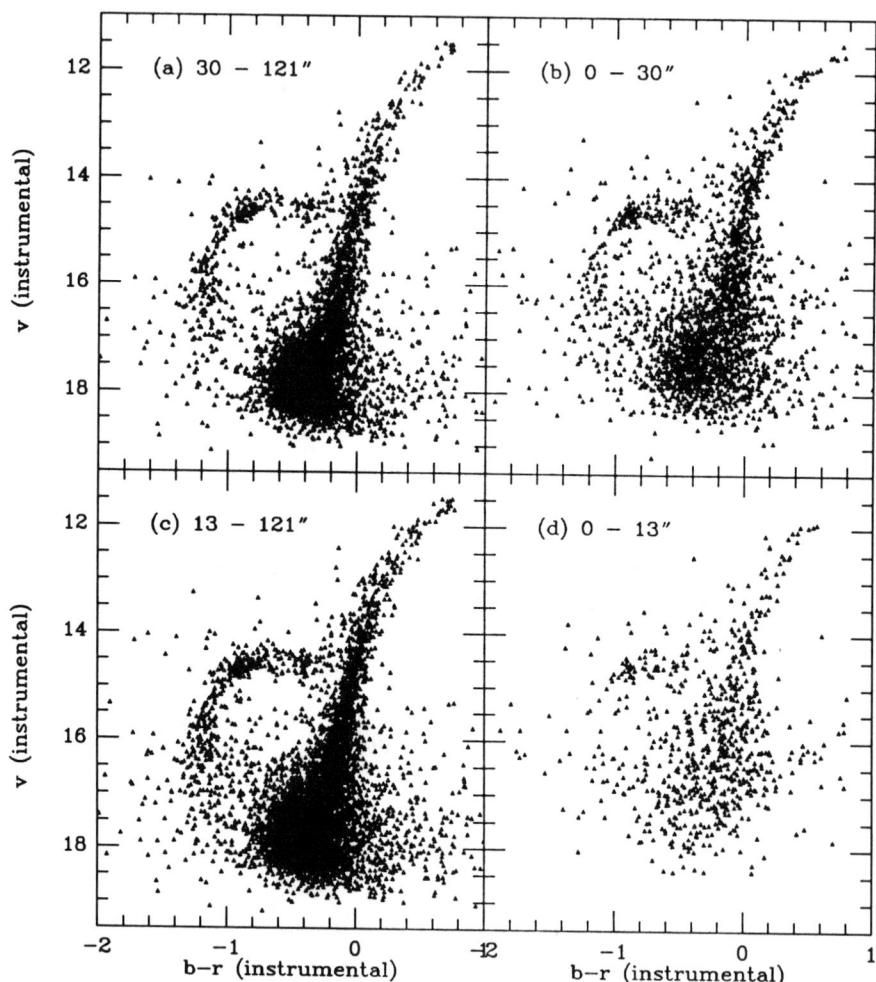

Fig. 3. The CCD color-magnitude data for the M 15 stars, divided into several radial zones: (a) 30" < radius < 121"; (b) 0" < radius < 30"; (c) 13" < radius < 121"; and (d) 0" < radius < 13".

blends of two or more images which, of course, will be far more common in the crowded cluster center. The reality of each of these apparent trends can be determined by artificial-star tests, which I have not yet carried out, or by HST observations. Until one of these happens, I am not prepared to make any definitive statements about the supra-HB stars or the stars below the level of the flat part of the HB. Still, if a deficiency of extended-BHB stars and excesses of blue stragglers and supra-HB stars *are* eventually confirmed in the center of M 15, remember: you heard it here first. Since the stars involved in analysis of the RHB:BHB and GB:HB ratios were all of the same apparent magnitude, I doubt that radius-dependent incompleteness could produce the population trends documented in Table 2.

Close examination of Fig. 3c and d reveals another potentially interesting anomaly: in the outer part of the cluster, the GB tip is half a magnitude brighter than it is in the center: outside a radius of 13" it reaches to v ~ 11.5, inside it only goes to v ~ 12. Is this real? To look into this question, I considered all stars in my data with instrumental v < 15, and sorted them by radius. I then plotted their apparent magnitudes as a function of radial *rank* - their sequential numbers in order from the center of the cluster out. This stratagem was adopted order to produce a plot where stars are uniformly distributed along the abscissa. If there is no change in the stellar population with radius, stars of *all* magnitudes will be evenly distributed across the diagram, while an absence of stars in any particular radial zone will produce an obvious void in the figure. The missing GB-tip stars are represented by the empty rectangle in the upper left corner of the plot. Outside a radius of 13", 17 stars out of the 694 brighter than v = 15 are within a half magnitude or so of the cluster's GB tip. Inside 13" there are 218 stars brighter than v = 15; by comparison with the outer zone, 5.3 of them should be in the top half magnitude. In fact, none are. Of course, this result is statistically corrupt, because I drew in the boundaries *after* I decided what I wanted to test. However, I have not been completely arbitrary because, after all, Piotto et al. (1988) and Djorgovski et al. (1990a, b) have found that giants avoid the center of M 30 and several other cusp clusters. If my result were a fluke it could just as easily have gone the other way - I could have found 10 bright giants in M 15's center, and in fact I didn't. Likewise, if the stars that avoided the center of the cluster produced a gap in the *middle* of the GB, or if the bright giants were missing from an *intermediate* radial zone, it would seem much more reasonable to call it an accident. I think that seeing the very brightest giants avoid the very center of the cluster is special enough that it is not completely *arbitrary*, whether you think it is *real* or not. If you look at Fig. 4, you might have the feeling that in the upper left corner where no stars are found, there should be maybe four or so. The Poisson probability of observing zero objects where four are expected is 0.018.

I can also make other not entirely arbitrary divisions of the data. There are 14 stars in the distinct clump that stands separate at the tip of the GB in Fig. 3c; these stars have $11.49 \leq v \leq 11.77$, i.e. one star per 0.02 mag, then there is a tenth of a magnitude gap between the 14th brightest star and the 15th, and from there on it averages ~ 0.01 mag per star. If I divide my sample at the median radius of 30", I find that five of these 14 stars are in the inner half of the sample, and nine are in outer half. The binomial probability of taking 14 objects and getting a split at least as asymmetric as 5:9 on even odds is 0.21. If I instead consider the brightest 20 stars in the cluster (a number adopted because it is the largest whose binomial probabilities are given in the CRC Standard Mathematical Tables), I find these divide into seven in the inner half and 13 in the outer half - a 7:13 or greater split has a probability of 0.13. Thus, I think there is evidence that the GB tip stars avoid the center of M 15, as compared with the average distribution of all stars on and above the flat part of the HB. The significance level of this finding is somewhere in the range 80 - 90 - 98%, depending on how badly corrupted by the expected result you think my statistical tests are. Unlike the situation in M 30 (Piotto et al. 1988) but like the situation in NGC 6397 (Djorgovski et al. 1990b), the central deficiency does *not* seem to extend to the fainter giants (cf. Table 2 and Fig. 4).

Discerning viewers may see in Fig. 4 a trend for stars with v ~ 14 to be concentrated toward the cluster center. This trend is in fact statistically significant at about the 2 σ level, but since it might be caused by optical doubles and the increase in random photometric errors, as discussed above, I'm not yet prepared to assert its reality.

3. SURFACE PHOTOMETRY

To investigate further the nature of the color and population gradients in M 15, I then performed simple aperture photoetry for annuli centered on the cluster's cusp (defined for this purpose as the centroid of the prominent, small equilateral triangle of bright stars near the cluster center). For each bandpass, I selected the five deepest FOCAM frames and used the geometric and photometric offsets derived during the construction of the CM-Diagrams to transform them to a common system, and averaged them together. I also used the photometry derived above to *subtract* all stars brighter than v = 15 from each of the original frames, and averaged these star-subtracted frames together to represent the distribution of the light of fainter stars in the cluster. The same was done with the available HRCam images. The radial distributions of the visual light of all M 15 stars, and of only those stars fainter than v = 15, are shown as the upper and lower sets of circles in Fig. 5. In each case, the closed circles are for the FOCAM data (which extend to larger radii), and the open circles are for the HRCam data. The v magnitude difference between the total light and the light of the faint stars alone, averaged over the radius range $2" \leq r \leq 40$ sec, is 0.30 mag, which implies that in this annular zone ~ 75% of the visual light comes from *below* the flat part of the HB. The corresponding figures for b and r are 0.25 mag (~ 80%) and 0.31 mag (~ 75%), respectively.

ON THE COLOR AND POPULATION GRADIENTS IN M 15

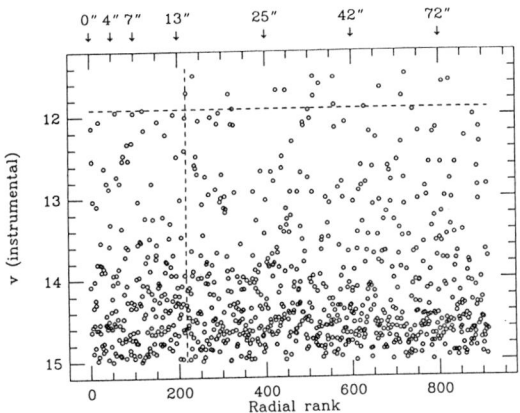

Fig. 4. The apparent, instrumental v brightness for stars in the central 2' of M 15, plotted as a function of radial rank, or percentile. A scale of arcseconds is given at the top. The dashed lines delimit the area of the diagram not occupied by bright giants at the center of the cluster.

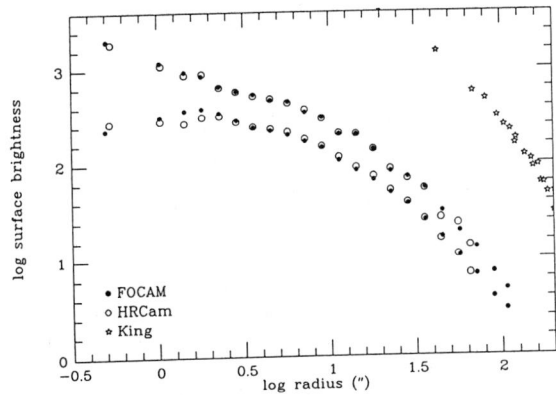

Fig. 5. Apparent, instrumental v brightness profiles for the central 2' of M 15. The upper sequence of open and closed circles is for the total integrated light of all cluster stars; the lower sequence is for the light which remains when all stars at and above the level of the flat portion of the horizontal branch have been subtracted. In each case, the closed circles are derived from the larger area, lower resolution FOCAM data, and the open circles are for the smaller area, higher resolution HRCam data. The stars at the upper right represent the photographic star counts of King et al. (1968).

Of course, before I could produce these light profiles, it was necessary to determine and subtract the diffuse sky brightness. In their analysis of some of these same data, Bailyn et al. (1989) determined a working sky value from the four corners of the frame: in each corner (~ 2' from cluster center), the median observed brightness value was determined from a box 50 pixels square. The least of these four medians was then adopted as the sky brightness for the frame. However, it remains true that the stars in each of these boxes will have skewed the pixel-brightness distribution toward higher values and, furthermore, two' arcminutes is not really far enough out to avoid entirely the light of unresolved cluster stars. Consequently, I think that it is still necessary to regard Bailyn et al.'s method as producing a firm upper limit to the diffuse sky brightness, rather than an unbiased estimate. I therefore attempted to determine the sky brightness for the FOCAM data in several other ways.

First, I made the assumption that, on average, stars brighter than v = 15 should have about the same radial distribution as the integrated light of fainter stars: the light of the stars fainter than v = 15 will be dominated by subgiants and main-sequence turnoff stars, which have masses very similar to the visible stars in more advanced stages of evolution. The length of time spent on the GB and HB is short enough that any mass lost in these phases should not have time to reveal itself in radial stratification. Thus, the profile representing the light of all M 15 stars and that representing just the light of stars fainter than v = 15 should be asymptotically parallel at the outer edge of the available data. There is only one sky value which will make this so.

Second, I assumed that the integrated light of M 15's stars - both that of all stars considered together and that of the fainter stars alone - should be parallel to the star counts of King et al. 1968; these counts are represented by the star symbols in Fig. 5. Again, for a given observed light profile, there is only one presumed sky value which will enforce this parallelism.

Third, once you believe that you have determined the correct sky brightness in one of the three photometric bandpasses, an *incorrect* sky value in one of the other bandpasses will produce a color gradient that diverges exponentially at the outer rim of the available data. I therefore made minor adjustments to the adopted sky values in b, v and r to produce zero net color gradients in the outermost couple of data points. The various sky values resulting from this effort are presented in Table 3a. As expected, the apparently best sky values are slightly smaller than the lowest of the four median values from the frame corners. Once these cluster-brightness zero points had been determined for the FOCAM data, sky values for the HRCam data were determined by requiring them to parallel the FOCAM curves at their own outermost points, both for all stars taken together and for the fainter stars alone; these results are given in Table 3b.

TABLE 3a

Sky Brightness Determinations - FOCAM Data

Filter	Median in Corner	Total Light vs Minus bright stars	Total Light vs King	Minus bright stars vs King	Zero color gradient - Adopted
b	6	5.8	1.2	5.6	3.8
v	3	3.4	1.5	2.7	2.5
r	2	1.9	0.7	1.4	1.3

TABLE 3b

Sky Brightness Determinations - HRCam Data

Filter	Total Light vs FOCAM Data	Minus Bright stars vs FOCAM Data	Adopted
b	13.3	12.2	12.5
v	13.9	11.7	12.0
r	-4.4	8.2	8.0

The total-light profiles derived from the FOCAM and HRCam data agree very well. However, there is a bug in the procedure which I used to transform each star's mean apparent magnitude, as determined from all available frames, back to the photometric system of each individual frame. This reverse transformation from the mean observed magnitude back to the instrumental magnitudes which *should* have been observed in the individual frames is required for two reasons: it ensures that exactly the same stars are subtracted from all frames, and it permits each star to be subtracted using the very best available estimate of its apparent brightness (except RR Lyraes, but any errors for them should cancel out to reasonable approximation when the star-subtracted frames are averaged). As a result of this bug, there has been some oversubtraction of the bright stars in the central 3" of the FOCAM data and in the central 1" of the HRCam data, as may be seen in the central dips in the lower profiles in Fig. 5. This must be kept in mind as the color curves below are considered.

Fig. 6a shows the observed mean annular values of instrumental (b-r) color as a function of log(radius) for the total integrated light of all M 15 stars. As before, closed circles represent FOCAM data and open circles represent HRCam data. The error bars are derived directly from the observed values of σ_b and σ_r in each annulus. These error bars and the point-to-point wiggles in the mean color curves are both comparatively large because of the stochastic distribution of the brightest cluster stars. It is clear, however, that the innermost arcsecond of the cluster, containing the three bright stars AC 214, 215 and 216 (Auriére et al. 1984) as well as an anonymous fuzzy object or blend inside the triangle, is redder than its immediate surroundings. Outside that central arcsecond there is a definite blueward inward color gradient. FOCAM and HRCam data parallel each other very well. Fig. 6b shows as solid curves the color profiles that would be obtained if changes of ± 1 ADU were made in the adopted sky brightness of the r frames - the more sensitive of the two bandpasses to a sky error. The dashed curve is my eyeball estimate of the net color gradient, +0.10 mag·dex^{-1}. Note that this perceived gradient extends more or less continuously from a radius of ~ 1" to > 30" - the change in stellar population, whatever it is, is not confined to the central dozen arcseconds, but rather seems to occur gradually out to the limit of the available data. While a change in one of the adopted sky values might alter or erase the perceived color gradient in the cluster outskirts, inside of 16" or so the observed effect is many times larger than any plausible sky error could produce.

Fig. 7 shows the color gradients that remain after stars brighter than v = 15 have been subtracted. The error bars here are much smaller than in Fig. 6 because, without the small-number statistics of the bright stars, M 15's light is much more smoothly distributed. I should reiterate that these star-subtracted data are corrupt inside a radius of 3" for the FOCAM data, and inside ~ 1" for the HRCam data. Outside this region, again a blueward inward color gradient is seen. However, in this case it is just half as strong as it was in the data for all stars taken together: here it is ~ +0.05 mag·dex^{-1}.

4. DISCUSSION

What do we now know about population gradients in M 15 and, by induction, in central-cusp clusters in general? Let me summarize the available factoids.

• Red HB stars are rare in M 15 outside 2' (Buonanno et al. and Bingham et al.), but not too rare inside 2'.

• The brightest giants *may* be missing from the center, but there is no evidence for an *overall* central depletion of giants relative to the HB (unlike M 30: Piotto et al. 1988; like NGC 6397: Djorgovski et al. 1990b).

• A blueward inward gradient in M 15 is confirmed. It is *not* confined

ON THE COLOR AND POPULATION GRADIENTS IN M 15

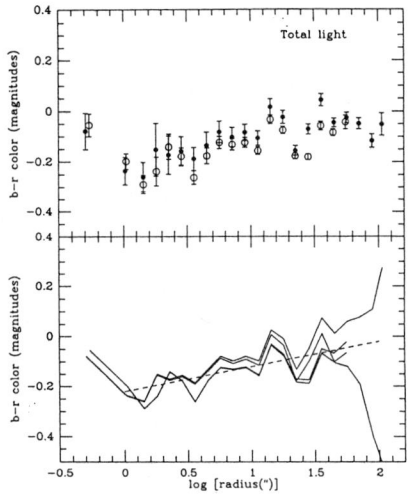

Fig. 6. Instrumental (b-r) color profile for M 15's total integrated light. (a) Individual radial data points and their standard errors based on (closed circles) the FOCAM images, and (open circles) the HRCam images. (b) The color profiles that would result if the adopted sky value in the r frames were wrong by $\pm$ 1 ADU in each set of data. The dashed curve represents a smooth gradient of +0.10 mag·dex^{-1}.

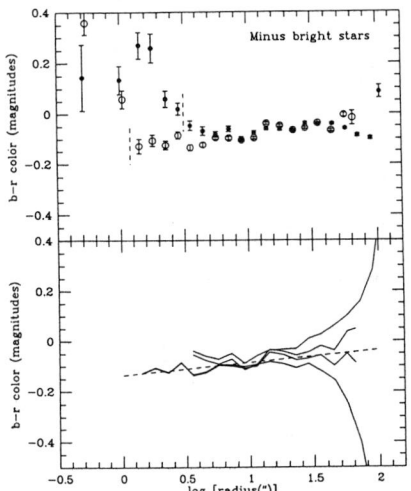

FIG. 7. Instrumental (b-r) color profile for M 15's integrated light, after all stars on and above the flat part of the horizontal branch ($v \leq 15$) have been subtracted. (a) Individual radial data points and their standard errors based on (closed circles) the FOCAM images, and (open circles) the HRCam images. Data points to the left of the vertical dashed lines were corrupted by incorrect subtraction of the bright stars within 3" of the cluster center in FOCAM images, and within ~ 1" in the HRCam images. (b) The color profiles that would result if the adopted sky value in the r frames were wrong by $\pm$ 1 ADU in each set of data. The corrupt inner points have been omitted. The dashed curve represents a smooth gradient of +0.05 mag·dex^{-1}.

to the innermost 6", but rather it extends ~ linearly with log(radius) from ~ 1" to ~ 1' (as in M 30: Piotto et al. as in many cusp clusters: Djorgovski et al. 1990a, b).

• The color gradient is, if anything, *stronger* in the integrated light of *all* stars than in the light of faint stars alone (as in M 30: Djorgovski et al. 1988).

• Recall: M 15 also shows gradients in (U-B) and Hα (Bailyn et al. 1989, Cederbloom et al. 1990): (U-B) bluer in the middle; Hα absorption stronger in the middle.

A centrally concentrated population of blue stragglers can explain some of these effects. If blue stragglers are produced by mass exchange in binaries, they can have masses up to somewhat less than twice the turnoff mass; if they are coalesced stars, they can have masses up to twice the turnoff mass; while if they are internally mixed stars, their masses could even be more than twice the turnoff mass. These stars would be concentrated to the center because that is where they are more easily produced by multiple-body encounters and collisions, and/or because of dynamical mass segregation. The other implicit assumption is that while on the main sequence these stars have luminosities and surface temperatures appropriate to their (current) masses. These assumptions taken together would imply a centrally concentrated population of A and F stars, which would produce stronger Hα absorption and bluer colors in (B-V), (V-R), (b-r), etc. in the cluster center than in the outskirts. Since these stars would still have a finite velocity dispersion, albeit smaller than that of the other visible stars, their fractional contribution to the cluster's light should be a gradually decreasing function of radius rather than an abrupt step, just as is observed. Furthermore, even blue stragglers must die sometime. In the process, they'll probably prefer to turn into red HB stars than blue ones since they are more massive than the ordinary stars that are now leaving the main sequence.

I do not think that blue stragglers can explain all the observations quite as simply, however. First, remember Bailyn et al.'s observation of a (U-B) gradient, and recall the Johnson ((U-B), (B-V)) two-color diagram. The solid curve here is the standard Population I, luminosity class V locus, reddened by E(B-V) = 0.10 mag and E(U-B) = 0.07 mag; the dashed curve indicates the reddened Pop. I locus shifted upward by M 15's mean color excess at the turnoff, Δ(U-B) ~ 0.20. M-15's integrated (U-B) and (B-V) colors (taken in this instance from Reed 1985) place it near the F-star peak of the main-sequence color-color locus. Blue stragglers which lie slightly above M 15's observed turnoff at (B-V) ~ 0.45, (U-B) ~ - 0.10 (the open circle in Fig. 8) would indeed be bluer than the average M 15 color, but a prodigious concentration of them to the cluster center would be required to produce the observed color gradient in the face of dilution by the remaining cluster stars. A centrally concentrated population of particularly hot and bright blue stragglers, as observed by Auriére et

ON THE COLOR AND POPULATION GRADIENTS IN M 15

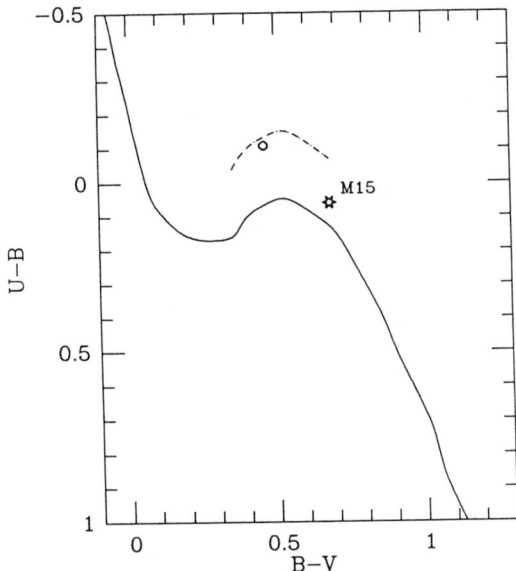

Fig. 8. Standard Johnson-system ((U-B), (B-V)) color-color diagram. The solid curve represents the luminosity-class V locus for Population I stars; the dashed curve suggests the unevolved main-sequence/ blue-straggler locus for stars of M 15's metallicity; the open circle represents the actual colors of stars at M 15's turnoff; and the star symbol represents the total, integrated colors of M 15 from the tabulation by Reed 1985.

No, I think the only way that all the available observations can be satisfied is to say that, in addition to a centrally concentrated population of AF-type stars below the level of the horizontal branch (still required to produce the Hα gradient), either (a) there is a major deficiency of red giants and subgiants in the center, or (b) the giant branch, subgiant branch, and/or horizontal branch are themselves bluer in the center. We have looked for (a) and not found it in M 15. While it's probably true that four or five bright giants are missing from the central 13" of the cluster, this fact by itself is not capable of producing the observations - recall that a continuous color gradient over the range 1 - 2" ~ radius ~ 30 - 60" is observed. I am therefore driven to the hypothesis that the giant and subgiant branches themselves may be shifted toward the blue in the center of M 15.

al. 1990 in NGC 6397, would produce *less* of a gradient in (U-B), both because of the increased Balmer discontinuity, and because the smaller amount of line blanketing expected in these hotter stars would decrease their Δ(U-B) color excess, moving them closer to the Pop. I sequence. Stars blue enough to make a big difference in the (U-B) color would have to be mid-B or earlier, but these would lie near or above the HB and likely wouldn't have enough Hα absorption. Finally, blue stragglers alone cannot produce the observation that the color gradient is *stronger* when the brighter stars are included than when they are removed.

This may not be as crazy as it sounds. If some large fraction of the evolving stars in the center of M 15 are of order twice as massive as the "normal" subgiants and giants, then their GB will indeed lie to the blue of the normal GB. If the relative numbers of these overmassive giants, and maybe also their degree of overmassiveness, increase toward the center of the cluster, then the average color of the GB will shift in proportion. This is an effect which I should be able to dig out of my data, when I perform a true calibration complete with color terms for each of the data sets.

How about the deficiency of GB tip stars in the center of M 15, and of giants in general in other central-cusp clusters? As discussed by Djorgovski et al. 1990b, it may have something to do with binary stars. Perhaps *most* of the stars in the centers of these clusters are binaries, whose semimajor axes have been driven by dynamical interactions to only a few A.U. or less. Could mass exchange inhibit the swelling of giant stars in such systems beyond a certain point? Maybe the stars at the tip of the GB in M 15's outer regions are really asymptotic giants, and for some binaries in the very center of the cluster enough mass is exchanged/lost that the helium core flash never occurs. Such stars would bypass the HB and AGB phases of evolution, becoming helium white dwarfs instead. Whatever is happening, the process seems to have gone much farther in M 30 than in M-15: in M 30 there is a deficiency of all giants, not just the tip stars; the color gradients are far stronger in M 30 than in M 15 - 0.28 mag·dex^{-1} in (B-R) for M 30's total integrated light (Piotto et al. 1988, Djorgovski et al. 1988), as compared to 0.10 mag·dex^{-1} in (b-r) (which should scale to ~ 0.12 mag·dex^{-1} in (B-R)) for M 15; and Rose et al. 1987 found that the Balmer absorption was far stronger in the center of M 30 than in their reference object - the center of M 15! Whatever the disease is, M 30 has it far worse than M 15, but M 15 and several other clusters have a significant dose as well.

I am grateful to the many coinvestigators on this project who have helped to acquire and interpret these data: Charles Bailyn, Steve Cederbloom, Haldan Cohn, Josh Grindlay, Jim Hesser, Phyllis Lugger, Bob McClure and Jim Rose. I hope that they aren't too humiliated by their association with this project after reading my speculations. I am also pleased to acknowledge helpful conversations with Pierre Demarque, Allen Sweigart and Don VandenBerg. George Djorgovski also kindly provided comments on this paper, as well as preliminary drafts of his and his collaborators' results. Thanks.

REFERENCES

Arp, H. C. 1955, AJ 60, 317
Auriére, M., Le Févre, O. and Terzan, A. 1984 A&A, 138, 415
Auriére, M., Ortolani, S. and Lauzeral, C. 1990 Nature 344, 638
Bailyn, C. D., Grindlay, J. E., Cohn, H., Lugger, P. M., Stetson, P. B. and Hesser, J. E. 1989 AJ 98, 882
Bingham, E. A., Cacciari, C., Dickens, R. J. and Fusi Pecci, F. 1984 MNRAS 209, 765
Buonanno, R., Buscema, G., Corsi, C. E., Iannicola, G., and Fusi Pecci, F. 1983 A&AS 51, 83
Cederbloom, S. E., Moss, M. J., Cohn, H., Lugger, P. M., Bailyn, C. D., Grindlay, J. E. and McClure, R. D. 1990 in The Formation and Evolution of Star Clusters, Astron. Soc. Pacific Conf. Series, No. 13, K. Janes, ed., Astron. Soc. Pacific, Chelsea, Mich., p. 246
Djorgovski, S., Piotto, G., and King, I. R. 1988 in the Proceedings of the CITA Workshop on Dynamics of Dense Stellar Systems, D. Merritt, ed., Cambridge Univ Press, p. 147
Djorgovski, S., Piotto, G. and Mallen-Ornelas, G. 1990a in The Formation and Evolution of Star Clusters, Astron. Soc. Pacific Conf. Series, No. 13, K. Janes, ed., Astron. Soc. Pacific, Chelsea, Mich., p. 262
Djorgovski, S., Piotto, G., Phinney, E. S. and Chernoff, D. F. 1990b ApJL, submitted
King, I. R., Hedemann, E., Jr., Hodge, S. M. and White, R. E. 1968 AJ 73, 456
Piotto, G., King, I. R. and Djorgovski, S. 1988 AJ 96, 1918
Reed, B. C. 1985 PASP 97, 120
Rose, J. A., Stetson, P. B. and Tripicco, M. J. 1987 AJ 94, 1202
Sandage, A. 1969 ApJ 157, 515
Sandage, A. 1970 ApJ 162, 841

DISCUSSION

KURUCZ: 1. How do you establish membership?
2. Do you need a coincidence correction?

STETSON: 1. It's not necessary. The number of field stars per square arcminute is much less than the number of cluster stars at the center of M 15.

2. In the final analysis, yes, you should. In fact, under these conditions, with 0.5-arcsecond seeing (FWHM) and 0.1-arcsecond pixels, there are of the order of 10^4 independent resolution elements within 13" of the cluster center, and only ~ 200 stars of horizontal-branch luminosity or higher. So, except for the central two or three arcseconds, the crowding isn't all that bad.

TWAROG: Since the blue stragglers lie brighter than the turnoff, is there hope that with state-of-the-art CCDs one could measure directly the ultraviolet magnitudes of the stars?

STETSON: Yes. Unfortunately the HR camera is rather opaque in U but excluding the central few arcseconds, it should be possible to study the rest of the cluster center to sufficient depth in U.

JANES: Regarding the RR Lyraes in M 15, T. Lutz and I have J and K data of a number of RR Lyraes in the cluster. With these data we will be able to get good absolute K magnitudes for them. If they are different from other RR Lyraes at this metallicity, we will be able to detect that.

CCD DETECTORS APPLIED TO GLOBULAR CLUSTER RESEARCH

Harvey B. Richer and Gregory G. Fahlman

University of British Columbia

1. INTRODUCTION

Because of their enhanced sensitivity and digital readout, CCDs permit measurement of fainter objects with higher accuracy and more convenience than was possible with photographic plates or early versions of other electronic detectors. Further, because it is possible to carry out Monte-Carlo simulations on the data frames themselves, a better estimate of the incompleteness and the errors involved in measuring objects can be obtained. We illustrate the improvements possible using these detectors by looking at two areas of research interest involving globular star clusters: ages, and luminosity (and mass) functions. In this discussion, we emphasize the new astrophysical insight obtained from the improved data.

2. GLOBULAR CLUSTER AGES

The ages of globular clusters are of interest because they are the oldest objects in the universe for which a reliable age can be determined. In this context, their absolute age is of cosmological interest as it provides a lower limit to the age of the universe. The relative ages of clusters at a given [M/H] as a function of Galactocentric distance may provide clues to the formation of the Milky Way Galaxy, or any age dispersion with metal abundance might indicate how the clusters themselves formed. As a minimum, we might expect that the relative ages of the globular clusters will be able to distinguish between the Eggen, Lynden-Bell and Sandage (1962) picture of Galaxy formation wherein a rapid collapse took place, and that of Searle and Zinn (1978) where the Galaxy accumulated its mass chaotically over a period of several Gyrs. In the first scenario, the globular cluster age dispersion should be less than about 1 Gyr, whereas in the latter picture, we might expect to see an age range of at least 2 or 3 Gyrs.

The art of determining cluster ages (in particular relative ages) has recently been advanced by a new technique developed by VandenBerg, Bolte, and Stetson (1990) and Sarajedini and Demarque (1990). This technique involves simply shifting cluster fiducial sequences (for clusters of similar metal abundance) horizontally to bring the bluest

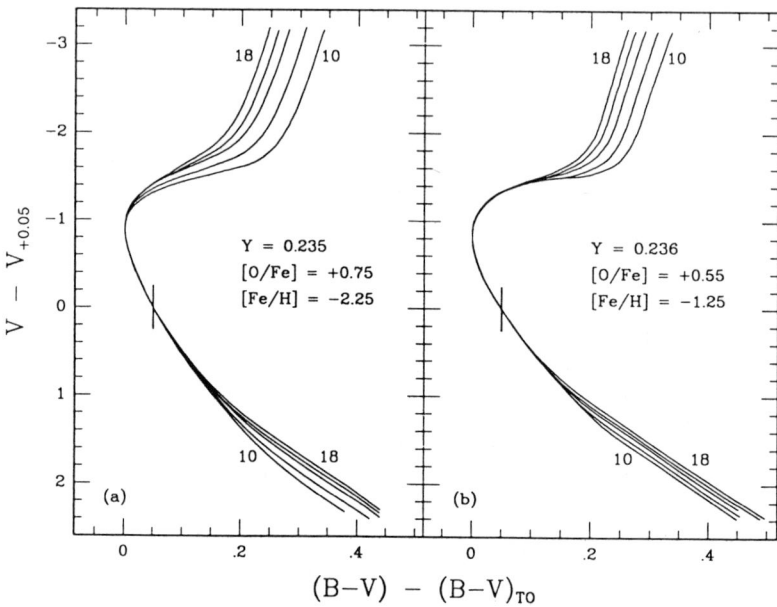

Fig. 1. Plot of theoretical isochrones for globular clusters for the abundances indicated at 10 to 18 Gyrs. The isochrones have been aligned as described in the text. The age dispersion in the clusters is apparent in the color difference between the turnoff region and the nearly vertical portion of the red giant branch.

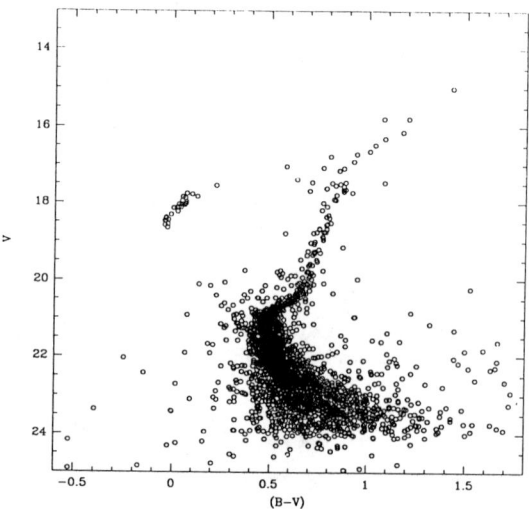

Fig. 2. Color-magnitude diagram for NGC 7492.

turnoff points to the same color, and then shifting vertically to force the upper main sequences to align at a color that is 0.05 mag redder than the turnoff. Any age differences in the clusters will then become apparent through the color differences between the turnoff points and the steeply rising red giant branch. This effect can be clearly seen in Fig. 1 which is taken from the VandenBerg, Bolte and Stetson (1990) paper.

VandenBerg, Bolte and Stetson examined a total of 22 globular clusters with this technique. They find that the most metal-poor clusters exhibit no age dispersion within the errors (however see Buonanno et al. 1990 who conclude that the metal-poor cluster Ruprecht 106 may be several Gyrs younger than M 68), but that clusters with [M/H] in the range of 1.3 exhibit a significant age dispersion; perhaps covering a total range of as much as 6 Gyrs. They also conclude that there is no evidence for an age-galactocentric distance relation in the outer halo of the Galaxy. This result rests primarily on the conclusion that their two most distant clusters from the galactic center (Pal 5 and NGC 7492; 16.5 and 18.7 kpc respectively) exhibit ages similar to the fiducial clusters. From their Fig. 14 however, there is some weak evidence that Pal 5 may be somewhat older than the fiducial cluster NGC 362. We present new data and analysis of the other distant cluster NGC 7492 and investigate whether it appears coeval with the fiducial cluster for its metallicity, NGC 6752.

The CCD imaging data for NGC 7492 were obtained at CFHT and consist of two 10 minute frames in each of B and V centered on the cluster center. The seeing was superb for these frames, averaging about 0.65" FWHM. The resulting calibrated color-magnitude diagram is shown in Fig. 2, and the comparison with the fiducial cluster for this metallicity (NGC 6752) is displayed in Fig. 3.

As with Pal 5, there is some evidence that NGC 7492 is about 2 Gyrs older than the fiducial cluster for its metal abundance. We conclude from this that the question of an age gradient in the galactic halo must still be considered an open one, and that there now seems to be added incentive to determine ages for clusters with distances in excess of about 20 kpc from the galactic center.

3. LUMINOSITY AND MASS FUNCTIONS

CCD detectors are superb devices for determining luminosity functions in globular clusters. The digital data that they provide can be manipulated to subtract out the increased scattered light near the cluster core, and images generated from the stellar point spread function can be added back into the frames with subsequent re-reduction yielding both incompleteness corrections and error estimates. This procedure was always difficult to do with analog photographic plates (King, Hedemann, Hodge and White 1968). The major deficiency with CCD detectors is their small format. This limitation is important as mass segregation in these systems will result in varying luminosity

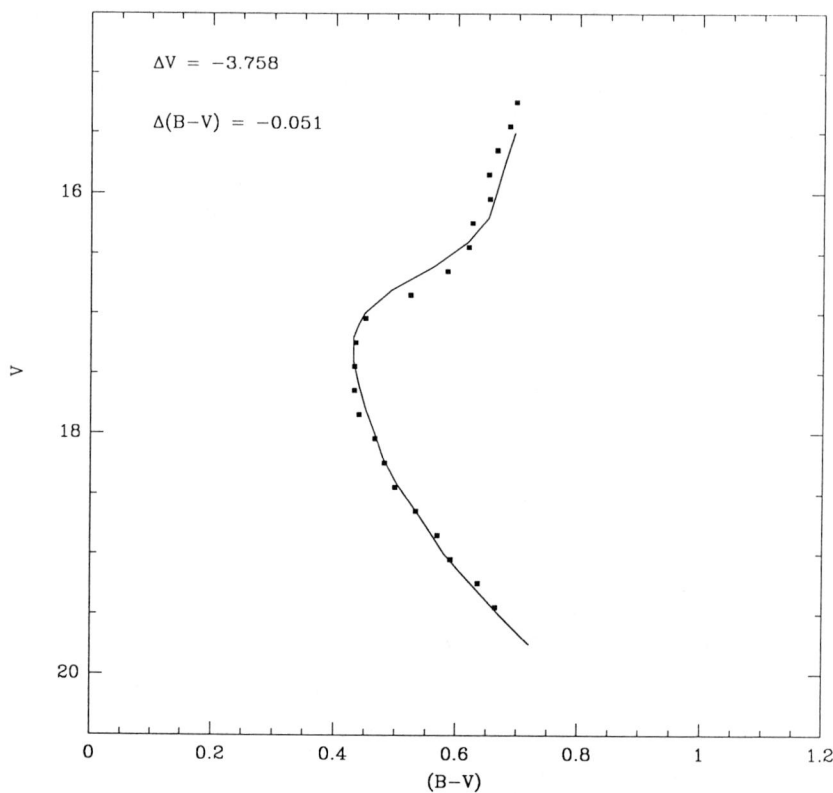

Fig. 3. Comparison of the fiducial sequences for NGC 7492 (squares) and NGC 6752 (line). The sequences were shifted as described in the text.

functions in different regions of the cluster. As a concrete example of what can be obtained regarding luminosity functions in these clusters, we consider the case of NGC 5053.

This particular cluster is of interest as it is very loose (c = 0.77) as well as being very metal poor with [M/H] = -2.02 (Webbink 1986). We originally observed it as a test of the McClure et al. (1986) relation between metal abundance and mass function slope, the idea being that complications due to mass segregation could be eliminated given the looseness of the system as it is practical to observe low mass main sequence stars right into the center of the cluster. This means that it is possible to obtain the global mass function of the cluster directly without any need for the somewhat uncertain mass segregation corrections. Of further interest is the fact that the central relaxation time of the cluster is in excess of 5 $\times$ 10^9 years (Peterson and King 1975), so that the cluster is dynamically young. Its observed mass function may then be similar to its initial mass function if extensive tidal stripping has not as yet occurred.

The data for this cluster consist of a strip beginning about two minutes east of the cluster center and extending westward to a radius of about 10.5'. In total 44 U, B and V CCD frames of five overlapping fields were secured at CFHT. Exposure times ranged from 3600 sec for the U frames in the central regions, through 900 sec for the B and V frames also in the cluster center. Where the crowding was not so severe, 1800 sec exposures in B and V were obtained. We discuss below only a few results from the V frames. More details of the work on NGC 5053 can be found in Fahlman, Richer and Nemec (1991).

In Fig. 4 we display the cluster luminosity function derived from the data discussed above. It is presented in three panels; the inner panel representing stars within 1.7' of the cluster center, the outer one consisting of all stars from 4.9' to 9.5' and the global luminosity function the sum of these two plus the stars at intermediate distance (1.7' to 4.9'). Several points about these diagrams should be mentioned. Firstly, all appropriate geometrical corrections were made in order to account for the portions of the cluster not observed (ie the numbers of stars observed were multiplied up by the ratio of the area in the cluster actually observed to the full area so the numbers represent the complete numbers in the cluster). Secondly, the outer CCD frames do not extend all the way to the cluster tidal radius which is at 13.6'. However, we have observed out to a radius which contains most of the cluster mass (see Fahlman, Richer and Nemec 1991 for more discussion on this point), consequently our global luminosity function should be quite complete except possibly for the lowest mass stars which may have been depleted in the inner parts of the cluster due to mass segregation. As can be seen in Fig. 4, the cluster does exhibit mass segregation. The amount observed is consistent with expectations based on multimass King models (Fahlman, Richer and Nemec 1991).

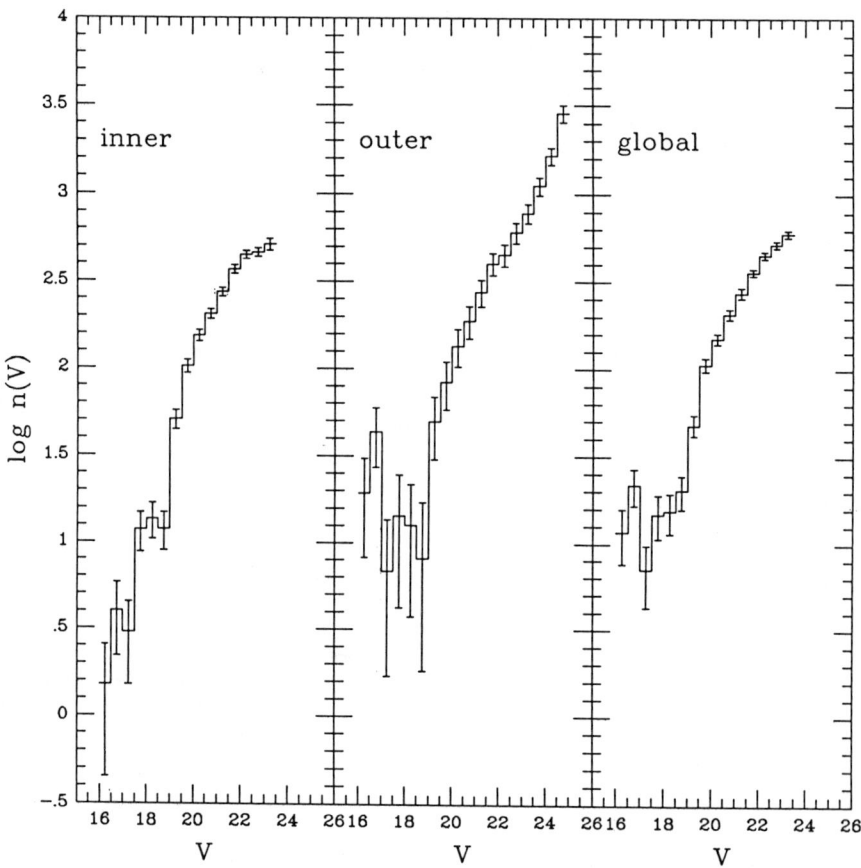

Fig. 4. Luminosity functions for NGC 5053. The inner function is for stars within 1.7' of the cluster center, while the outer one covers the area from 4.9' to 9.5'. The global function is the sum of these two together with the luminosity function for the stars at intermediate cluster distances.

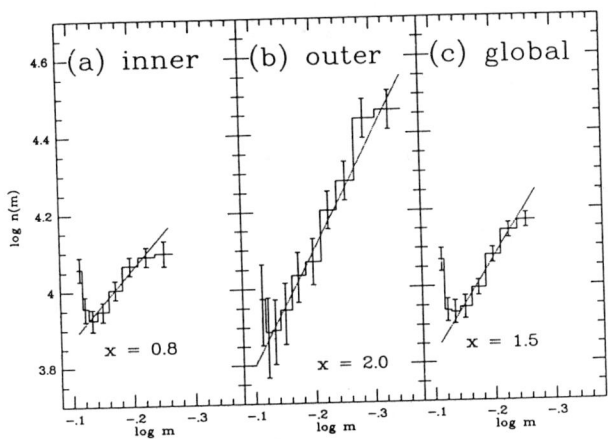

Fig. 5. Mass functions for the inner and outer regions of NGC 5053, as well as the global mass function. Slopes indicated are eyeball estimates only.

The luminosity functions derived above were converted into mass functions using the 16 Gyr oxygen enhanced isochrone employed in the analysis of M 68 (McClure et al 1987). These functions are displayed in Fig. 5. The lines drawn through the calculated mass functions are eyeball estimates only, intended to demonstrate that there are obviously discernable differences in the functions. It seems clear from Fig. 5. that the global mass function for NGC 5053 is well represented by a power law with slope x = 1.5 over the mass range sampled ($0.54 < m/M_o < 0.78$). As mentioned above, this may be close to the initial mass function slope for this cluster, and within this context it is noteworthy that it is so steep. Curiously enough, there are really only two globular clusters for which mass functions have determined globally; NGC 5053 and M 71 (Richer and Fahlman 1989). This latter cluster is metal-rich and has a mass function slope (for stars more massive than $0.4 M_o$) that is very flat. As we have seen, that for the metal-poor cluster NGC 5053 is very steep.

4. A FEW COMMENTS

That CCDs have revolutionized research in globular star clusters is undeniable. Almost solely due to their development, progress on the following important topics has been made.

(1) Metallicity dispersion within most clusters is found to be very small (less than a few tenths of a dex).

(2) Clear evidence of dynamical relaxation is seen.

(3) Relative ages are accurate to better than 1 Gyr. Absolute ages accurate to about 2 Gyrs.

(4) Main sequence mass functions are measured to very low masses.

(5) Time resolved photometry is used to discover low-luminosity variables including photometric binaries.

(6) Discovery of extensive populations of globular clusters surrounding external galaxies.

The interested reader can doubtless provide several additional important items to this list. Future breakthroughs will likely come from large format chips wherein entire clusters can be measured (or at least a large fraction of them), or from observations in new spectral regions such as the near ultraviolet from the Hubble Space Telescope. Of particular interest will be CCD observations in conjunction with high resolution telescopes allowing a probe of the stellar content and structure of cluster cores.

The authors are indebted to several colleagues and students who have been involved with some of the work described here. In particular mention should be made of J. Nemec who was involved in the work on NGC 5053, and P. Cote who carried much of the analysis relating to NGC 7492. This research has been supported by grants from the Natural Sciences and Engineering Research Council of Canada.

REFERENCES

Buonanno, R., Buscema, G., Fusi Pecci, F., Richer, H.B. and Fahlman, G. G. 1990 AJ, in press
Eggen, O. J., Lynden-Bell, D. and Sandage, A. 1962 ApJ 136, 748
Fahlman, G. G., Richer, H. B. and Nemec, J. M. 1991, in preparation
King, I. R., Hedemann. E., Jr., Hodge, S. M. and White, R. E. 1968 AJ 73, 456
Landolt, A. U. 1983 AJ 88, 439
McClure, R. D., VandenBerg, D. A., Bell, R. A., Hesser, J. E. and Stetson, P. B. 1987 AJ 93, 1144
McClure, R. D., Vandenberg, D. A., Smith, G. H., Fahlman, G. G., Richer, H. B., Hesser, J. E., Harris, W. E., Stetson, P. B. and Bell, R. A. 1986, ApJL 307, L 49
Peterson, C. J. and King, I. R. 1975 AJ 80, 427
Richer, H. B. and Fahlman, G. G. 1989 ApJ 339, 178
Sarajedini, A. and Demarque, P. 1990, preprint
Searle, L. and Zinn, R. 1978 ApJ 225, 357
VandenBerg, D. A., Stetson, P. B. and Bolte, M. 1990 AJ 100, 445
Webbink, R. F. 1988 in IAU Symposium No. 113, Dynamics of Star Clusters, J. Goodman and P. Hut, eds., Reidel, Dordrecht, p. 541

DISCUSSION

STETSON: A minor technical point regarding your CM-Diagram for NGC 7492. It seems to me that the color scatter on the main sequence is large enough that you can't tell whether your normal points might be biased by the presence of a binary sequence. If there are a large number of binaries, the age might be less by a couple of gigayears.

RICHER: Yes, this is certainly possible. The presence of binaries will move the fiducial to the right, decreasing $\Delta(B-V)$ and thus decreasing the age. In theory, knowing our errors we can do Monte Carlo simulation to check if the width is broader than expected.

DEMARQUE: You mentioned at the beginning of your talk the need to shift isochrones in the CM-Diagram both in (B-V) and magnitude to fit the observations. Shifting in (B-V), while preserving the isochrone shape makes little physical sense. Shifting in magnitude is definitely incorrect. How do you justify these procedures?

RICHER: The reason why isochrones might have to be shifted is simple. If the theoretical color-temperature relation is not quite correct, or the cluster reddening is slightly in error a small shift in (B-V) might be required. Similarly if the cluster distance is slightly in error a small shift in magnitude might be required. This is clearly not a very satisfactory situation but remains an important fact of life.

THE α(16)Λ(9)-PHOTOMETRIC SYSTEM REVISITED

Eugenio E. Mendoza V.

Institute of Astronomy, UNAM

ABSTRACT: This work is based on α(16)Λ(9)-photometry of normal and peculiar stars of the MK system of spectral classification. Photometric characteristics are briefly described through selected stars of both groups. The peculiar stars discussed are; Wolf-Rayet stars, classical Be stars, Herbig Ae/Be stars, upper main-sequence stars with anomalous abundances, metallic-line stars, T Tauri and related stars. Normal stars are used to illustrate the calibrations of the photometry versus equivalent widths, and absolute magnitudes. New observations and future work are included in the paper.

The α(16) and Λ(9) indices are defined by:

$$m_1 - 0.5*(m_{awc} + m_{lwc}) = \begin{matrix} \alpha(16) + 0.941 \\ \Lambda(9) + 0.268 \end{matrix}$$

where m_1, m_{awc} and m_{lwc} are the measurements (in magnitudes) through three interference filters. The characteristics of these filters are given by Mendoza (1990). The constants on the right-hand of equation (1) are such that a vanishing index corresponds to a vanishing equivalent width (Mendoza 1989). The indices measure the total absorptions of the lines of hydrogen, Hα, and O I-triplet at λ 7774 Å, respectively.

The α(16)Λ(9)-photometric system was originally designed to separate supergiant stars from other luminosity class stars. The results (Mendoza 1989) indicate that this system neatly separates luminosity class I stars later than B2, but only the most luminous stars with spectral type later than G1, because the luminosity effect in O I λ 7774 is very strong through G1, and falls off rapidly after that (see Keenan and Hynek 1950).

This paper is divided into three parts: the first is concerned with "normal" stars; the second deals with some groups of stars whose spectra do not fit the two-dimensional spectral classification of the MK system, herein they will be called "peculiar" stars; the third gives an account of future work.

1. NORMAL STARS

We have selected as an example of "normal" stars the revised list of fundamental MK standards for the O4 - G2 spectral range (Morgan and Keenan 1973, Table 1). The $\alpha(16)\Lambda(9)$-photometry for these stars is listed in Table 1; this photometry has been partly published before (Mendoza 1979, Mendoza et al. 1983).

A few remarks are in order; most stars have been observed at least during three different nights. Thus, this photometry has a high internal precision; mean errors are, in $\alpha(16)$ ± 0.003 mag., and in $\Lambda(9)$ ± 0.004 mag. Measurements with standard deviations larger than three σ are an indication that the star is a variable star or it is suspected of variability, at these wavelengths. For more details on this subject see Mendoza (1978, 1983), and Mendoza et al. (1983). ρ Orionis is suspected of variability in the $\alpha(16)$-index because its standard deviation is ± 0.012 mag. (on four different nights). α Cygni is a variable star in the hydrogen Hα-line (see Mendoza and Johnson 1979, and Johnson and Mendoza 1980); the standard deviation of the $\alpha(16)$-index measurements is ± 0.098 mag. (on 15 different nights); both stars are listed in Table 1. However, their $\Lambda(9)$-index shows no variations larger than 0.01 mag.

The photometry corresponding to η Tauri is quite different from many B7 III type stars. Its $\alpha(16)$-index discloses a value typical of Hα emission-line stars (see below), its $\Lambda(9)$-index corresponds to an O I-line somewhat contaminated by emission, probably since it is slightly fainter than the mean value of several "normal" B7 III type stars. Finally three stars listed in Table 1 have a lower quality photometry, because the only night they were observed the photometer was unstable. They have been included in Table 1 for the sake of completeness.

The photometry contained in Table 1 is well suited for calibration with equivalent widths. The results yield:

$$W(H\alpha) = 21.91\alpha(16) \pm 0.6$$

where $W(H\alpha)$ is in Ångstroms, and $\alpha(16)$ in magnitudes

$$W(7774) = 9548.3\lambda(9) \pm 163$$

where $W(7774)$ is in milli-Ångstroms and $\Lambda(9)$ in magnitudes.

We are not going to describe the calibrations of the $\alpha(16)$-index, because in principle the method to derive the calibrations of the β-index can be applied to our index (Crawford 1975, Breger 1975). It suffices to say, for example, that for a small range of spectral type (say F-type stars) the calibrations in terms of

Table 1

α(16)Λ(9)-Photometry of the MK System of Standard Stars in magnitudes

BS	MK	α(16)	Λ(9)	BS	MK	α(16)	Λ(9)
HD46223	O4	+0.109	+0.016	1713	B8 Ia	−0.017	0.145
HD46150	O5	+0.127	+0.021	7906	B9 IV	+0.341	0.070
3165	O5f	−0.067	+0.010	7001	A0 V	+0.435	0.078
2456	O7	+0.140	+0.013	4554	A0 V	+0.406	0.068
8622	O9 V	+0.161	+0.011	7924	A2 Ia	+0.037	0.222
1899	O9 III	+0.109	+0.006	8728	A3 V	+0.430	0.087
2782	O9 Ib	0.041	0.017	6081	A5 II	0.337	0.109
1931	O9.5 V	0.161	0.015	3569	A7 IV	0.375	0.063
6165	B0 V	0.122	0.023	1412	A7 III	0.365	0.083
1855	B0 V	0.147	0.008	7876	A9 II	0.282	0.089
1903	B0 Ia	−0.031	0.008	1351	F0 IV	0.314	0.081
5953	B0.3 IV	0.138	0.008	4031	F0 III	0.263	0.079
1220	B0.5 III	0.134	0.022	292	F0 II	0.275	0.116
2004	B0.5 Ia	0.002	0.009	1865	F0 Ib	0.283	0.122
5993	B1 V	0.175	0.039	4931	F2 V	0.277	0.052
1892	B1 V	0.158	0.022	2107	F2 IV	0.284	0.051
1131	B1 III	0.141	0.029	21	F2 III-IV	0.260	0.083
1203	B1 Ib	0.089	0.027	1279	F3 V	0.252	0.037
6141	B2 V	0.235	0.031	HD27524	F5 V	0.224	0.034
39	B2 IV	0.159	0.040	856	F5 III	0.228	0.058
153	B2 IV	0.163	0.043	7495	F5 II	0.223	0.059
1790	B2 III	0.157	0.043	1017	F5 Ib	0.217	0.127
2135	B2 Ia	−0.071	0.043	1543	F6 V	0.215	0.033
7121	B2.5 V	0.168	0.040	544	F6 IV	0.201	0.044
5191	B3 V	0.212	0.038	6577	F6 III	0.198	0.045
1641	B3 V	0.213	0.048	HD27808	F8 V	0.192	0.013
2653	B3 Ia	−0.020	0.091	HD27383	F9 V	0.178	0.022
1749	B5 V	0.250	0.048	4540	F9 V	0.180	0.013
6092	B5 IV	0.237	0.063	4983	G0 V	0.156	0.018
2827	B5 Ia	−0.017	0.123	4883	G0 III	0.149	0.041
1145	B6 IV	0.232	0.052	339		0.133	0.044
1165	B7 III	−0.270	0.016	HD27836		0.146	0.017
1144	B8 V	0.289	0.049	6212		0.142	0.016
1178	B8 III	0.239	0.044	SUN		0.141	0.016

Notes: BS 1931 has a possibly variable α(16)-index. BS 7924 has a definitely variable α(16)-index. BS 6212, 6577 and 7876 have lower quality photometry.

temperature are linear $[T_{eff} = a + b\alpha(16)]$.

On the other hand the $\Lambda(9)$-index for stars later than B2 can be calibrated in terms of absolute magnitudes,

$$M_v = -25.16\Lambda(9) + 4.820 \pm 0.5 \quad (A0 - G1)$$

$$M_v = -25.99\Lambda(9) = 2.878\ 0\pm 0.4 \quad (B3 - B8;\ G2 - G4)$$

The $\alpha(16)\Lambda(9)$-array for high and low luminosity stars listed in Table 2 is illustrated in Fig. 1 to show how these two groups are separated by this photometric system.

2. PECULIAR STARS

There exist several well-known groups of stars whose spectra do not fit the MK scheme of two-dimensional classification (see Keenan and Morgan 1951). We have observed a few of them, namely

a. Two sequences of Wolf-Rayet stars

b. Early-type emission line stars (Classical Be stars)

c. Herbig Ae/Be stars

d. Upper main-sequence stars with anomalous abundances (Bp and Ap)

e. Metallic-line stars (Am)

f. T Tauri stars

The $\alpha(16)\Lambda(9)$-photometry for a few selected members (no space for more) is presented in Table 2. A brief description of the main photometric characteristics follows:

2.1 Wolf-Rayet Stars

The photometric data listed in Table 2 for Wolf-Rayet (W-R) stars have been selected from Mendoza (1990); they comprise four WC and four WN stars. Let us recall that (W-R) stars have abundant helium and perhaps only traces of hydrogen. Therefore, the $\alpha(16)$-index is most likely a measure of the He II line at λ 6560 Å. An important characteristic of (W-R) stars is broad emission lines (1000 km/s are not uncommon); thus, $\alpha(16)$ will be a lower limit, since the width of this helium line is in general much broader than the width of the corresponding central filter (Mendoza 1990). Hence, the $\alpha(16)$-index should not be used in calibrated relationships.

On the other hand the $\Lambda(9)$-index for WN-stars shows an exceedingly small range in values, close to that of a normal B1-type

Table 2

$\alpha(16)\Lambda(9)$-Photometry of Stars Which do not fit the MK
Two-Dimensional Classification (mean values in magnitudes)

Star	Sp	$\alpha(16)$	$\Lambda(9)$	Star	Sp	$\alpha(16)$	$\Lambda(9)$
HD165763	WC5	-0.841	0.461	HD190918	WN4.5+O	-0.262	0.026
HD156327	WC7	-0.529	0.251	EZ CMa	WN5	-1.271	0.034
BS 6265	WC7 + O	-0.602	0.176	HD191765	WN6	-1.087	0.035
γ^2 Vel	WC8 + O	-0.567	-0.106	HD211853	WN6 + O	-0.410	0.032
γ Cas	B0 IVe	-0.786	-0.031	BS 1445	B9 Hg	0.297	0.049
ω CMa	B2 IV-Ve	-0.599	0.002	14 Hya	B9 Hg	0.280	0.051
ϕ Per	B2 Ve	-1.377	-0.047	112 Her	B9 Hg	0.304	0.050
48 Per	B3 Ve	-0.774	0.002	GY And	B9 Cr	0.427	0.026
ς Tau	B4 IIIpe	-0.701	0.167	84 Uma	B9 Cr	0.387	0.016
o Cas	B5 IIIe	-0.413	0.012	45 Her	B9 Cr	0.373	0.026
48 Lib	B5 IIIpe	-0.664	0.305	41 Tau	B9 Si	0.251	0.021
ψ Per	B5 Ve	-1.166	-0.023	Cu Vir	B9 Si	0.301	0.018
28 Tau	B8 Vpe	-0.416	0.247	108 Aqr	B9 Si	0.298	0.024
V594 Cas	B2-B5eq	-1.570	-0.245	RW Aur	dG5:e	-1.677	0.049
HD200775	B3+shell	-1.968	-0.076	CO Ori	G5pe	-0.154	0.114
V380 Ori	A1:e	-1.783	-0.317	T Tau	K0 IV-V	-1.490	-0.065
LkHα 234	Aeβ	-1.448	+0.224	RY Tau	K1 IV-Ve	-0.649	0.095
R CrA	A5e	-2.155	-0.067	HD283447	K3 Ve	-0.173	-0.001
R Mon	Ae	-1.858	-0.071	GW Ori	dK3e	-1.191	-0.047
63 Tau	A1m	0.338	0.033	BS 1519	Am	0.384	0.051
60 Tau	A3m	0.316	0.038	τ UMa	Am	0.314	0.048

Notes: Hg stands for Hg-Mn, Cr stands for Cr-Eu-Sr.
Photometry sources: WC and WN, Mendoza 1990. Classical Be stars, Mendoza 1982. Herbig Ae/Be stars, Mendoza 1987 and this paper. Upper main-sequence stars with anomalous abundances (B9p), Mendoza 1986b, Metallic-line stars (Am), Mendoza 1976, 1986a. T Tauri objects, this paper.

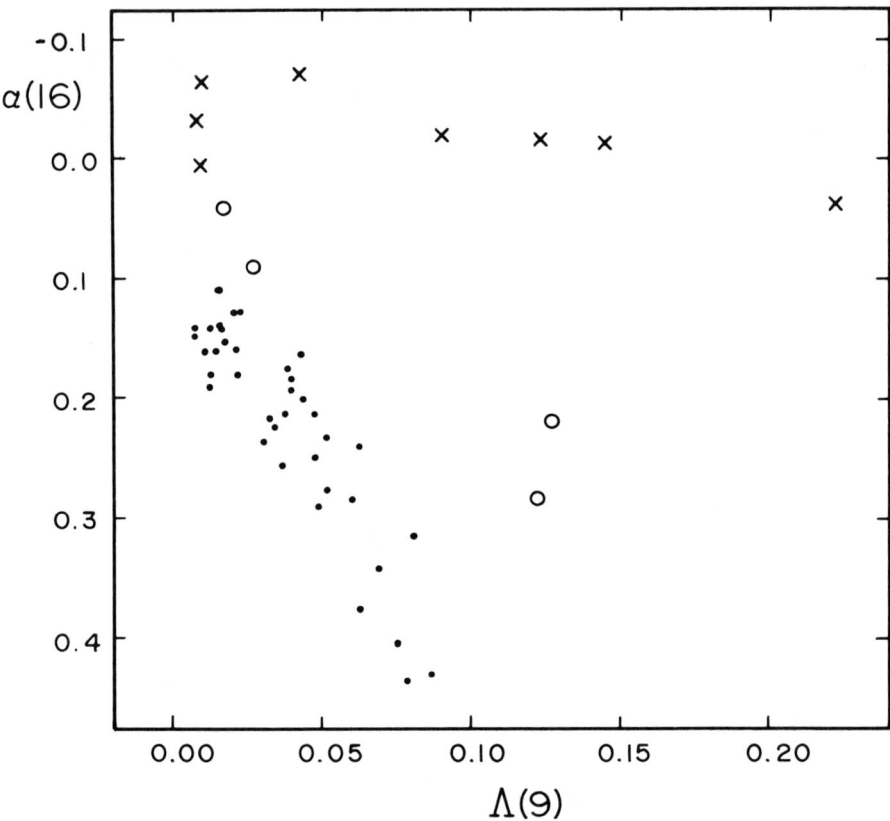

Fig. 1. $\alpha(16)\Lambda(9)$-plane. Coding: crosses, supergiant stars of luminosity class Ia, light circles, supergiants of luminosity class Ib, dots, stars of luminosity class IV and V. All data are from Table 1.

star (see Tables 1 and 2). However, the Λ(9)-index for WC-stars is quite peculiar probably, because this index is affected by spectral features which fall in the wavelength range covered by the filter that measures the short wavelength continuum (swc) around the O I-line as manifested by the raw data (Mendoza 1990). If this is the case, then the continuum will be artificially raised up, furnishing a somewhat large Λ(9)-index, that most probably has nothing to do with luminosity, since at these high temperatures the total absorption of the O I-line at λ 7774 Å is not a well defined luminosity criterion.

In spite of these facts, most WN and WC stars are separated from each other in the $\alpha(16)\Lambda(9)$-diagram (Mendoza 1990).

2.2 Classical Be Stars

The photometric data listed in Table 2 for early-type emission line stars (classical Be stars) have been selected from Mendoza (1982). They have been chosen to represent Be stars with and without shell characteristics; with the Λ(9)-index clearly contaminated by emission or with a very small equivalent width. Altogether, we have collected nine Be stars (see Table 2), three of each kind. The range in the $\alpha(16)$-index for these stars is similar to those in the sample given by us before (Mendoza 1982 and Mendoza et al. 1983). The Λ(9)-index for Be stars with shell characteristics is larger than most supergiants of luminosity class Ia (see Tables 1 and 2). Other classical Be stars disclose a Λ(9)-index which ranges from faint, if any, to a clear O I-line contaminated by emission. It seems that the stronger the emission in Hα-line is, the stronger the emission in the O I-line will be, whenever it exists. It should be pointed out that the emission in this hydrogen line is much stronger in classical Be stars than in supergiants.

2.3 Herbig Ae/Be Stars

We have selected six Herbig Ae/Be stars from 49 stars we have observed (see Mendoza 1987). The $\alpha(16)\Lambda(9)$-photometry for these stars denotes that most Herbig Ae/Be stars have stronger emission in the Hα-line than do classical Be stars. When the O I-line is in emission, it is also stronger in Herbig Ae/Be stars than in classical Be stars. The same is true when the O I-line is in absorption.

A comparison among supergiant, classical, and Herbig Ae/Be stars shows that, on the average, the Hα-line is stronger in Herbig Ae/Be, and faintest in supergiant stars. In general, these photometric characteristics are extreme in Herbig Ae/Be stars, intermediate in classical Be stars, and feeble in supergiant stars. For example, in a sample of 60 supergiant stars (Mendoza 1989), there is no indication that the Λ(9)-index is contaminated by emission. However, it is polluted for 11 out of 45 classical Be stars; and in 22 out of 49 Herbig Ae/Be stars.

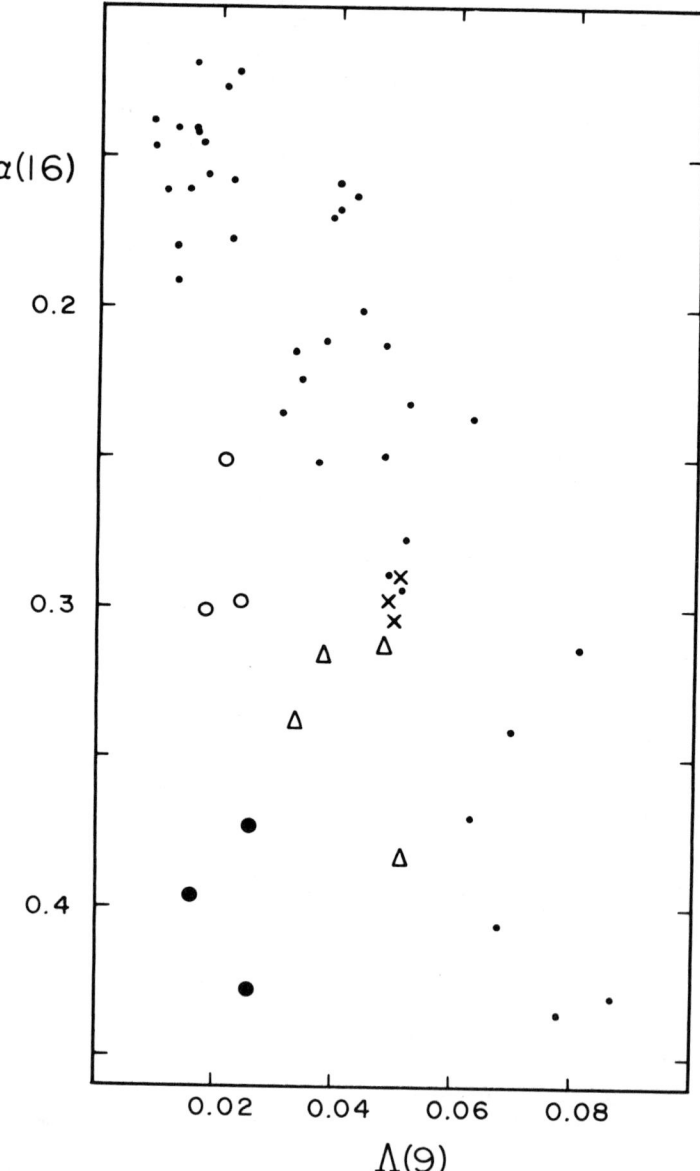

Fig. 2. $\alpha(16)\Lambda(9)$-array. Coding: dots, stars of luminosity classes IV and V (from Table 1), crosses, B9p-Hg-Mn stars, light circles B9p-Si stars, dark circles, B9p-Cr-Eu-Sr stars, and light triangles, classical Am stars (from Table 2).

Most Herbig Ae/Be stars are well separated from classical Be and supergiant stars in the $\alpha(16)\Lambda(9)$-plane.

2.4 Upper Main-Sequence Stars with Anomalous Abundances

We have selected from Mendoza (1977, 1986b) nine B9p stars, three of each one of the groups, Hg - Mn, Si, and Cr - Eu - Sr which have the following photometric features;

Hg - Mn stars are quasi-normal B9 low luminosity class stars; they nevertheless seem to define a boundary between these two kinds of stars.

Si stars have a quasi normal Hα line strength for their spectral types. However, the O I-line is extremely weak for these spectral types.

Cr - Eu - Sr stars have definitely stronger hydrogen line and a very weak neutral oxygen line, as compared with normal B9 low luminosity class stars.

The upper main sequence stars with anomalous abundances form three different photometric groups; two of them are exceedingly well separated from normal stars in the $\alpha(16)\Lambda(9)$-array (Mendoza 1977, 1986b and Fig. 2).

2.5 Metallic-line Stars

We have selected only four classical Am stars from Mendoza (1976, 1986a). Three belong to the Hyades cluster and one is a very well known Am star (see Table 2). Metallic-line stars are separated from upper main-sequence stars with anomalous abundances, and from normal A-type stars in the $\alpha(16)\Lambda(9)$-diagram (Mendoza 1977, 1986b). The $\alpha(16)$-index for Am stars is quite similar to normal B9, A5 - F0 low-luminosity class stars. However, the $\Lambda(9)$-index is smaller than the corresponding index for these normal stars.

Thus, according to the $\Lambda(9)$-index strength there are three well defined groups among A-type stars of low luminosity class, namely; 1. normal stars have a "strong" O I-line, 2. Am stars, intermediate and 3. Ap stars, weak (see Fig. 2).

2.6 T Tauri Stars

We have listed five classical T Tauri stars and one related star in Table 2. They lie, as a group, between classical Be and Herbig Ae/Be stars in the $\alpha(16)\Lambda(9)$-plane with little overlapping.

Classical Be, Herbig Ae/Be, T Tauri and related stars are also variable stars in this photometric system. The largest amount of variation in the $\alpha(16)$ and $\Lambda(9)$-indices corresponds to Herbig Ae/Be stars, and the smallest to classical Be stars. T Tauri stars are in the middle.

Some supergiant stars of luminosity class Ia also show light variations, which on the average, are smaller than those of classical Be stars (Mendoza 1977, 1986b).

3. FUTURE WORK

The brief account of the $\alpha16)\Lambda(9)$-photometric system given above indicates that it will be worth while to do the following;

1. To improve and extend the present calibrations through a larger and selected sample of adequate stars.

2. To investigate the behavior of Population II stars, in particular white dwarfs.

3. To use CCD techniques to be able to observe fainter and/or distant stars. This method may also be employed to study selected fields, such as cluster and stellar associations, in and out of the Galaxy. In particular distant and/or highly reddened yellow supergiant stars will be very easily identified through this photometry. Upper main-sequence stars with the anomalous abundances of the Si and the Cr - Eu - Sr groups will also be found by $\alpha(16)\Lambda(9)$-CCD-photometry.

It is interesting to mention that the $\alpha(16)\Lambda(9)$-photometric system may be combined with other photometric systems to improve the stellar classification. This has been done with the 13-color system (Mendoza et al. 1983), with infrared photometry (Mendoza 1982). Strömgren-photometry looks very promising as indicated by the work of Mendoza et al. (1990).

REFERENCES

Breger, M. 1975 in Multicolor Photometry and the Theoretical HR Diagram, A. G. D. Philip and D. S. Hayes, eds., Dudley Obs. Repts. 9, p. 31
Crawford, D. L. in Multicolor Photometry and the Theoretical HR Diagram, A. G. D. Philip and D. S. Hayes, eds., Dudley Obs. Repts. 9, p. 17
Keenan, P. C. and Hynek, J. C. 1950 ApJ 111, 1
Keenan, P. C. and Morgan, W. W. 1951 in Astrophysics, A Topical Symposium, J. C. Hynek, ed., McGraw-Hill Book Co, p. 12
Johnson, H. L and Mendoza, E. E. 1980 SPIE 264, 230
Mendoza, E. E. 1976 Rev Mexicana Astron. Af. 2, 29
Mendoza, E. E. 1977 Rev Mexicana Astron. Af. 2, 259
Mendoza, E. E. 1978 BAAS 10, 615
Mendoza, E. E. 1979 A&A 71, 147
Mendoza, E. E. 1982, in IAU Symposium 98, Be Stars, M. Jaschek and H. G. Groth, eds., Reidel, Dordrecht, Reidel, p. 3

Mendoza, E. E.   1983   IAU Inf. Bull. Var. Stars, number 2444
Mendoza, E. E.   1986a   Rev Mexicana Astron. Af. 12, 193
Mendoza, E. E.   1986b, in IAU Colloquium 90, Upper Main Sequence Stars with Anomalous Abundances, C. R Cowley, M. M. Dworetsky and C. Mégessier, eds., Reidel, Dordrecht, p. 195
Mendoza, E. E.   1987 in IAU Colloquium 92, The Physics of Be Stars, A. Slettebak and T. P. Snow, eds., Reidel, Dordrecht, p. 529
Mendoza, E. E.   1989   AJ 97, 1147
Mendoza, E. E.   1990   A&A 233, 137
Mendoza, E. E., Gomez, T., Ortega, R. and Quintero, A. 1983   PASP 95, 48
Mendoza, E. E. and Johnson, H. L. 1979 PASP 91, 465
Mendoza, E. E., Rolland, A. and Rodr guez, E. 1990 A&AS 84, 29
Morgan, W. W. and Keenan, P. C. 1973 ARAA 11, 29

DISCUSSION

GARRISON: 1) The band passes are very narrow. How faint can you observe with a normal single-channel photometer?

2) Have you observed any of the high latitude F-G supergiants, like HD 161796? If not, I hope that you will be able to observe some of them, since this system should give some interesting results.

MENDOZA: 1) I have observed as faint as 15 mag with an 1.5 m telescope.

2) Not yet. So far I have observed only MK supergiant standard stars (cf. Mendoza 1989).

V, (B-V) AND DDO COLORS EXTRACTED FROM GRISM SPECTRA

David J. Bell and Kenneth M. Yoss

University of Illinois

1. INTRODUCTION

Single-channel photometry becomes less accurate as fainter stars are observed, due to decreased signal-to-noise, increased errors due to sky brightness and increased interval of time between first and last observations, allowing sky brightness and extinction to change. Direct-image CCD observations for faint stars reduce the sky subtraction uncertainty, and represent a decided improvement over the standard single-channel approach, provided the stars in question are in a relatively small region, such as in a cluster. However, for all-sky photometry, or in cases where the stars are in an extended region, with only a few recorded per exposure, it also suffers from the time factor needed for recording the sequence of filter observations.

We describe here a multichannel procedure which permits simultaneous star and sky exposures, and which permits rapid acquisition of data over the entire sky. Low-resolution grism spectra are recorded, from which V, (B-V), and DDO colors are extracted. This procedure is much faster than the single-channel approach, allows observations past 16th magnitude, and produces accurate colors even on slightly subphotometric nights.

2. THE INSTRUMENT

The focal reducer/grism spectrograph instrument, a copy of a system developed at the University of Hawaii, was built by Dr. Laird Thompson for the Illinois 1-m reflector at Mount Laguna Observatory. Two remotely controlled motors allow both the grism and an aperture assembly to be moved in and out of the field as needed. The spectra are focused with a 50 mm Zeiss camera lens onto a TI 800 x 800 CCD detector with a readout noise of about 8 to 10 e⁻, which has been UV flooded for increased blue sensitivity (Leach and Lesser 1987). The system is quite efficient, with a throughput of approximately 50% at 4500 Å, but the focal-reducer lens system drops off in transmission toward the blue and is opaque shortward of 3900 Å.

The 400 line/mm 5000 Å blaze grism produces a reciprocal

dispersion of approximately 410 Å/mm corresponding to 6.2 Å/pixel. The resolution of stellar spectra can vary from about 10 to 20 Å since it is usually limited by the seeing instead of by the detector. A typical example of an early-type star is shown in Fig. 1. Background skylight is suppressed with a 1.0 mm wide slot corresponding to 15 arcseconds at f/13.5.

3. PROCEDURE

3.1 Data Collection

Several short direct exposures are usually necessary to center the target star in the aperture before moving the grism into the beam and recording a spectrum. Exposure times of late-type stars vary from 20 seconds at V = 8 to 30 minutes at V = 16, yielding a S/N of over 100 for the extracted DDO counts for most observed stars. The use of an off-axis guider is necessary for exposures longer than a few minutes.

Despite the overhead time required for centering, guiding, and CCD readout, data acquisition is faster than the single-channel approach at all magnitudes, due to the high quantum efficiency of the detector and the simultaneous aspect of the instrument. At V = 14, the practical limit for traditional single-channel photometry with our 1-meter reflector, the speed is about 10 times as great and yields considerably better results for the reasons outlined in the introduction.

A dozen or more standards of each photometric system (BV standards of Landolt (1973) and Landolt (1983), and DDO standards of McClure (1976) and Clark and McClure (1979)) are observed throughout the night, as is done with traditional photometry, to establish extinction and transformation coefficients.

3.2 Data Reduction

One-dimensional spectra are extracted from the CCD images with standard IRAF routines. The two-dimensional images are bias subtracted, flat-fielded, aperture summed and sky subtracted in a manner similar to those found in many Cassegrain-spectrograph cookbooks.

The one-dimensional spectra are then aligned through a cross-correlation process over a small window containing the atmospheric A˜band. The A-band is ideal for this step as it is visible in all stars and is relatively uncontaminated by stellar features for stars earlier than K5. Aligning the spectra is essential in order to remove small errors in guiding and instrumental flexure, which would significantly affect the extracted colors. This process aligns the images with an accuracy of about 1 - 2 Å. A dispersion solution is then determined by identifying absorption lines in an A-type stellar spectrum, and is applied to all exposures.

V, (B-V) AND DDO COLORS EXTRACTED FROM GRISM SPECTRA   113

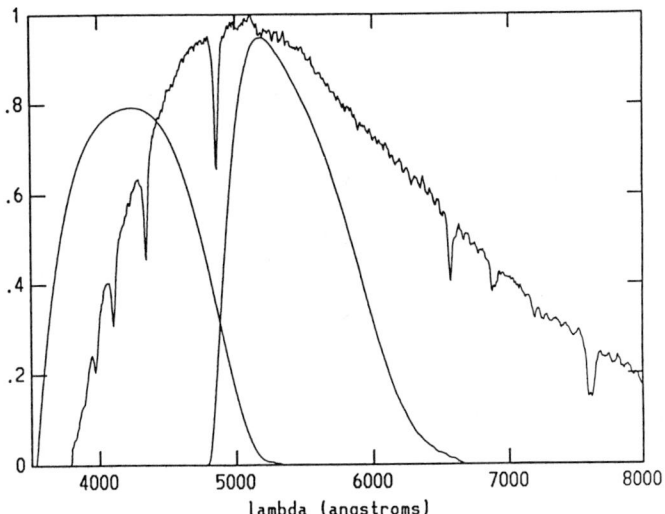

Fig. 1. The B and V passbands superimposed on a grism spectrum of an A star.

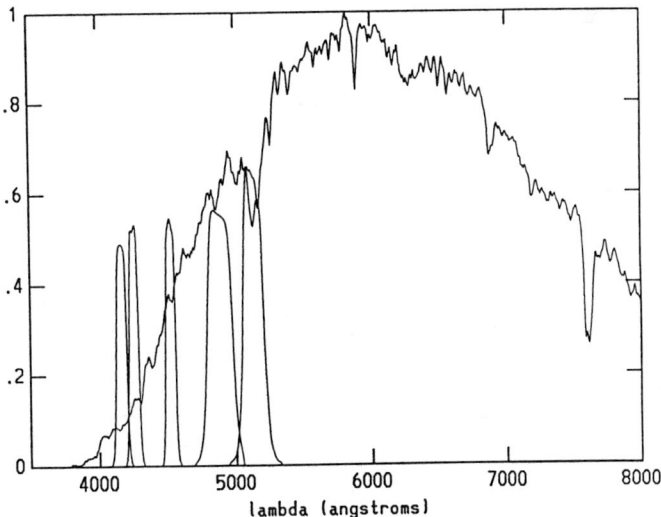

Fig. 2. The DDO passbands, including the Mgb + MgH passband, superimposed on a grism spectrum of a K star.

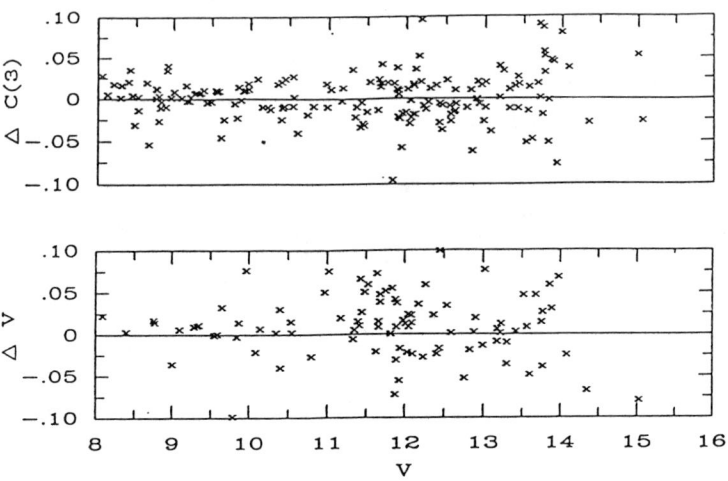

Fig. 3. The differences, photoelectric minus grism vs V mag, for C(41-42).

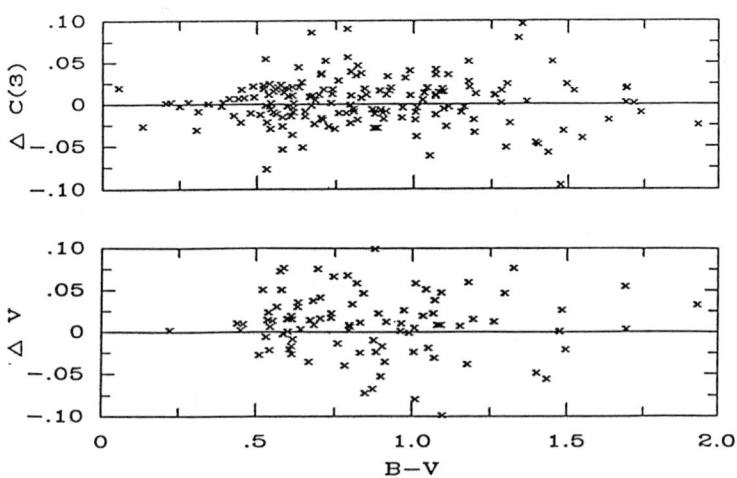

Fig. 4. The differences, photoelectric minus grism vs (B-V) color, for C(41-42) and V.

The resulting spectra are then operated on with the passbands of the DDO system and the V and B system. See Figs. 1 and 2. It is evident from the response of the Zeiss camera lens that the derived B is only an approximation of the standard B. We might thus expect a lower accuracy for (B-V).

The extracted flux data are then reduced with the same programs we use for our photoelectric work, yielding DDO colors and B and V. Exposures known to be taken in poor sky conditions are reduced in the DDO system only.

4. RESULTS

We have made observations on stars to V ~ 16, and find the system to be more accurate and faster than the conventional single-channel multifilter system for stars fainter than V ~ 10. Comparison of multiple grism observations yield rms values for single observations of V and (B-V) of ±0.010 and ±0.007 mag respectively, based on 61 stars; and comparison of 131 stars for the three DDO colors C(45-48), C(42-45), and C(41-42) of ±0.004, ±0.006 and ±0.008 mag respectively.

Comparison of grism and photoelectric colors for stars observed by both methods gives mean errors of ±0.025 and ±0.017 mag for V and (B-V) respectively (N = 94), and ±0.024, ±0.027 and ±0.028 mag for the three DDO colors C(45-48), C(42-45) and C(41-42) respectively (N = 176). Only a small number of photoelectric observations have been made for the Mgb + MgH index C(48-51), and it therefore is not discussed here. No systematic trends with V or (B-V) are evident, and much of the observed scatter is due to the photoelectric data, particularly at the faint end. Fig. 3 shows the differences of C(41-42) and V versus V, and Fig. 4 shows the differences versus (B-V).

5. CONCLUSION

The grism method for deriving colors and magnitudes is shown to be comparable to conventional photometry in terms of accuracy, to the practical limit of conventional single channel photometry. With our 1-meter reflector, the grism method is faster and in the case of fainter stars, more accurate.

Additional colors can, of course, be extracted from the spectra, for example (V-R). Narrow-band photometry, such as Hβ, should be possible with the instrument set up for higher resolution spectra. Unfortunately, full four-color photometry cannot be emulated, due to the UV cutoff of the instrument, although the metal index, m_1 = (v-b) - (b-y), can be obtained. Highly time-resolved photometry of variable stars should be possible, and this application will be tested. One-dimensional spectral classification, to ~ 0.5 class, also is possible, which gives an independent comparison of the DDO classification.

The present approach is particularly useful to us in our on-going program, in which we are pushing our observations to V ~ 16 for complete coverage within a two square degree region near the NGP. The objective of the project is to derive $D(z)$ and $\Sigma(\infty)$ as functions of composition, and $K_z(z)$. Preliminary results are contained in Yoss, Neese and Bell (1989).

REFERENCES

Clark, J. P. A. and McClure, R. D. 1979 PASP 91, 507
Landolt, A. U. 1973 AJ 78, 959
Landolt, A. U. 1983 AJ 88, 439
Leach, R. W. and Lesser, M. P. 1987 PASP 99, 668
McClure, R. D. 1976 AJ 81, 182
Yoss, K. M., Neese, C. L. and Bell, D. J. 1989 in The Gravitational Force Perpendicular to the Galactic Plane, A. G. Davis Philip and P. K. Lu, eds., L. Davis Press, Schenectady, p. 123

DISCUSSION

TWAROG: Rose and Aghostino have been using objective prism, spectrophotometric indices to study the SGP and have found a thick disk population of intermediate age and metallicity. Can you make use of your prism CCD spectra to reproduce his indices and place your data on the same system?

YOSS: Yes, and we will try it. Their work is excellent and can be duplicated for the NGP with the Burrell Schmidt at KPNO.

GARRISON: I'm concerned about transformations from your internal system, which may be self-consistent, to a standard system. You may be able to do it for normal stars, if you work hard at it, but any weird star may be difficult to transform in any meaningful way (e.g. a supernova, nova, or even maybe an extreme Pop II star). In other words, several people are developing filter systems with different CCD's and the total combined systems may be so difficult to transform that we won't be able to talk with one another. More standardization is needed (filters, CCD transformers, etc.).

YOSS: We hope to improve our wavelength calibration by using more lines than the A-band for our dispersion and zero point. This will help to improve our results. Furthermore, for any interesting stars we will get follow-up slit spectra. Since we have the full spectra, we can spot unusual stars. We also will use the Mgb + MgH index, which stands out for Pop. II stars.

PHOTOMETRY OF STARS IN SEVERAL SOURCE CATALOGUES

A. R. Upgren and E. W. Weis

Van Vleck Observatory

In the last few years we have obtained photoelectric photometry of many stars, mostly at KPNO, to deduce the distances and physical properties of stars belonging to one of several source catalogues and to delineate their extent on color-luminosity planes. An invited review outlining this and other related work has been given recently (Upgren 1991) and this report emphasizes in more detail some of the points raised there. In that review recent observations of stars in catalogues limited by either distance, apparent magnitude or proper motion were summarized, making them more useful for further analysis.

The most significant of these developments to be considered here are the completion of the photometry of stars in one of the major catalogues in each of the three categories. As has been reported elsewhere, broad-band photometry has been completed (Upgren and Weis 1991) for all remaining stars in the catalogues of dwarf K and M stars of Vyssotsky (1963), still the premier magnitude-limited source of the most representative stars in the galactic disk. The photometry obtained by Weis (1991) and 20 other investigators has been intercompared in order to obtain for each star or separable component of each of the 895 stellar systems the most precise values of the magnitude and (B-V) and (R-I) colors. The photometric parallaxes allow distances, absolute magnitudes and transverse velocities to be derived from ,which the probability of each star's membership in the young or thick disk has been derived by Ratnatunga and Upgren (1991). They use a maximum likelihood analysis developed by Casertano, Ratnatunga and Bahcall (1990) which overcomes incompletenesses in the data (trigonometric parallax, proper motion and radial velocity) using conditional probabilities.

The source limited in distance is that of Gliese (1969) for which Bessel (1990) has published similar photometric data for 937 stars, which represent the great majority of the stars in the 1969 or second edition of Gliese's catalogue of nearby stars. As for proper motion limited sources, Weis has just completed and is preparing for publication photometry of the stars without photometry of Eggen in the LHS Catalogue, which lists all stars known to have proper motions in excess of 0".5/yr. Most of the newly-added stars are north of +30° in

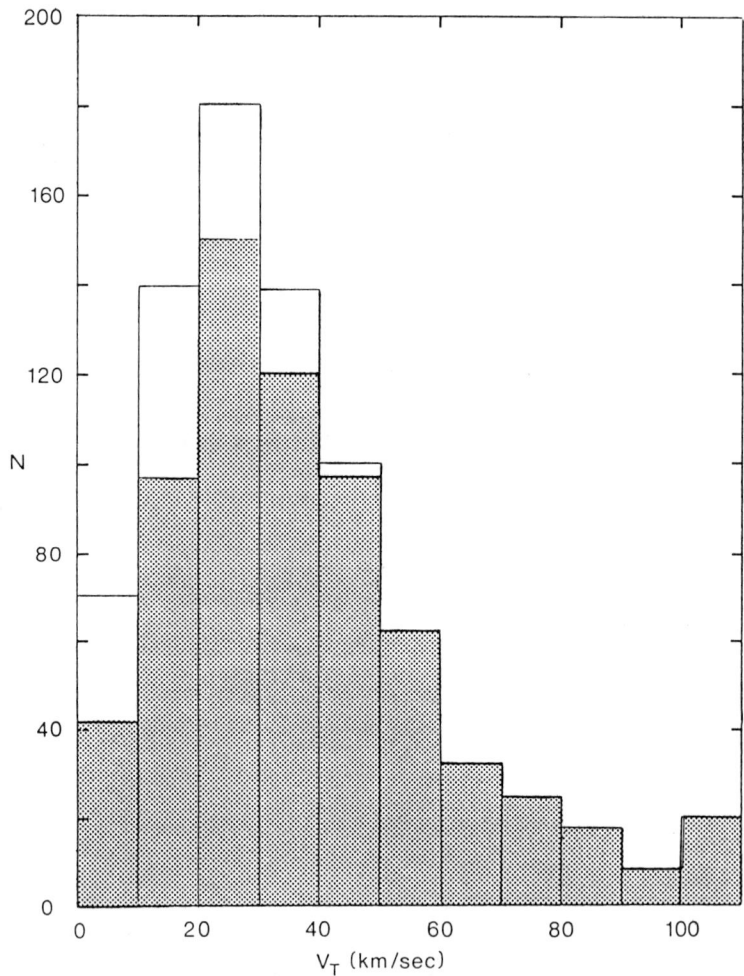

Fig. 1. Distribution of stars by transverse velocity. The shaded and unshaded portions represent stars with trigonometric parallaxes and those with only photometric parallaxes. The bar at the far right represents all stars with space motions in excess of 100 km/sec.

declination and are therefore inaccessible to Eggen.

The dangers of using incomplete data for any purpose can be illustrated in any number of ways but we will do so in Fig. 1. This diagram shows the number of stars or stellar systems listed as nearby stars by Gliese (1969) with (B-V) between 0.4 and 1.4 (corresponding to MK spectral types for main sequence stars of about F5 to M0). There are 782 such stars of which 656 or 84 % have trigonometric parallaxes. (In addition to these the 77 brighter main-sequence stars and the 86 evolved stars (giants and subgiants) all have trigonometric parallaxes.) Despite the seemingly overwhelming preponderance of parallaxes among the stars in this group, a strong bias towards high velocity is still present as is shown in the figure. The average transverse velocities among the stars with and without parallaxes are 31 and 15 km/sec, despite the fact that virtually all of the stars are brighter than an apparent visual magnitude of 11. Omission of the one star in six without a parallax determination leads to a sample very deficient in slow-moving and presumably young stars. The completeness of the photometry at least, goes a long way towards providing a sample free of such selection effects.

REFERENCES

Bessel, M. S. 1990 A&AS 83, 357
Casertano, S., Ratnatunga, K. U. and Bahcall, J. N. 1990 ApJ 357, 435
Gliese, W. 1969 Veröff Astron Inst Heidelberg, No. 22
Ratnatunga, K. U. and Upgren, A. R. 1991, preprint
Upgren, A. R. 1991 Robotic Observatories: Research, Education and Global Networking, S. Baliunas, ed, Fairborn Press (in press)
Upgren, A. R. and Weis, E. W. 1991, submitted, ApJ
Vyssotsky, A. N. 1963 Stars and Stellar Systems, Vol. III, Basic Astronomical Data, K. Aa. Strand, ed., University of Chicago, Chicago, P. 192
Weis, E. W. 1991, in preparation

THE STROMGREN uvby SYSTEM: WHY?

David L. Crawford

Kitt Peak National Observatory*
National Optical Astronomy Observatories

ABSTRACT: Strömgren designed the uvby system so as to maximize the information content possible in intermediate band photometry for a wide range of spectral types. It has now been in use for precision photometry for about three decades and appears to be meeting his objectives well. This paper reviews the concept of the system, its advantages for precision photometry, and some of its pitfalls.

It was a great pleasure to work with Bengt Strömgren during many of the years of the development of the uvby system. Professor Strömgren was a person with a remarkable breadth of interests and abilities, as well as a most personable human being. I consider it a great privilege to have known him well, both professionally and personally.

I was there at Yerkes Observatory, as a graduate student, when we used the l, c system mentioned by Knude, and during the later developments. My thesis concerned the definition and use of the $H\beta$ system, which supplements the uvby system well. I was also on hand during much of the observing at McDonald, and for the runs at Lick Observatory and at Palomar that were mentioned by Knude. Strömgren was also one of the early users (and supporters) of KPNO, and one of the early versions of the four-channel systems was used at KPNO for many years. I must say that he had remarkably good luck with the weather during his runs. It generally was clear until the last half of the last night. I wonder how it did it, but he did have remarkable talents.

---

*Operated by the Association of Universities for Research in Astronomy, Inc. under cooperative agreement with the National Science Foundation.

I want to share with you some of my memories about the early days of the system and of my conversations with Strömgren during those days. Certainly, he had a great deal of knowledge and experience with earlier work, by others and by himself. This included the use of color gradients, in different spectral regions, research which was most useful for understanding of what could be accomplished by photometric systems, any systems. He had a great deal of experience with stellar atmosphere models, and, of course, wanted to develop a system that would be efficient in helping obtain the maximum useful observational information to compare to theoretical models. He had published what is still one of the best writeups about photometric principles, in the Handbook of Experimental Physics volume, which Knude also referred to. In addition to these theoretical and observational talents and experiences, he had an excellent understanding of optics and of astronomical instruments. All in all, an excellent background to be a developer of one of the most used, and useful, photometric systems.

The uvby system was set up to address the issue of getting the maximum (intermediate band) information resolution from the stellar spectral output in the region that the common photomutliplier detectors were sensitive. The UBV system uses three filters to cover the entire region (in fact, they overlap each other to some extent) between the atmospheric cut off in the near ultraviolet and the fall off of PMT sensitivity in the red. The uvby system uses four, and they are well located relative to the information that is contained in this spectral region. They do not overlap. The u filter is above the atmospheric cutoff, and so the u band is filter defined, not as in the case of U. The red cutoff of the y filter is below the region where the 1P21 sensitivity dies, and so the y band (or V, if transformed to a V magnitude) is also filter defined, unlike the V of the UBV system.

The u is located in the region below the Balmer discontinuity, and does not reach above it, as does U, due to its greater width. The v filter is located entirely above the discontinuity, and is in a region of the spectrum where "metallic line blanketing" is present. The b band is located above this latter region. So the color index (b-y) is rather free of blanketing effects, unlike (B-V), while the color difference m_1, which is defined (v-b) - (b-y), is a good measure of the blanketing. Since all the filters are above the Balmer discontinuity, the index is rather free of any luminosity effects, unlike (U-B), often used to estimate blanketing.

The goal was to separate the parameters of interest, while still being quite sensitive to the feature being measured. I think that the vast amount of data obtained shows that his goal was achieved.

While the band widths are narrower than the ones defining the UBV system, the key issue, for most problems, is the information resolution, not the spectral resolution. In general, if you increase the spectral resolution you also increase the information resolution,

THE STROMGREN uvby SYSTEM: WHY?

but not always. For example, if one wants to obtain an apparent magnitude, then V is a good choice. Here one can use a rather wide bandwidth and still get as good an information resolution as with a narrower band, which would naturally have less photons. So one should always tailor the band widths to the scientific job: common sense, but often ignored.

The uvby system does an excellent job of giving much more information resolution than does the UBV system. The power of the latter is for the very faintest stars (or such) where there are just not enough photons to use narrower bands. One just has to give up information resolution to get anything at all. The uvby system covers the ordinary optical region with four intermediate bandwidth filters, and it gives us information about 1) apparent magnitude, 2) a color index (temperature parameter, generally) free (to first order) of adverse blanketing effects, 3) an index measuring blanketing differences, free (to first order) of luminosity effects, and 4) an index measuring the Balmer discontinuity, free (to first order) of blanketing effects. The parameters are well separated: they are rather clean measures of the underlying astrophysical parameters we are trying to estimate. The band widths are as wide as one can go and still have the advantages. They are still wide enough to be of use for all stars except the faintest. The information resolution is excellent.

In addition, the narrower bands widths (than for the UBV system) means that some of the observational problems are minimized, such as the effect of second-order terms in the reductions for atmospheric extinction and in the color transformations to standard system. There are still many problems left, to all systems, of course, that must be well understood and handled well in order to get accurate and precise photometric data.

For stars of spectral type O to G, when combined with the Hβ parameter, one can then measure the apparent magnitude, the apparent color, the blanketing, and estimate both the intrinsic color and the absolute magnitude. These then give us an estimate of the interstellar reddening and absorption, of the temperature (or spectral type), of the luminosity, and of the chemical composition, all basic parameters for understanding and classifying the stars, and their location in the Galaxy. We also can use the stars as probes of the interstellar matter. For some stars, those above the zero-age main sequence, we can also estimate the age.

All of these things were of interest to Strömgren (and many more of us, of course). The system was designed to get this basic information so as to use it for interesting astronomical problems. Strömgren, and many others, have done so. Many more will as time goes on. We are all in his debt for a good job.

The system is most useful on telescopes of any size (I am most pleased that Strömgren is one astronomer who never caught the aperture fever disease), and with detectors other than the 1P21, including even the new use of CCDs for imaging photometry (see several papers at the present meeting, for example). It has been done with filters and also with photometers using slots to define the spectral regions. I am sure that it will be done with spectrophotometric instruments using CCDs too.

I think that the keys to the success of the system are as follows:

1. It was well thought out and well implemented.
2. It had a generous choice of standard stars selected and well observed early in the game. More are needed, of course.
3. It is scientifically most useful.
4. It has been used a lot, on many useful programs. Most of this work has been carefully done.

There are other most interesting systems out there, now and in the past, that did not achieve nearly the success that the uvby system has. I think it is because they did not have all the above four items.

Let me review some of the questions he addressed, and which anyone defining or selecting a system must address:

1. Why the central wavelength for each filter?
2. Why the band width for each filter?
3. What color indices and/or color differences to use?
4. Are these clean estimates of the parameters?
5. Are the parameters well separated?
6. Impact of interstellar reddening?
7. Stable system?
8. Standards adequate?
9. Precision and accuracy potential?

I think the uvby system has also helped us understand the question of precision photometric systems much better. It used to be that 0.02 mag was "high precision"; now we can do nearly ten times better, but only if great care is used, of course! Photometry is still evolving, and still most useful. (See my review article on issues in photometry, Crawford 1988 for more of this philosophical stuff.)

I don't plan to review any of the uses of the system here; the papers at this meeting, and the astronomical literature, give ample examples of the power of the system. It has proved to be excellent in determining luminosities and temperatures for stars of spectral type O to early G, both in the field and in clusters. It is excellent in separating "normal" stars from "peculiar" ones, such as Am and Ap stars. It has also been excellent in studies of the interstellar medium. Let me just add that there is lots more to come: later type stars, supergiants, special stars of all kinds, considerable more work on open and globular clusters (with imaging photometers), and so forth.

THE STROMGREN uvby SYSTEM: WHY?

Certainly Strömgren would be the first to show that there are other useful photometric systems as well, and he would urge observers to use what they need. The goal is to maximize the information resolution and the efficiency and accuracy.

Certainly he would also caution about the pitfalls, both in the use of the uvby system or any other system. Most of these arise because the observer (or user) is not aware of the issues involved in doing accurate and precise photometry, being out of touch with the sky and with the system and with the data. Let me just mention the "problem" of m_1 transformations to the standard system, where the range in the index is quite small, and where interstellar reddening decreases the value of the index as well. One must be very careful in using any system, and must insure that one is using a careful selection and an adequate number of standard stars. In all cases, one must "interpolate"; extrapolations always cause problems. That is a universal truism, but one often ignored.

I should perhaps make a few remarks on the potentials of the system in the future. As I noted above, the extensions to later spectral types and to supergiants is already underway. I think we will see increasing use of automation (one hopes the observer will still stay in close touch with the data, however), in imaging photometry (with CCDs), better and more effective use of both large and small telescopes, with better and more efficient photometers, the development of efficient spectrophotometers to do photometry with varying information resolution (bandwidths), better understanding of the system itself, and accuracies even better than those now being obtained.

One can do an excellent job, using the uvby system. Many observers are doing so now, and accuracy and precision of close to ± 0.003 have been obtained with the most careful work. We all have Bengt Strömgren to thank for the establishment of this most excellent and useful system. His insight will be of value to us for many years in the future. Thanks, Bengt!

REFERENCE

Crawford, D. L. 1988 PASP 100, 887

CCD PHOTOMETRY ON THE uvby SYSTEM

Barbara J. Anthony-Twarog and Bruce A. Twarog

University of Kansas

ABSTRACT: The Strömgren uvby photometric system has become the dominant photometric tool for studying stellar populations and Galactic structure. Despite this, application of CCDs to the system has been limited due to the uncertainty regarding the reliability of the approach and to the paucity of standard fields for calibration. Results of the last five years and work in progress are used to illustrate the feasibility and the value of this approach. For open clusters (e.g., M 67, IC 4651 and NGC 3680) one can derive the fundamental cluster parameters, as well as identify probable binaries and field stars. For globular clusters (e.g., NGC 6397 and Omega Cen), estimation of the cluster abundance from stars at the turnoff is possible while permitting the isolation of field stars from the cluster sample. Recent and ongoing improvements include the establishment of a set of approximately 30 standard fields for use with a CCD, extension of the system to globular cluster red giants, and the expansion of the uvby system to include a fifth filter centered on Ca II H and K, leading to the most sensitive metallicity index currently available for halo giants. Analysis of field star data indicates that there is weak metallicity sensitivity of the c_1 - (b-y) relation for metal-poor giants; when combined with the Ca index, it leads to the possibility of reliable reddening and abundance estimation for metal-poor globular clusters.

1. Introduction

Since its introduction by Strömgren (1966), the uvby system has shown itself to be an exceptional tool for the determination of the fundamental stellar parameters of luminosity, temperature, and metal content for stars of spectral type B through early G, especially when used in conjunction with Hβ (Crawford 1975, 1978, 1979). This ability has helped to make it the predominant photometric approach for the study of stellar populations (Twarog 1986). The extensive growth of the data set, due primarily to Olsen (1983) and his coworkers, coupled with the extension of the system to giants and cool dwarfs (Ardeberg and Lindgren 1985) makes it probable that the importance of the system will only grow in the near future.

Despite this optimistic projection, the one area where the system has lagged sadly behind is in the application of CCDs. Over the last five years, we have been engaged in a program to test and develop the use of CCDs with the uvby and, more recently, Hβ systems. The purpose of this paper is to summarize what we have learned to date, to discuss the most recent results, and to explain future planned uses of the system, including the addition of a new filter for application to metal-poor systems.

2. UNCERTAINTIES AND APPLICATIONS

Given that CCDs can be used with broad-band photometric systems to derive colors with a frame to frame uncertainty of 0.5%, is there any reason why such systems should fail to produce comparable accuracy for intermediate-band systems such as the uvby system? Clearly, NO. Taking into account the need for longer integrations to compensate for the

TABLE 1

Comparison of CCD and Photoelectric uvby Photometry

		$(m-M_o)$	s.e.m.	N	[Fe/H]	s.e.m.	N	<V>	Tel
M 67									
	PE	9.61	0.04	44	-0.07	0.02	36	13.0	1.5 m
	CCD	9.53	0.14	24	-0.07	0.03	27	13.5	4 m
IC 4651									
	PE	9.80	0.05	30	0.22	0.03	36	12.4	1.5 m
	CCD	9.87	0.05	29	0.25	0.03	31	13.2	4 m
NGC 3680									
	PE	9.78	0.06	24	0.10	0.02	24	12.8	1.5 m
	CCD	9.68	0.05	21	0.10	0.02	21	13.1	4 m

the smaller bandpass, CCD photometry on the uvby system accurate to 1% should be obtainable to some limiting magnitude dependent upon the chip, the filter, and the telescope aperture, leading to indices reliable to between 1.5 and 2.0%. Since this wasn't definite when the project first began, our initial observations concentrated upon systems where internal checks on the transformations were readily available and on stars where the calibrations of the indices were most applicable, i.e., F stars in intermediate-age open clusters, specifically M 67, IC 4651 and NGC 3680. In a series of papers (Anthony-Twarog 1987a, Anthony-Twarog and Twarog 1987, and Anthony-Twarog, Twarog and Shodhan

1989), we have discussed the analysis of the photometry in these clusters and the comparison of the results to the photoelectric data within the same object. The critical information is shown in Table 1, where we compare the mean values for the metallicity and distance and, more importantly, the standard error of the mean for each. Also noted is the typical V magnitude of the stars in the sample and the telescope used to obtain the data.

The PE data for M 67 are from Nissen, Twarog and Crawford (1987), while the PE data for NGC 3680 are from Nissen (1988). The table clearly demonstrates that the CCD results are as good as those derived from high quality photoelectric data. For M 67, the scatter is larger for the CCD distance modulus because potential binaries were not excluded from the CCD sample while they were from the PE sample. The gain from the use of CCDs is more dramatic if one takes into account the limiting magnitude of the surveys and the integration times. The CCD surveys extend typically three magnitudes below <V>, while the PE data reaches about one mag fainter. The total integration time for a typical cluster series on a 1.5 m telescope to provide the same accuracy at <V> = 13.0 for three fields is about three hours. In that time frame, PE photometry would provide a single observation at the same <V> for approximately 15 to 20 stars, with an average of three - four observations needed per star to obtain the same final accuracy. Thus, CCDs lead to a gain of at least a factor of five over standard PE work, with significant additional improvement occurring as larger format and uv sensitive chips become readily available.

In addition to obtaining the fundamental properties of open clusters, what additional uses can one make of the CCD approach? Within open clusters, Twarog (1983) in NGC 752 and Nissen et al. (1987) in M 67 have demonstrated that accurate uvby data can be used to identify probable binaries. In both clusters, every radial velocity binary for which uvby data are available had been tagged photometrically as a binary based upon anomalously small distance moduli; some photometric binaries, about 20%, exhibit no evidence of variations and may be photometric scatter, long-period variables, and/or inclined systems. Such an application could greatly improve the definition of the main sequence for more distant open clusters.

A more dramatic application of CCDs and uvby photometry is the observation of globular clusters. The preliminary study of NGC 6397 (Anthony-Twarog 1987b) has been expanded from one to six fields, concentrating on vby alone (Anthony-Twarog, Twarog and Suntzeff 1991). The expanded sample allows definition of the turnoff and main sequence mean relations with an internal accuracy of 0.01 in (b-y). The m_1 index provides a highly successful approach to eliminating field stars which have distinctly higher metallicities, both for the main sequence and the red giant stars. Fig. 1a shows the CM-diagram for all stars in the sample; in Fig. 1b we have removed the stars which do not fall along the cluster m_1 - (b-y) relation convolved with the errors at each

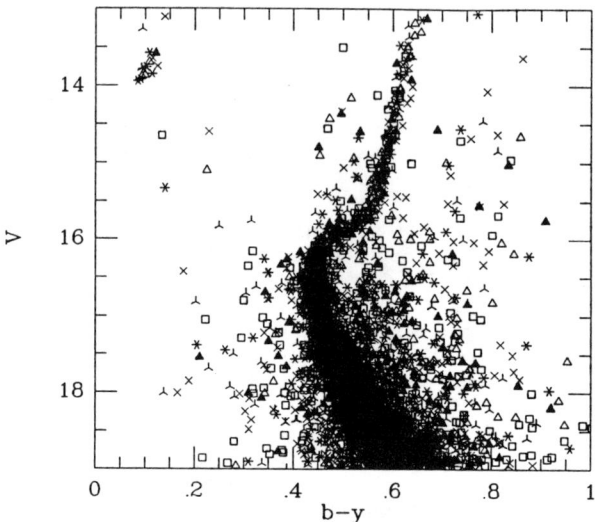

Fig. 1a. CM-diagram of NGC 6397, all stars.

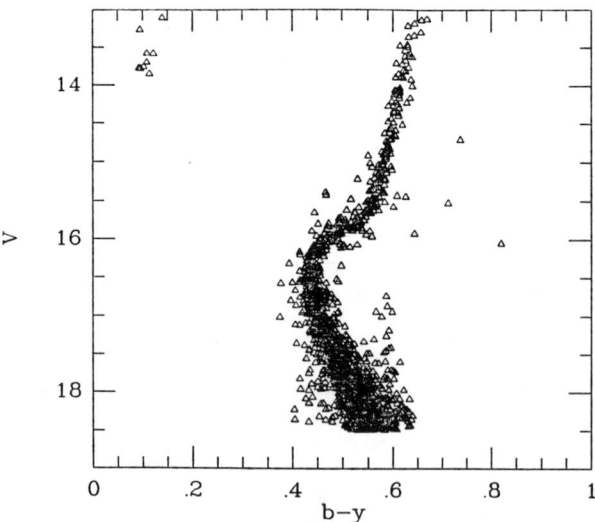

Fig. 1b. CM-diagram of NGC 6397, stars with discrepant m_1 values removed.

magnitude level. Fig. 1b also illustrates one of the primary interests in this cluster, the possibility of binaries evidenced by the asymmetry in the color distribution of the main sequence. Since we can eliminate the scatter due to field star contamination, the red tail in the color sample must be real. The only question is whether the composite systems which populate this tail are true binary systems or optical binaries caused by image confusion. Fig. 2 shows the radial variation of the fraction of "cmd" binaries. The concentration near the cluster center could be explained by either binary model, but the apparent increase at larger radii is inconsistent with the optical binary scenario. A better statistical sample at large radii would be valuable.

A second globular cluster studied to well below the level of the turnoff is Omega Cen (Mukherjee 1989). The goal was to test if the main sequence exhibited the same spread in color shown by the giants, and if the metallicity as measured by m_1 was consistent with an intrinsic metallicity range. The CM-diagram of the cluster is shown in Fig. 3, and it is apparent that in comparison with Fig. 1b, the scatter near the turnoff is excessive. From the m_1 data, the intrinsic spread in the index was significantly larger than predicted from the photometric errors, leading to an intrinsic range of 0.8 dex in [Fe/H] among the turnoff stars based upon the calibrations of Schuster and Nissen (1989).

3. WORK IN PROGRESS

Beyond the obvious need to use CCD data on both open and globular clusters to measure intrinsic properties of turnoff stars, what additional work can be done? Though hardly exciting, a fundamental need within the uvby system is for a set of well-defined CCD standard fields with a broad range in color. Over the last three years we have collected multiple frames on approximately 30 fields around the celestial equator in the hope of alleviating this deficiency. To ensure that they are on a common system, we have tied them to cluster fields which contain reliable data and have tried to observe photoelectrically at least one star in each field. The latter approach is needed because most of the cluster fields do not contain a uvby sample with a large color range; these clusters are now being observed to place some of their giants on the standard system defined by the work of Olsen (1984). The need for better standardization is best exemplified by the early results on NGC 6397 (Anthony-Twarog 1987b). The zero-points of the CCD photometry were defined by a set of four stars observed by Ardeberg, Lindgren and Nissen (1983). When compared to the field star data of Schuster and Nissen (1988), it is apparent that the turnoff in NGC 6397 is too blue in (b-y) and c_1 is too large, implying a significantly younger age. Our recent CCD and PE observations to renormalize the data show that the discrepancy lies with the photometry of Ardeberg et al. (1983), not the intrinsic properties of the cluster.

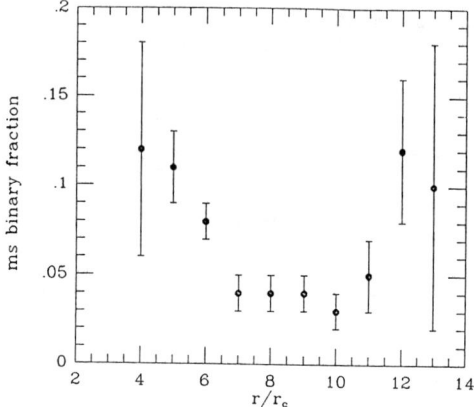

Fig. 2. Binary fraction as a function of distance from the cluster center in NGC 6397. Units are core radii.

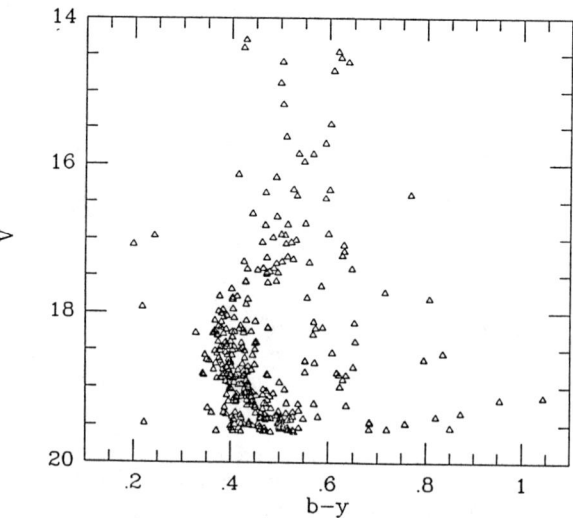

Fig. 3. CM-diagram for Omega Cen, all stars.

CCD PHOTOMETRY ON THE uvby SYSTEM

Two areas where the uvby system has been underutilized are in observations of metal-poor stars and of cool stars. The recent work of Schuster and Nissen (1988, 1989) has remedied the former problem for dwarfs, demonstrating conclusively the declining sensitivity of the system as one nears [Fe/H] = -3. For the latter problem, the program of Ardeberg and Lindgren (1985) should be the solution, but progress in this area has been lacking in recent years. To close out our discussion of CCD photometry, we will report on a pair of ongoing, linked projects to illustrate the value of expanding and extending the uvby system.

Though uvby photometry of turnoff stars in globular clusters can help put them on the same metallicity scale as the field stars, the faintness of most turnoffs and the declining sensitivity combine to make this impractical for many clusters. The obvious alternative is to use the red giants, which are brighter and do not suffer the declining sensitivity found among hotter dwarfs (Gustafsson and Bell 1979). Using the data of Bond (1980) and Twarog and Anthony-Twarog (1991), we have recalibrated the m_1 - (b-y) diagram for metal-poor giants (Fig. 4). Near (b-y) = 0.6, the giant branch just above the horizontal branch, the Δ[Fe/H]/Δm_1 slope is about 12 at [Fe/H] = -3 , the same as for F dwarfs of solar composition, and grows smaller at redder (b-y). CCD photometry of red giants in globular clusters will allow us easily to define the cluster metallicity to better than 0.1 dex.

In an effort to overcome the declining sensitivity of the m_1 index at low [Fe/H], for the last seven years we have been testing and calibrating a new filter for addition to the standard uvby system. The filter is approximately 100 Å wide, centered on Ca II H and K, a common region for abundance estimation at low [Fe/H]. We now have observations of over 2200 stars, including about 150 primary standards (Anthony-Twarog, Laird, Payne and Twarog 1991). Our observations confirm that at a given temperature, the Ca index is about three times more sensitive than m_1, meaning that it should work well for hot dwarfs, even at [Fe/H] = -3. This trend continues for red giants, as illustrated by the calibration in Fig. 5, and has produced a unexpected result for the most metal-poor red giant known, CD -38 245. Observations on five different photometric systems, including Ca, show that CD -38 245 has an ultraviolet deficiency which leads to a metallicity greater than -3, while BD -18 5550, supposedly at [Fe/H] = -2.7, has a uv excess which leads to [Fe/H] = -3.4. Though the uv excess can be explained by a blue companion, the discrepancy for CD -38 245 cannot be explained in this way. Even stranger is the result that over the 2.5 year span of observations, the Ca index for CD -38 245 has grown larger, implying higher metallicity (Twarog and Anthony-Twarog 1991).

The final contact point between uvby photometry and globular clusters is reddening estimation. The c_1 - (b-y) diagram for field

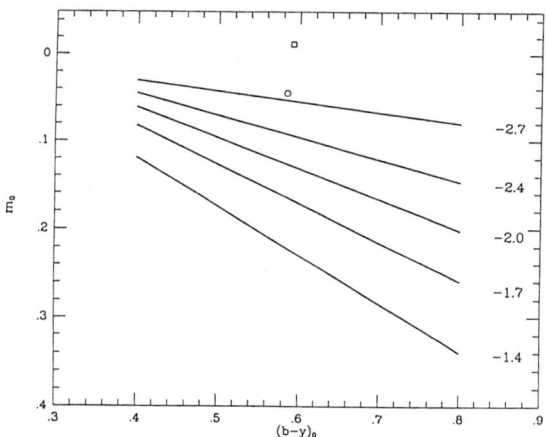

Fig. 4. Metallicity calibration for red giants in the m_1, (b-y) plane. The position of BD -18 5550 is noted with a square, CD -38 245 with a circle.

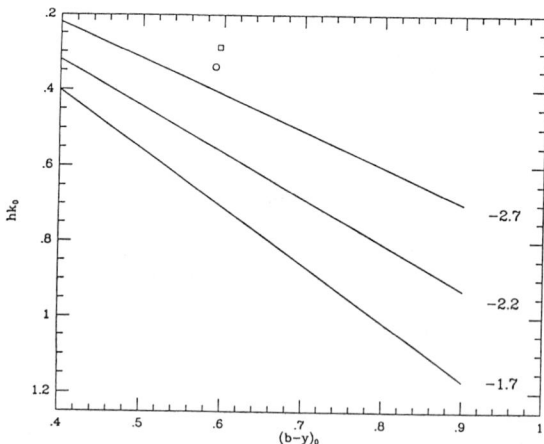

Fig. 5. Metallicity calibration for red giants using the Ca index, hk. Symbols for -18 5550 and -38 245 are the same as in Fig. 4.

halo giants in Bond (1980) shows considerable scatter, not unexpected given the range in [Fe/H] from -1.2 to -3.0. However, our recent observations show that the scatter is somewhat exaggerated by the photometric errors coupled with the use of a csc(b) law to correct for reddening. If we plot our photometry combined with that of Eggen (private communication) and reddening-corrected colors, using the maps of Burstein and Heiles (1982), the resulting diagram (Fig. 6) shows well-defined relations with modest scatter. Without taking [Fe/H] into account, at a given c_1, the dispersion in (b-y) is less than 0.03, slightly less than the errors expected from photometry and reddening corrections alone. To show that the metallicity has little effect, we can divide the sample into two groups, metal-poor (circles) and metal-rich (plusses, [Fe/H] > -2.0). It is apparent that there is little significant separation of the data points. This implies that if one defines a field star c_1 - (b-y) relation, for clusters with [Fe/H] < -1.5 one can derive reddenings by forcing the cluster relation to match the standard. Differentially, one should be able to set reddening values whose accuracy is only limited by the photometric uncertainty of the red giant branch.

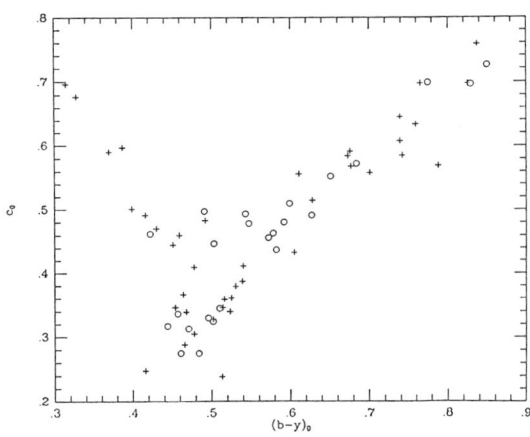

Fig. 6. Reddening-corrected c_1, (b-y) relation for metal-poor giants. Stars with [Fe/H] $\geq$ 2.0 are noted with plus signs, stars with [Fe/H] < -2.0 with circles.

REFERENCES

Anthony-Twarog, B. J. 1987a AJ 93, 647
Anthony-Twarog, B. J. 1987b AJ 93, 1454
Anthony-Twarog, B. J. and Twarog, B. A. 1987 AJ 94, 1222
Anthony-Twarog, B. J., Twarog, B. A. and Shodhan, S. 1989 AJ 98, 1634

Anthony-Twarog, B. J., Twarog, B. A. and Suntzeff, N. B. 1991, in prep
Anthony-Twarog, B. J., Laird, J. B., Payne, D. and Twarog, B. A. 1991, in preparation
Ardeberg, A. and Lindgren, H. 1985 in IAU Colloquium No. 88, Stellar Radial Velocities, A. G. D. Philip and D. W. Latham, eds., L. Davis Press, Schenectady, p. 151
Ardeberg, A., Lindgren, H. and Nissen, P. E. 1983 A&A 128, 194
Bond, H. E. 1980 ApJS 44, 517
Burstein, D. and Heiles, C. 1982 AJ 87, 1165
Crawford, D. L. 1975 AJ 80, 955
Crawford, D. L. 1978 AJ 83, 48
Crawford, D. L. 1979 AJ 84, 1858
Eggen, O. 1990, private communication
Gustafsson, B. and Bell, R. A. 1979 A&A 74, 313
Mukherjee, K. 1989 M.S. Thesis, University of Kansas
Nissen, P. E. 1988 A&A 199, 146
Nissen, P. E., Twarog, B. A. and Crawford, D. L. 1987 AJ 93, 634
Olsen, E. H. 1983 A&AS 54, 55
Olsen, E. H. 1984 A&AS 57, 443
Schuster, W. and Nissen, P. E. 1988 A&AS 73, 225
Schuster, W. and Nissen, P. E. 1989 A&A 221, 65
Strömgren, B. 1966 ARAA 4, 433
Twarog, B. A. 1983 ApJ 267, 207
Twarog, B. A. 1986 in IAU Symposium No. 118, Instrumentation and Research Programmes for Small Telescopes, J. B. Hearnshaw and P. L. Cottrell, eds., Reidel, Dordrecht, p. 135
Twarog, B. A. and Anthony-Twarog, B. J. 1991 AJ, in press

DISCUSSION

RICH: Considering NGC 6397 as a system of unknown abundance, what is the δ[Fe/H] in the cluster, based on the observation?

TWAROG: The standard error of the mean, due to internal error is less than 0.003 in m_1 and therefore less than 0.05 dex. The real limit to the [Fe/H] accuracy is the uncertainty in the m_1 zeropoint, which is no better than 0.015, or 0.30 dex in [Fe/H].

STETSON: Would you care to characterize the metallicity distribution in Omega Cen? Square? Gaussian? Bimodal?

TWAROG: Because of the intrinsic errors, it is better to use the intrinsic scatter in (b-y) rather than m_1, assuming the (b-y) spread is due solely to [Fe/H]. This distribution is broad and continuous, rather than Gaussian or bimodal.

STETSON: I realize that you have chosen and begun to observe your 30 equatorial standard fields. I would strongly urge you to consider including the open cluster M 11. It is conveniently equatorial and contains stars over a broad range of colors over a broad range of magnitudes. Have you explicitly included Pop II stars or equatorial globulars among your standards?

TWAROG: We have Pop II stars within the primary standards for Ca. Within the uvby CCD standards we have a range of open clusters along the equator. There are some standard fields within current globular cluster studies, but these will be expanded as our red giant program continues.

STETSON: When dealing with field Pop II stars of unknown distance, can you distinguish horizontal-branch stars from subgiants from main-sequence stars at the same (b-y)?

TWAROG: One can use c_1 to easily separate HB stars from less evolved stars at the same (b-y).

RATNATUNGA: What is the practical limiting magnitude of the system, exposures times and accuracy at faint magnitudes?

TWAROG: With a 36" and CCD, one can work comfortably at V = 15. With a single-channel photometer, Ca can be observed to V~ 12 for a metal-rich giant, fainter for metal-poor giants and hotter dwarfs. With a CCD, the gain should be at least 2 mag.

CRAWFORD: uvby is useful also for work in young open clusters, as m_1 gets smaller with reddening, and the separation from unreddened A and F field stars increases. One can easily separate members from non-members, therefore, unlike with the UBV system.

TWAROG: Agreed. CCD work on a large format detector or a Schmidt with a standard detector would be an invaluable way to isolate members of young open clusters, especially nearby clusters.

GARRISON: I have a little trouble with your nomenclature w.r.t. $-38°245$. Did you mean it was "metal-rich" or did you really mean "it was not not as metal-weak" as you thought?

TWAROG: Not as metal-poor as originally thought.

GARRISON: Regarding transformations to a "standard" system--How do you cope with the different response curves for different CCDs? I presume you carry your own filters around with you. e.g. Johnson defined UBV by a filter set <u>and</u> a 1P21 Tube (at 6000 ft alt !!) and if you didn't use exactly those conditions, it was difficult to transform.

TWAROG: We have our own Ca filter, but we have always used the uvby set provided by C.T.I.O. Our transformations are done over separate color ranges and normally include color terms for V, m_1, and c_1.

GRAY: In your transformation equations, what size are your color coefficients? Do you have problems in transforming highly reddened stars to the standard system?

TWAROG: Typically the color terms are at the 10% level, though the exact value varies with the type of chip. We have, as yet, not observed significant numbers of highly reddened stars, so the problem has not appeared.

RICHER: You showed a color-magnitude diagram on the Stromgren system for ω Cen. It shows a rather broad main sequence as expected because of the metallicity variations in the cluster. However, the number of stars in the diagram is still small. Is there some chance of increasing this sample size (do you have more data on hand?) so that you can look at the diagram in different metallicity regions.

TWAROG: Yes. All that is required is additional observation of more fields, rather than just the one used to define the color-magnitude diagram.

DEMARQUE: Do you have any information on the ω Cen metallicity spread from the giants?

TWAROG: The information should be derivable once the photometry is reliably transformed to the standard system.

WARREN: Is your calcium index sensitive across the complete abundance range of red giants?

TWAROG: No, as [Fe/H] increases, the Ca line saturates and changes only occur in the wings. Thus, the sensitivity declines.

SYSTEMATIC COLOR TRANSFORMATION EFFECTS IN STROMGREN PHOTOMETRY

J. Manfroid

Institute of Astrophysics, University of Liège

C. Sterken

Astrophysical Institute, University of Brussels

ABSTRACT: A major cause of systematic errors in photometry can be attributed to the color transformation. A thorough analysis of numerous observing runs performed with various photoelectric equipment in the Strömgren system shows how reliable the transformations in the different indices are. Conclusions are drawn about the choice of the standard stars defining the system.

1. INTRODUCTION

The aim of color transformations in photometry is to convert data obtained with a given spectral curve into a standard system having only slightly different spectral characteristics. Mathematically the problem is to find a relation between integrals involving various functions: spectral response of the equipment, spectral distribution of the stellar flux and atmospheric extinction. If the stellar spectra could be described by a well-defined class of functions, as is the case with black bodies, it would be possible to find such a relation. But stellar energy distributions have great diversity and the problem has no general solution.

Hence the color transformations used are only approximations and the accuracy of the transformation procedure depends on several factors involving the width of the bandpasses and the variety of stellar types observed. The degree of conformity of the instrumental system to the standard one is of course essential. A perfect match could eliminate the need for any color transformation.

Young (1974, 1988) has shown that if bandpasses are chosen in such a way that the selection corresponds to a dense sampling of the wavelength domain, one would obtain accurate results. Unfortunately that condition is never met when photometric systems are being designed. The range of stellar types is the other major factor affecting the accuracy for a given photometric system. If a transformation is valid for some stars (say main-sequence stars), it may not be valid when other stars (e.g. higher luminosity classes) are included. There is a general consensus (based on careful analysis of observational data) that different color transformations should be applied for different sets of stars. This approach however makes the reduction procedure very cumbersome.

Even when a single transformation is used for a given set of stars, it may be difficult to determine the right color transformation, especially for systems having three or more indices. This is because the number of parameters to be fixed is rather large, even when simple linear transformations are used. Moreover the standard stars used to derive the transformations may not have the optimal characteristics for that purpose, so that measurement errors of a few millimagnitudes and inaccuracies of the same order in the standard values, will cause much larger uncertainties for the indices of many stars.

In the following we give some evidence concerning this latter point, based on observations made with various equipment in the Strömgren uvby system.

2. OBSERVATIONS AND REDUCTIONS

We consider data which were obtained during a program aimed at studying long-term variations of variable stars. Since October 1982, a considerable amount of photometric observing time at the European Southern Observatory has been allotted to the Long-Term Photometry of Variables (LTPV) program (Sterken 1983, 1986). This represents an average of six months per year at one of the small photometric telescopes at La Silla Observatory (the ESO 50-cm, the University of Bochum's 61-cm and the University of Copenhagen's 50-cm telescopes), and the measurements so far yielded a total data set of more than 600 nights. This data bank, with more than 100,000 entries encompassing 10 Mb, is very well suited for statistical studies.

In uvby reduction schemes, the color transformations are generally linear and can be described by incomplete matrices, i.e. wherein only the diagonal and the first column (corresponding to (b-y)) are not zero. The linear transformation is then written as:

$$U_s = M U_o + K$$

where U is the vector of indices:

$$U = \begin{pmatrix} (b-y) \\ y \\ m_1 \\ c_1 \end{pmatrix}$$

The suffixes s and o respectively denote the standard and instrumental values. K is the vector of zero-points. The color transformation

SYSTEMATIC COLOR TRANSFORMATION EFFECTS IN STROMGREN PHOTOMETRY

matrix M is written as

$$M = \begin{pmatrix} m_{11} & 0 & 0 & 0 \\ m_{21} & 1 & 0 & 0 \\ m_{31} & 0 & m_{33} & 0 \\ m_{41} & 0 & 0 & m_{44} \end{pmatrix}$$

The standard stars are taken from the list of Olsen (1983). They have been supplemented by a few stars from Olsen (1984) which have been used as comparison stars. Manfroid (1985) has emphasized the role that constant stars, with known uvby colors, may play in securing a consistent solution.

It is not unusual to see photometrists calculate color transformations with as few as six or ten measurements of standard stars. But serious observers use many more standards and this was certainly the case during the LTPV measurements. We shall show below how crucial it may be.

In a first step the reduction of the data obtained during periods of roughly one month has been carried out with an improved version of the program PHOT2 (Manfroid 1985). This program uses every measurement of every constant star and of every standard star. Since the LTPV project involves a large number of measurements of comparison stars, the major advantages of PHOT2 are obvious. The implementation of this reduction procedure equally facilitates the task of the observer who does not have to carry out a tedious and complicated schedule of extinction measurements.

The color transformation coefficients were calculated from this relatively small number of nights. Although they could be considered accurate by general standards, we were of the opinion that more accurate values are needed because stars with extreme indices lying outside the standard range, are very sensitive to small variations of those transformation coefficients.

We consequently used the preliminary derived matrices only for checking the stability of a given instrumental system. We found that when the same filters and the same type of photocathodes were used, the reproducibility of the color equations was excellent, even in the case of using two different telescopes and photometers. A statistical analysis of the color matrices obtained for each period shows that the rms values of the m_{ij} (in 0.0001 mag) (given in Table 1 and graphically represented in Fig. 1) for any of the five instrumental configurations stay within a few percent.

The final data are then obtained through a second stage of the reduction process (for a description, see Manfroid et al. 1990). The underlying hypothesis is that the true color matrices are stable over a time scale of several years. This is a straightforward extrapolation

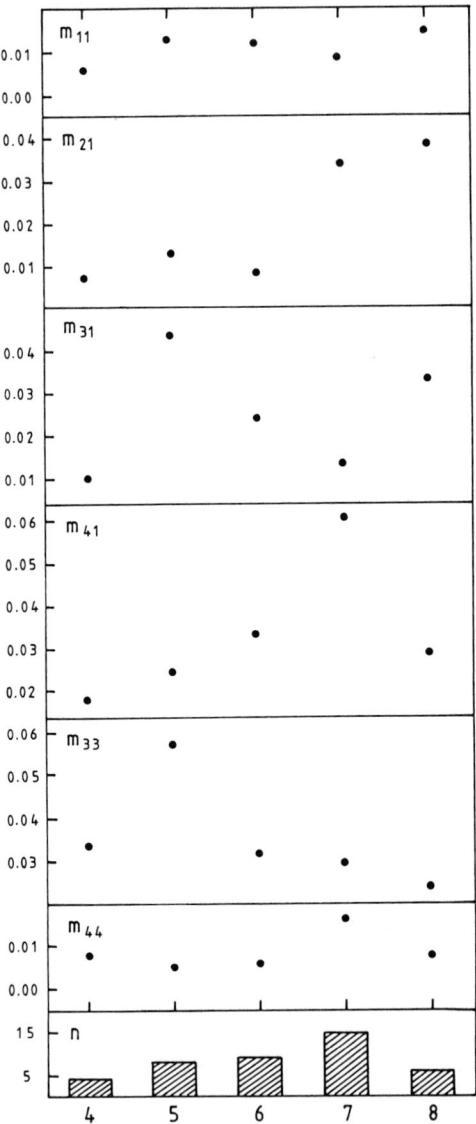

Fig. 1. Mean errors of the transformation coefficients $m_{i,j}$ for each instrumental configuration. n denotes the number of observing runs during which each configuration was used.

of one of the fundamental assumptions in the multinight algorithm. The adopted procedure allows a continuous updating of the data sets. Every time additional measurements are obtained in one of the instrumental systems, the complete set corresponding to this system is reprocessed.

TABLE 1.

rms values of the m_{ij} (in 0.0001 mag) for each of the five instrumental configurations.

Run #	$\sigma(m_{11})$	$\sigma(m_{21})$	$\sigma(m_{31})$	$\sigma(m_{41})$	$\sigma(m_{33})$	$\sigma(m_{44})$	number of periods
4	57	70	102	178	336	79	4
5	131	129	440	245	573	51	8
6	119	84	238	334	317	61	9
7	86	334	144	611	296	164	15
8	150	387	335	291	245	81	6

3. DISCUSSION

The month-to-month variations of the computed color coefficients of the instrumental systems do not follow a clear trend, but appear to be random fluctuations mainly caused by the particular distribution of the standard stars, by inaccuracies in the standard values, and by measurement errors.

Table 1 (and Fig. 1) show that some parameters are better defined than others. It is well known for instance that the m_{11} coefficient representing the (b-y) transformation can be easily determined. This is also the case for m_{44} corresponding to c_1. However the c_1 transformation is not so well determined because of the relatively large uncertainty affecting the non-diagonal coefficient m_{41}. Both other non-diagonal coefficients show a similar instability affecting y and m_1. But the m_1 index is the most sensitive because the uncertainty on the diagonal term m_{33} is rather large. Let us recall that by definition $m_{22} = 1$ and shows no variance. Since measurement errors are comparable in each band we conclude that errors on the m_1 index mainly come from a bad distribution of the standard values. The range of m_1 is smaller than for other indices. There are few stars with extreme values, they change according to the epoch of the observations, and they have a strong weight in the calculation of the solution. All this contributes to increased inter-period variations. The instability of the non-diagonal coefficients points toward a less than perfect justification of the transformation equation. One of the reasons is that a single transformation is used for all stars, although the range of indices is large and could need distinct relations. The opportunity of those non-diagonal terms may also be questioned. Data relative to system 7, which has sharp and rectangular passbands, show an abnormally large $\sigma(m_{41})$ and a rather large $\sigma(m_{21})$. This may indicate that the classical relations may suit some equipment better than others. This would confirm the need of smooth passbands, in spite of the advantage of higher transmission of the rectangular profiles. It also shows that more appropriate, and more general color transformations should be worked out. However this goal can only be achieved when a wider set of standard stars, including many more stars with peculiar or extreme indices becomes available. Several carefully derived extensions of the primary uvby standards have been published, but some of these lists have been obtained with the use of rectangular passband photometers. In the latter case it is impossible to determine what values the indices of these additional stars would have if they had been measured in the original instrumental system. Those generalizations of the uvby standards are not directly based on an instrumental system. They result from analytical transformations applied to original data obtained in non-standard systems. It can even be stated that those standard data do not represent any physically possible instrumental system. Mathematically the integral equations describing the photometric indices would not accept any solution for the instrumental functions. The very nature of those equations, which prevents general color transformations, also prevents such reconstruction.

REFERENCES

Manfroid, J. 1985 Thesis, University of Liège
Manfroid, J., Sterken, C., Bruch, A., Burger, M., de Groot, M., Duerbeck, H. W., Duemmler, R., Figer, A., Hageman, T., Hensberge, H., Jorissen, A., Madejsky, R., Mandel, H., Ott, H. -A., Reitermann, A., Schulte-Ladbeck, R. E., Stahl, O., Steenman, H., vander Linden, D. and Zickgraf, F. -J. 1990 A&AS, in press
Manfroid, J. and Sterken, C. 1987 A&AS 71, 539
Olsen, E. H. 1983 A&AS 54, 55
Olsen, E. H. 1984 A&AS 57, 443
Sterken, C. 1983 The Messenger 33, 10
Sterken, C. 1986 in The Study of Variable Stars Using Small Telescopes, J. R. Percy, ed., Cambridge University Press, p. 165
Young, A. C. 1974 in Methods of Experimental Physics, Vol 12A, Academic Press
Young, A. C. 1988 in Second Workshop on Improvements to Photometry, NASA Conf Publ 10015, p. 215

DISCUSSION

CRAWFORD: It is easy to show that the amount of down time of telescopes is strongly correlated with the number of instrument changes. It is least on a telescope with no changes. This fact appears to be independent of observatory.

STERKEN: I agree. Dedicated instruments which remain attached to a telescope are less likely to produce surprises; furthermore any changes made to the instrument are more consistently recorded in the log books than it's done in the case of non-dedicated instruments.

RICHER: You have mentioned that you are planning on placing an order for 59 sets of Stromgren filters. I have 2 questions. (a) Is it possible that some of these could be 2-inch filters instead of the 1-inch ones that you mentioned and (b) are these filters going to be perfect imaging filters?

STERKEN: I have been quoted prices for both imaging and non-imaging quality filters. I'm considering buying a large set of standard 1" filters, and a smaller sample of larger imaging filters.

STETSON: Is it possible that some of the filter sets could be two inches square?

STERKEN: Yes, in case we have enough candidates for placing an order.

WARREN: How will you ensure that the transmission curves for your fifty filter sets are not square and that they are acceptably uniform over all the sets?

STERKEN: The manufacturer guarantees uniformity within each batch of filters produced. For what concerns the shape of the transmission curve, they should be within the tolerances specified by the manufacturer. So we have control over the shape.

GARRISON: As you may know, the University of Toronto operates a well-equipped 60 cm telescope at Las Campanas, with the idea of carrying out projects that do not require a large telescope, but do require the kind of instrumentation available on large telescopes, such as a good CCD. The telescope has been very successful in filling this niche; an assessment of the productivity for this small telescope is published in the August 1990 Journal of the Royal Astronomical Society of Canada.

GARRISON: Is Erik Olson willing to use this filter system? If not, how well can the data be transferred to his system, since his has more stars observed than anyone else? We would be interested in buying a set of filters for Las Campanas.

STERKEN: If filters are used which reasonably match the original set, transformation may be possible to some extent, though it will take a large observational effort to observe a sample that is representative of his data. If deviant filter sets are used, transformation will not be possible.

BURKI: By performing differential photometric measurements, it can be difficult to control the eventual trends in the instrumentation over the years. For this reason, we prefer at Geneva to perform absolute photometry of variable stars. Depending on the accuracy we want, some additional standard stars are added during the night.

STERKEN: In principle we do the same: we observe many standard stars, and transform the measurements to the standard system, so we obtain all-sky photometry and we publish all variable and comparison star measurements *in extenso*. But every program star measurement is sandwiched between comparison stars. By using the differenced between variable and comparison stars, we eliminate the largest uncertainties in the extinction correlation, and we obtain a higher precision than if we would use non-differential measurements in the standard system.

E(b-y) AS THE PRIMARY PARAMETER FOR POLAR δm_1, δc_1 AND Z(pc) RELATIONS

Jens Knude

Copenhagen University Observatory

ABSTRACT: In a complete magnitude limited sample of A5 - (F9 - G0) stars at the NGP significant correlations are found between the extrinsic parameter E(b-y) and the intrinsic parameters δm_1 and δc_1.

The correlations' origin is E(b-y)'s variation with distance in a magnitude limited sample and are the results of a metallicity gradient perpendicular to the Galactic plane and a luminosity bias coming from the use of a magnitude-limited sample.

The luminosity selection with distance may cause a chemical gradient in any magnitude limited sample - for narrow spectral ranges at least.

1. INTRODUCTION

All A and F stars above b = +70° and brighter than B = 11.5 mag have been observed in the uvbyβ system. For the ≈ 5000 stars fulfilling the calibration criteria from the standard system to the intrinsic stellar properties δm_1, δc_1 and the extrinsic parameters distance ≈ Z(pc) and E(b-y) have been computed. It is the intention to compute the density variation $\nu(Z)$ for various sub populations and to understand possible selection effects wherefore δm_1 and δc_1 were plotted versus E(b-y), Fig. 1 and 2 respectively. Rather astonishingly these parameters seem to correlate tightly with E(b-y). A trend indicating a similar variation may have been noticed by Marsakov and Shevelev (1989) in catalogs from the literature, Philip and Egret (1980), Twarog (1980), Olsen (1984), Olsen and Perry (1984). Stars with negative and positive color excess were found to have significantly different average metallicities and state of evolution. Marsakov and Shevelev's explanation of the <δc_1> - E(b-y) variation is based on the assumed uncertainty in the measurement of particularly virtually zero color excesses and they suggest that the two-point relation entirely may be due to random errors in the uvbyβ photometry. It is mainly the errors from β observations that seem to be important. A similar scheme does, however, not work out for the δm_1 trend.

A. G. Davis Philip, A. R.
Upgren and K. A. Janes (eds.)
PRECISION PHOTOMETRY
147 - 151 © 1991
L. Davis Press

Jens Knude
University Observatory
Oster Volgade 3
DK 1350 Copenhagen
Denmark

As a consequence of their conclusion, Marsakov and Shevelev suggest that reddening corrections of nearby slightly reddened stars introduce spurious effects in the intrinsic stellar parameters and thus should be avoided.

With the advent of the IRAS satellite it has almost been accepted that interstellar dust is present even rather close to the Sun so reddening corrections seem required also in the solar vicinity. Otherwise the stars actually reddened will appear more metal poor than they are and a chemical gradient could thus be blurred.

An alternative explanation of the observed δm_1 - δc_1 - $E(b-y)$ relations is offered:

1. The color excess range that can be observed depends on distance.
2. The minimum stellar evolution that can be measured, for a given color, is also a function of distance.

2. OBSERVATIONS

In Figs. 1 and 2 are shown the observed variations of δm_1 and δc_1 with $E(b-y)$. The ordinates are averages over 0.010 mag intervals in $E(b-y)$. The diagrams seem to show that unreddened stars are metal poor (large δm_1) and old (large δc_1) and conversely that reddened ones are metal rich and young. The old/metal poor and young/metal rich trend fits the canonical concept of the Galaxy's chemical evolution but why should metallicity and age correlate with interstellar reddening? For very young stars a generic relation between dust and age is to be expected but probably not for the youngest stars in the present sample, ≈ 1 billion years.

In Fig. 3 we show the diagram where distances have been averaged over the same 0.010 mag $E(b-y)$ bins. Apparently smaller excesses are measured when the distance is increased. Such an effect is always present in reddening data from magnitude limited samples, see e.g. Knude (1987). A more detailed diagram would show that the small excesses are measured for the whole distance range whereas the larger ones only can be measured for the nearby stars. Teerikorpi (1990) has recently considered analogous problems from a statistical point of view.

Finally Fig. 4 shows how δc_1 varies with distance, but with a sample confined to the specific color $(b-y)_0$ from 0.325 to 0.350 mag. The diagram shows data for 1082 stars in this color bin and we notice that at a given distance δc_1 must exceed a certain minimum value to have been included. The complete sample shows the same effect which, however, is more clearly demonstrated for stars in a narrow color interval. The solid curve in Fig. 4 indicates the minimum δc_1 required for stars with $(b-y)_0$ = 0.338 mag and M_v = 4.11 mag.

E(b-y) AS THE PRIMARY PARAMETER

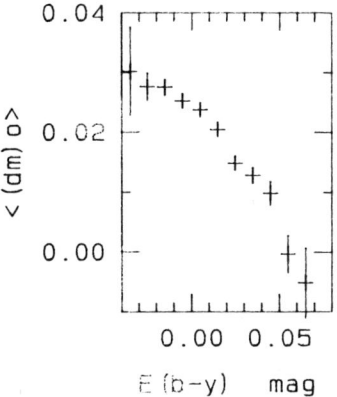

Fig. 1. $<(\delta m)_o>$ versus E(b-y) in 0.010 mag bins for ~ 5000 NGP stars. The error of the mean is also indicated.

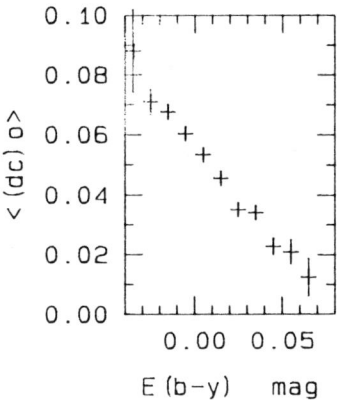

Fig. 2. $<(\delta c)_o>$ versus E(b-y). The error of the mean is also indicated.

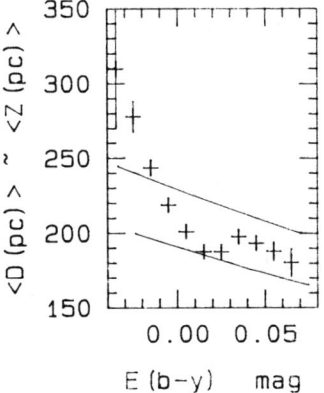

Fig. 3. Average distance from the plane plotted as a function of the color excess. The error of the mean $<Z(pc)>$ also is indicated. The solid curves show maximum distances where G0 (lower) and F8 (upper) stars may be observed as a function of reddening when B(lim) = 11.5.

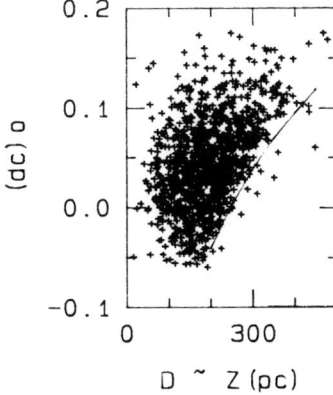

Fig. 4. $<(\delta c)_o>$ versus distance. The solid line indicates the minimum observable $<(\delta c)_o>$ in a sample with the B(lim) = 11.5 mag and M(V) = 4.11, valid for F7 stars.

3. INTERPRETATION

We propose that the variation of the intrinsic parameters with E(b-y) may be understood from the variation of color index with distance. If δm_1 and δc_1 do vary with distance from the Sun they will also depend on E(b-y) because of the "homogeneous" dependence of color excess on distance. The true variation of δm_1 with distance, in our case with Z(pc), is the chemical gradient. Fig. 4 shows that in the magnitude limited sample the average δc_1 will increase with distance because only the evolved stars can be included, the less evolved ones are too faint to have been included. The quantity δm_1 could increase with Z(pc) because there is an abundance gradient. This is probably true for if we plot δm_1 versus E(b-y) for stars in narrow distance intervals, where a range of color excesses have been observed; we won't see any variation with E(b-y). The Z(pc) variation is, however, difficult to separate from the age effect introduced with the δc_1 - Z(pc) dependence.

Furthermore a chemical dependence on distance in any direction is almost an inherent effect in magnitude limited samples. If the Galaxy is chemically enriched during its lifetime a true δm_1 - δc_1 relation is to be expected. Old stars are probably metal poor. Due to the sampling bias the most distant stars will have the largest δc_1 values; they will be among the relatively oldest stars in the sample and then probably also among the metal poorest ones. Any chemical gradient reported may therefore partly have been caused by luminosity selection.

REFERENCES

Knude, J. 1987 A&A 171, 289
Marsakov, V. A., Shevelov, Yu. G. 1988 AZh 65, 918
Olsen, E. H. 1984 A&AS 54, 55
Olsen, E. H., and Perry, C. L. 1984 A&AS 56, 229
Philip, A. G. D. and Egret, D. 1980 A&AS 40, 199
Teerikorpi, P. 1990 A&A 235, 362
Twarog, B. A. 1980 ApJS 44, 1

DISCUSSION

YOSS: How long did it take to observe your 5458 stars and how many observations per star?

KNUDE: The stars were observed during ~250 nights but due to the NGP's narrow right ascension interval another ~4000 stars in selected areas were included. Most stars are only observed once in uvby and once in Hβ but ~1200 stars were reobserved twice because of interesting δm_1 indices.

GRAY: I was interested to see the correlation which you have obtained between δm_1 and δc_1 for your stars. I would like to point out that in the A dwarf stars, rotation will produce a similar correlation. That is, rotation will increase δc_1 and slightly increase δm_1.

KNUDE: If such an effect also exists for the F dwarfs it must certainly be taken into account when chemical evaluation and gradients are considered.

MAJEWSKI: Can you give an average value, in E(B-V), of the total reddening at the NGP (i.e. for stars well out of the galactic plane)?

KNUDE: For these stars $<E(b-y)> = 0.01^m$ but this number may not be representative for the NGP. 10% of the lines of sight have $E(b-y) > 0.050^m$.

MAJEWSKI: How does your work compare to that of Hilditch, Hill & Barnes (1983)?

KNUDE: There are ~650 stars in common to Hill, Hilditch and Barnes program for stars above b ~ 75° and $<E(b-y)_{HHB} - E(b-y)_K> = -0.005^m$ with rms error = $\pm 0.018^m$.

CCD FOUR-COLOR PHOTOMETRY OF GLOBULAR CLUSTER STARS

A. G. Davis Philip

Union College and Van Vleck Observatory

ABSTRACT: Four-color CCD frames have been taken in the globular clusters M 4, M 5, M 22, M 55 and M 92 using 1-m telescopes at Kitt Peak National Observatory and Cerro Tololo Inter-American Observatory. The data have been reduced at the Dominion Astrophysical Observatory using Peter Stetson's program, DAOPHOT. Three frames in each color were obtained in each field studied. In a field of 2 by 3.5 minutes of arc from 5 to 22 BHB stars were measured as well as several hundred other stars. The internal probable errors, computed by comparing the magnitudes of the BHB stars in each of the three frames, were under ± 0.01 mag for magnitudes down to V = 15. In M 22 and M 55 external probable errors were calculated from comparisons with measures made by Arp and Melbourne (1959) and by Lee (1977). These came out to be ± 0.02 mag, the probable errors quoted in two papers cited. A CM-Diagram has been constructed for all the stars in the frame, as well as an additional CM-Diagram of just the BHB stars.

1. INTRODUCTION

The major part of the paper given at the meeting has been published in Philip (1990) and readers are referred to that article. The main conclusion was that precision photometry of BHB stars in several globular clusters yielded information concerning the state of their evolution. The BHB stars could be divided into two groups. One group fell, closely grouped, along a curved line in a y, (b-y) diagram; the second group scattered above this line with Δy values of 0.2 mag or more. This appearance matched the distribution of evolutionary tracks of horizontal-branch stars from Sweigart (1987). BHB stars of different masses have evolutionary tracks that initially fall on the Zero-Age Horizontal Branch (ZAHB) and then evolve to the blue. Depending on the stellar mass, the tracks turn upward and bend to the right, starting to evolve back up along the asymptotic red giant branch. If one imagines a group of BHB stars, of slightly different masses, evolving along BHB tracks, the distribution of points representing the BHB stars in a y, (b-y) diagram would seem to be just what is obtained by the plots of the CCD four-color data obtained here. Since the initial tracks, whatever the mass of the star, fall on the ZAHB, one would expect that stars in this stage of evolution would follow a common line in the y, (b-y) diagram. Stars of different masses have different turnoffs and one would expect that stars past the turnoff would exhibit a greater scatter and fall above the ZAHB. Figs. 8 and 9 in Philip (1990) show a combined y, (b-y) diagram for BHB stars in four globular clusters and compares the distribution with some evolutionary tracks from Sweigart (1987). In an article in these proceedings, Sweigart (1991) points out that the width of the horizontal branch can be used to estimate the helium abundance. The more accurate CCD four-color photometry, if applied to a greater number of HB stars in different globular clusters, would allow better estimates of the width of horizontal branches to be made.

CCD frames have been obtained of BHB stars in the globular clusters NGC 362 and NGC 4833 and these tapes will be reduced at the Dominion Astrophysical Observatory, using Stetson's program DAOPHOT on future visits there. Frames have also been obtained of FHB stars that have been set up as secondary four-color standards and these frames will be reduced using aperture photometry routines at DAO.

2. CM-DIAGRAM OF A FRAME IN M 5.

As part of a senior research project at Union College, Rebecca Koopmann (now at Yale University) worked on data brought back from DAO concerning a set of CCD frames taken at KPNO of the globular cluster M 5. Three CCD frames (y, b and v) had been reduced at DAO by AGDP using DAOPHOT and the data were transferred to a floppy disk and loaded into a microcomputer at the Physics Dept at Union College. Using the program "123" the data for the three frames in each color were compiled and merged. Stars were identified by their x, y coordinates and the three sets of photometric data were combined to produce a catalog of stellar photometry in M 5. Fig. 1 shows a y CCD frame of the area chosen for investigation. BHB stars located in this frame are identified by numbers and their identifications and photometry are presented in Table 1. Stars, identified in Arp (1955), are marked with a prefix, A and the Arp number. There are eleven BHB stars that were not identified on the Arp (1955) charts and these stars are identified by the number on the CCD frame (without a prefix). CCD photometry allows measures to be made of stars much closer to neighbor stars than can be done in single channel photometry and many of the additional BHB stars were found close to the cluster center.

Kurucz (1991), in an article in these proceedings, has presented his new atmospheric models and there are tables and figures relating four-color indices to T_{eff} and log g for Pop I stars. There are changes in these relations since the relations published by Philip and Relyea (1979) which used the older Kurucz (1979) models. For example if one investigates the change in the predicted T_{eff} and log g values for a Pop I A star with a c_1 index of 1.1 and (b-y) indices of 0.04, 0.08 and 0.12 differences (taken in the sense of new - old) of +0.05, -0.03 and +0.08 in log g and -25, +110 and +180 degrees in the effective

CCD FOUR-COLOR PHOTOMETRY OF GLOBULAR CLUSTER STARS

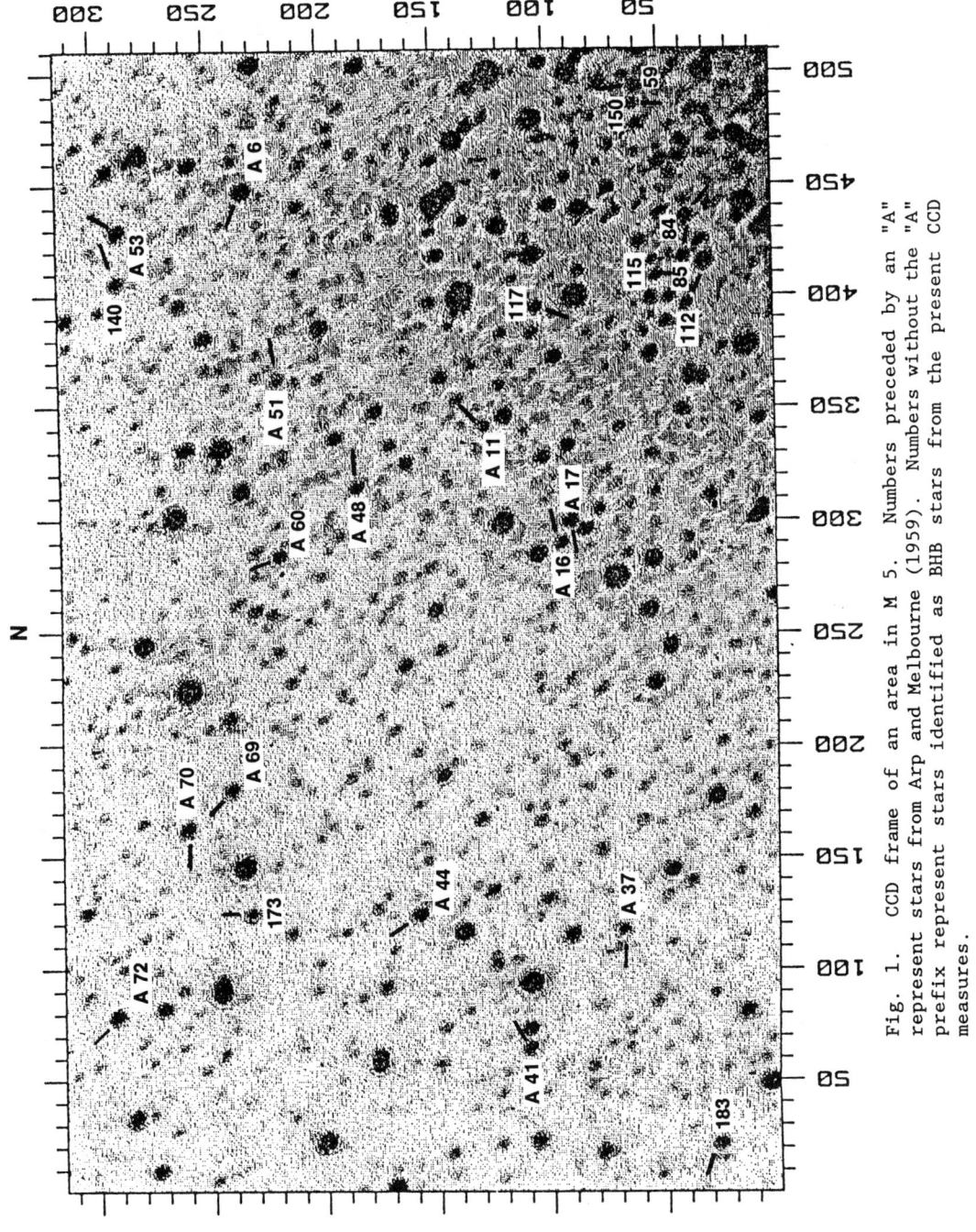

Fig. 1. CCD frame of an area in M 5. Numbers preceded by an "A" represent stars from Arp and Melbourne (1959). Numbers without the "A" prefix represent stars identified as BHB stars from the present CCD measures.

temperature can be found. In the 1979 grid relating color indices to T_{eff} and log g the rms errors were quoted as $\pm 160°$ and ± 0.2, so these differences are well within the calculated errors of the first grid for A-type stars.

Fig. 2 presents the CM-Diagram for all the stars measured in the b and y frames taken in M 5. On the left the blue horizontal branch can be seen. To the right there are a few red horizontal-branch stars and then, extending to the upper right one finds the giant branch and the asymptotic giant branch stars. On the bottom right the upper part of the main sequence is found.

Fig. 3 presents the CM-Diagram for just the BHB stars in the frame. The stars identified as BHB stars in Arp and Melbourne (1959) are plotted as triangles. The BHB stars, not in Arp and Melbourne, are plotted as crosses. The distribution of points on the lower left match the distribution of stars on the lower envelope in the combined CM-Diagram in Philip (1990). If this match is made, then the points in the upper middle and to the right in Fig. 3 fall on the "upper" track in the combined CM-Diagram. The y magnitudes and (b-y) colors of the stars plotted in Fig. 3 are tabulated in Table 1.

ACKNOWLEDGMENTS

The Kitt Peak National Observatory is thanked for the assignments of time for this project. Some of the observations were made with D. S. Hayes.

REFERENCES

Arp, H. C. 1955 AJ 60, 317
Arp, H. C. and Melbourne, W. C. 1959 AJ, 64, 28
Kurucz, R. L. 1979 ApJS 40, 1
Kurucz, R. L. 1991 in Precision Photometry, Astrophysics of the Galaxy, A. G. D. Philip, A. R. Upgren and K. A. Janes, eds, L. Davis Press, Schenectady, p. 25
Lee, S. W. 1977 A&AS, 29, 1
Philip, A. G. D. 1990 in CCDs in Astronomy. II., A. G. D. Philip, D. S. Hayes and S. J. Adelman, eds, L. Davis Press, Schenectady, p. 107
Philip, A. G. D. and Relyea, L. J. 1979 AJ 84, 1743
Sweigart, A. V. 1987 ApJS 65, 95
Sweigart, A. V. 1991 in Precision Photometry: Astrophysics of the Galaxy, A. G. D. Philip, A. R. Upgren and K. A. Janes, eds, L. Davis Press, Schenectady, p. 11

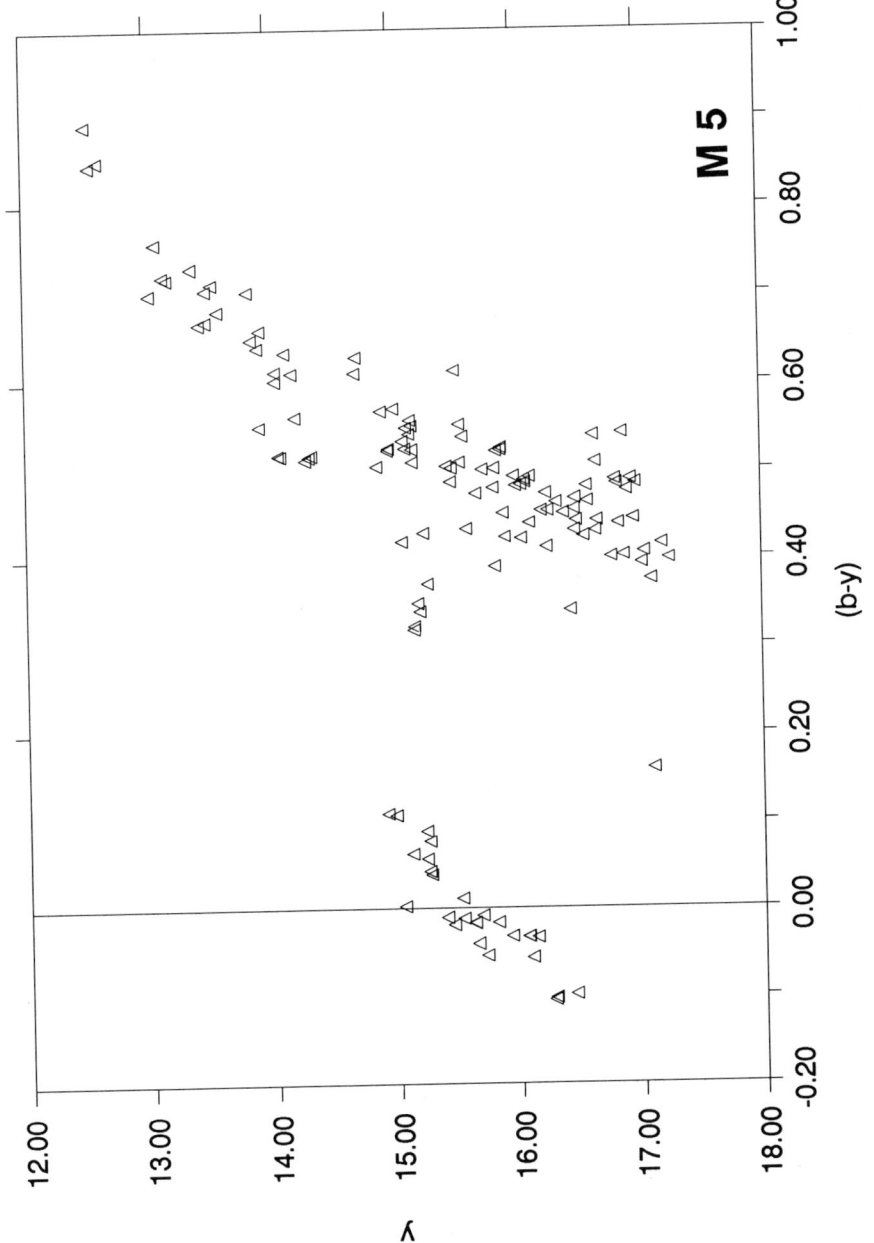

Fig. 2. CM-Diagram of the stars measured in CCD frame.

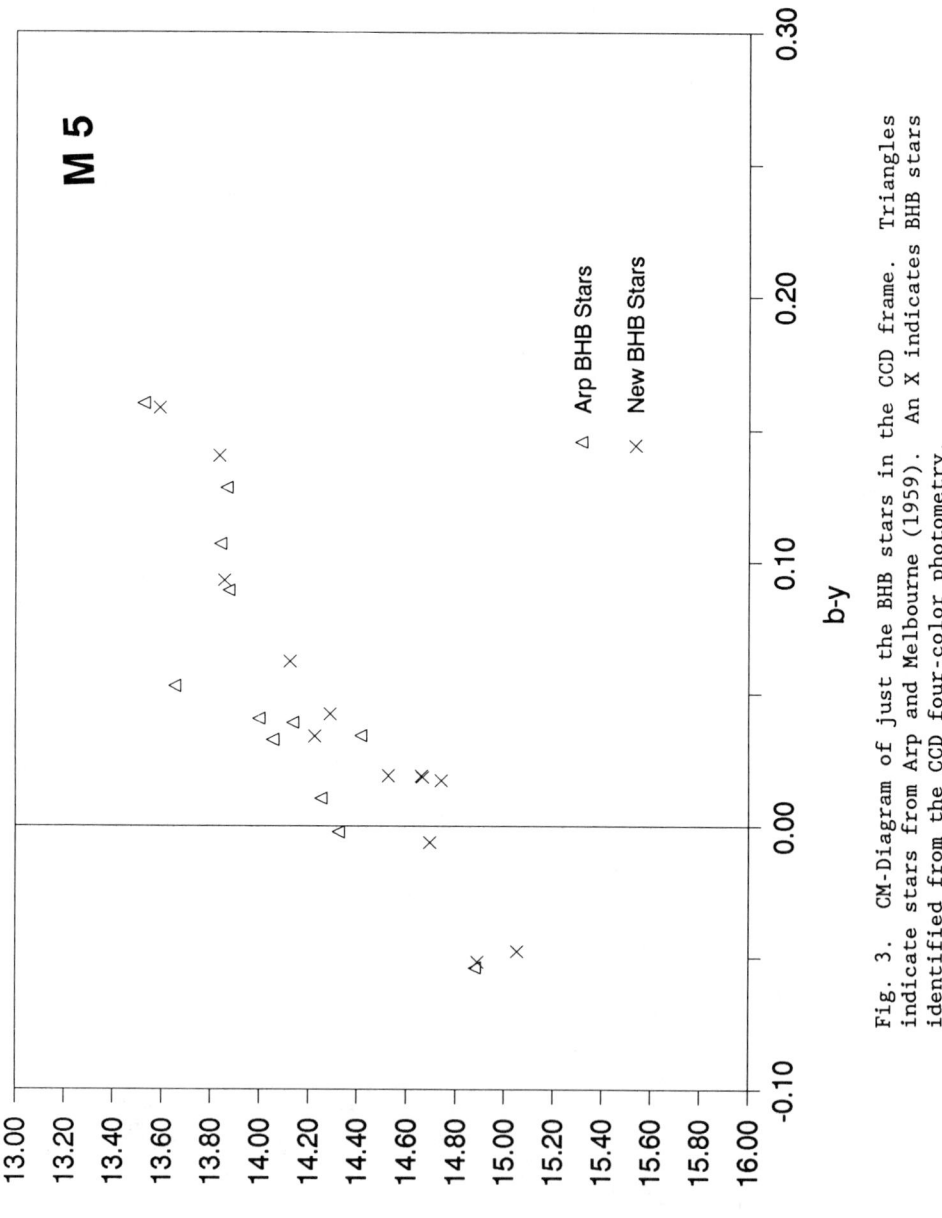

Fig. 3. CM-Diagram of just the BHB stars in the CCD frame. Triangles indicate stars from Arp and Melbourne (1959). An X indicates BHB stars identified from the CCD four-color photometry.

CCD FOUR-COLOR PHOTOMETRY OF GLOBULAR CLUSTER STARS 159

Table 1.

BHB stars measured in M 5

Arp #	# on Frame	y	(b-y)
6	157	14.696	-0.006
11	153	14.662	0.019
16	111	14.134	0.039
17	56	13.524	0.160
37	170	14.880	-0.054
41	97	13.998	0.041
44	65	13.652	0.053
48	88	13.865	0.128
53	87	13.840	0.107
60	91	13.875	0.089
69	128	14.323	-0.002
70	109	14.054	0.033
72	122	14.250	0.010
	59	13.588	0.159
	84	13.857	0.093
	85	13.836	0.140
	112	14.125	0.062
	115	14.288	0.043
	117	14.227	0.034
	140	14.527	0.019
	144	14.665	0.019
	150	14.741	0.017
	173	14.889	-0.051
	183	15.051	-0.047

The Arp numbers are from Arp (1955).

DISCUSSION

YOSS: As a historical question, isn't it true that you were the first to find A-type stars in the halo with a high velocity dispersion. I tell everyone that you were the first.

PHILIP: I know I was one of the early investigators but I don't know that I am the first.

JANES: I was somewhat struck by the metallicities of the stars from the Cayrel Catalog. They seem to be rather more metal-rich than the typical Pop II star. Could there be anything in this?

PHILIP: I think it is a statistical effect. There are not so many early-type Pop II stars in the compilation and we are dependent on the choices of many investigators who select stars for observation at high dispersion. If there were ten times as many stars of Pop II in the list we might find a different distribution with metallicity.

GARRISON: A comment on the [Fe/H] values in the literature for FHB stars. We looked at that catalog in more detail and the spread which you seem worried about is certainly reasonable, even for normal standard stars! Of course, the [Fe/H] value depends critically on the choice of log g and T_{eff}, which themselves; cover quite a large range for any given stars. As with photometric observations and reductions, you have to choose whom you want to trust.

PHILIP: It would be an improvement if one person would reobserve all the early-type Pop II stars in the Cayrel catalog and reduce the data. This new data set, even if it is off in the absolute [Fe/H] values should better represent the relative [Fe/H] determinations among stars in the group.

KURUCZ: When I compute U, I include the wavelength dependence of a 1p21 photomultiplier, aluminum reflectivity and atmospheric transmission. These effects are not negligible at the 0.001 mag level. A CCD should have large errors if the UV sensitivity is small and rapidly decreasing.

PHILIP: Yes, I agree. When I get good U measures the U response of the CCD will have to be modeled.

RICHER: Were you able to measure the m_1 index and have metal abundances for the globular clusters that you observed? This should be possible as some of the clusters have red horizontal branches and ought to be cool enough to use the index.

PHILIP: For spectral types A0 - A5 the δm_1 index has a small range and is not a good measure of [Fe/H]. Nissen has derived relationships between δm_1 and [Fe/H] for A5 - F0 and F-type stars. In these spectral ranges the difference is large enough to be used as a measuring tool. A complication is that the Hδ line appears in the v filter and this luminosity effect also acts to change the m_1 measure.

STROMGREN PHOTOMETRY OF REDDENED B-TYPE STARS

A. J. Delgado and E. J. Alfaro

Institute of Astrophysics of Andalucia

We have carried out an analysis of the uvby standard sequences for 17 young open clusters (main sequence between O9 V and A0 V), to determine whether the standard m_1, (b-y) and c_1 Strömgren color indices of B-type stars contain systematic erroneous trends and to what extent these trends would be correlated with the value of the color excess E(b-y). We refer mainly to photometric sequences based on observations obtained with filter defined instrumental systems. Two unpublished photometric studies (NGC 1502 and NGC 2169) are also included, performed with a Danish type photometer (slot-defined bands). The results show that the sequences of standard values are indeed affected by systematic trends, which can be formulated as a linear correlation of the observed reddening slopes with spectral type. The tendencies found affect the c_1 values mainly and are suggested to be not real but rather introduced in the standard sequences by the neglecting, in the transformation, of an actual dependence of the coefficients on the color excess, E(b-y). A complete account and discussion of this analysis is in the process of publication (Alfaro and Delgado 1990). We plan a systematic investigation of the four-color photometric system to establish the reality of these effects and the possible presence of such trends for reddened stars of later spectral types (A0 to G0).

The main steps in our planning would be: 1) Determination of the instrumental system response functions in the four uvby passbands. 2) Observation of highly reddened clusters. Clusters with E(b-y) larger than 0.5 and ages around and above 10^8 years are important in this connection. 3) Synthetic colors of spectrophotometric standards with and without artificial reddening. We aim to determine, reliably, in this way how and how much reddening affects the intrinsic Strömgren colors. In particular, estimates of the possible errors affecting the cluster parameters (age and distance modulus) are expected.

REFERENCES

Alfaro, E. J. and Delgado, A. J. 1990 A&A, in press

A COMPARISON OF THE SYSTEMATIC ACCURACY IN FOUR PHOTOMETRIC SYSTEMS*

J. W. Pel

Kapteyn Astronomical Institute

1. INTRODUCTION

After the Dutch 90-cm telescope and the Walraven five-channel photometer were moved in 1979 from the Leiden Southern Station in South-Africa to ESO, La Silla, a considerable effort was made to obtain an entirely new and very accurate calibration for the Walraven VBLUW photometric system. First of all, the five passbands were carefully redetermined, both from calculations and from spectrum-scans in the spectrophotometer, and a new standard star system was set up by J. Lub (Leiden Obs.) and the author. The internal accuracy of this system can be characterized by the mean standard deviations of the least squares "closure solution" for a network of standard-to-standard links over the whole southern sky, which are well below 0.001 mag in all indices:

	V	(V-B)	(B-L)	(B-U)	(U-W)
mean sigma (mag):	0.00045	0.00025	0.00031	0.00031	0.00027

Secondly, numerous calibrating stars of various kinds were observed by many observers in a wide variety of programs. By selecting and merging these data from the entire VBLUW database of the last 10 years, I have recently completed a "master list" of calibrating stars. High quality criteria were set for the selection of these data, and in many cases the observations (partly already published) were re-reduced to ensure strict homogeneity of the reduction procedures. The present list contains 1902 stars, all of which have been observed on at least two different nights. The average number of observations per star is 5.7. Two-dimensional MK spectral types are available for 95% of the stars. The catalog contains objects such as: photometric standards (for VBLUW and other systems), stars with spectrum scans and flux calibration standards, members of clusters and associations (Hyades, Pleiades, Ori OB-I, Sco-Cen, etc.), stars from the [Fe/H]-catalog, O-stars over a large range in reddening, etc. The typical accuracy (average rms error for a single measurement) is:

*Based on observations obtained at the European Southern Observatory at La Silla, Chile.

	V	(V-B)	(B-L)	(B-U)	(U-W)
mean sigma (mag)	0.0055	0.0031	0.0034	0.0038	0.0041

The above errors are very satisfactory, but they refer only to the internal accuracy, of course. The question remains whether the external accuracy of the VBLUW photometry can be trusted to the same level. My interest in this question is particularly strong because for some of my own programs (determination of metallicity differences between galactic, LMC and SMC Cepheids; calibration of the Cepheids PLC relation via photometry of Hyades, Pleiades and clusters with Cepheid members; photometric determination of T_{eff}, log g, [Fe/H] for F - G stars in fields of the Basel halo program) the systematic accuracy of the VBLUW system is of vital importance. In programs of this kind systematic photometric errors as small as 0.01 mag, e.g. as a function of position in the sky, can cause significant (and correlated!) errors between the derived physical quantities.

To check the external reliability of the VBLUW system I decided to make use of the large overlap of our calibration star list with photometry in other systems. Of the 1902 stars in the catalog 1637 have UBV data (Nicolet 1978), 1538 occur in the fourth catalog of Geneva photometry (Rufener 1988), and 881 have Strömgren photometry from the "southern" (La Silla) uvby catalogs (Grønbech and Olsen, 1976; Olsen, 1983; Schuster and Nissen, 1988). By comparing those indices that are transformable between all four of these systems, it should be possible to assess the corresponding systematic accuracies. The Walraven L, U and W bands unfortunately have no sufficiently close counterparts in the other systems to give useful transformations, so the comparison had to be limited to the Walraven V_W and (V-B) indices and their equivalents (V_J, (B-V) in the UBV system; V_G, (B_2-V_1) in the Geneva system; V_S, (b-y) in the uvby system).

2. COMPARISON OF THE PHOTOMETRIC DATA

Figs. 1 - 3 show the results of this comparison, in Fig. 1a, 2a and 3a for the visual magnitudes, and in Fig. 1b, 2b and 3b for the blue-visual colors of the four systems. Ordinates in all 21 panels are the residuals (quantity transformed from VBLUW) minus (quantity as observed in other system) in units of 0.01 mag. They are plotted against (V-B) color, visual magnitude, R.A. and Dec. The panels on the top side of each figure (residuals vs. (V-B)) are checks on the quality of the transformations. It should be noted that VBLUW colors are usually expressed in log units instead of magnitudes. The (V-B) zeropoint corresponds to about A0 V; unreddened O5 stars have (V-B) = -0.11, K0 V stars lie at about (V-B) = 0.40 dex.

We will not go into the details of the transformations here, but a few general remarks are in order. For the visual magnitudes simple relations with linear (V-B)-terms can be used in all cases. They are valid for all spectral types from O5 to late K, regardless of surface

A COMPARISON OF THE ACCURACY IN FOUR PHOTOMETRIC SYSTEMS

167

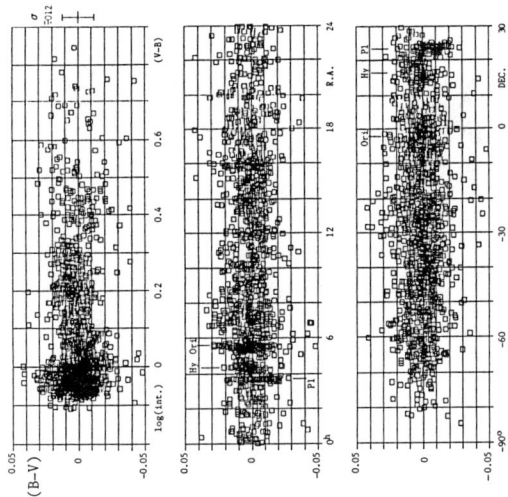

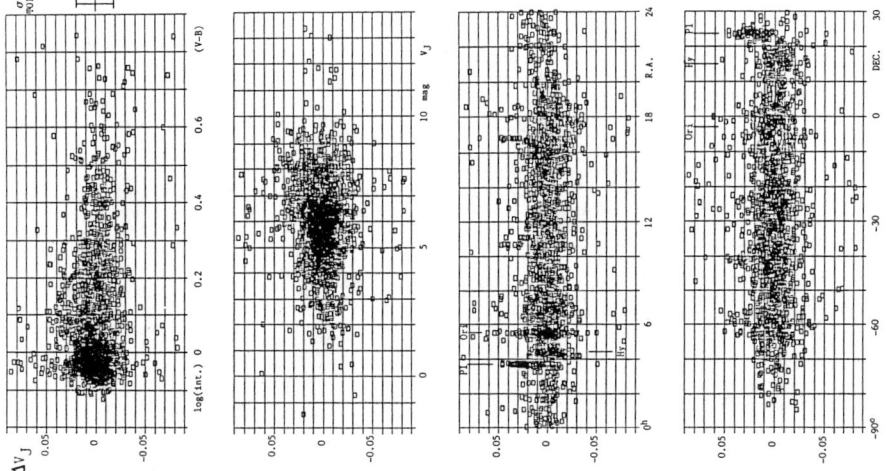

Fig. 1a. (Left) ΔV_J as a function of (V-B), V_j, R.A. and Dec.

Fig. 1b. (Above) Δ(B-V) versus (V-B), R.A. and Dec.

168 J. W. Pel

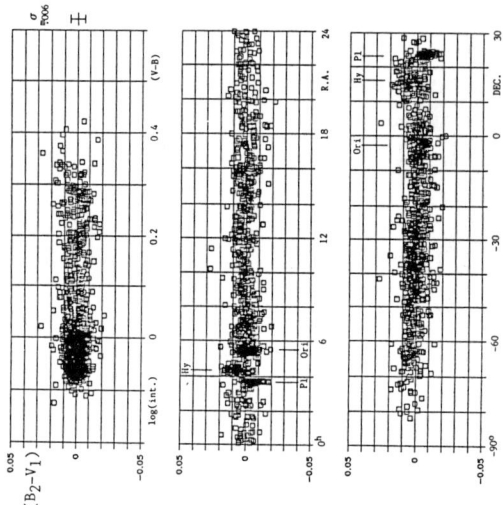

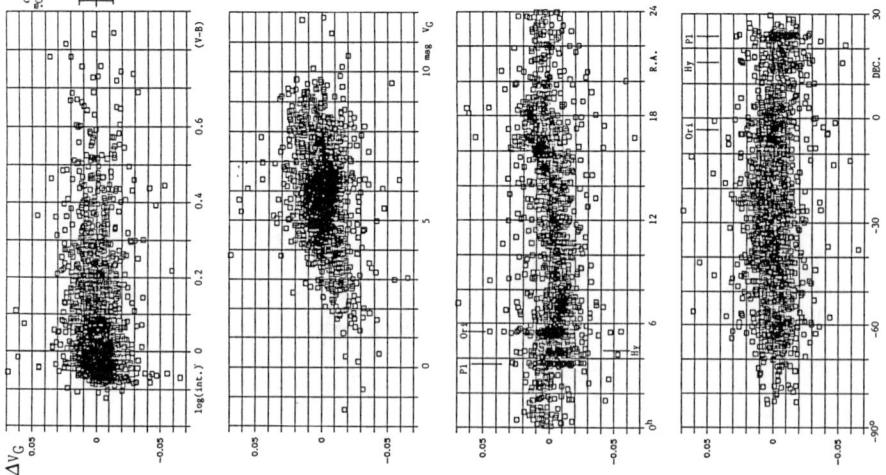

Fig. 2a. (Left) ΔV_G as a function of (V-B), V_G, R.A. and Dec.

Fig. 2b. (Above) $\Delta(B_2-V_1)$ versus (V-B), R.A. and Dec.

A COMPARISON OF THE ACCURACY IN FOUR PHOTOMETRIC SYSTEMS 169

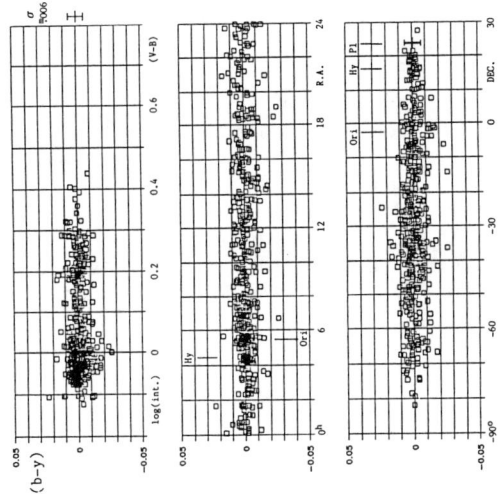

Fig. 3a. (Left) ΔV_S as a function of (V-B), V_S, R.A. and Dec.

Fig. 3b. (Above) Δ(b-y) versus (V-B), R.A. and Dec.

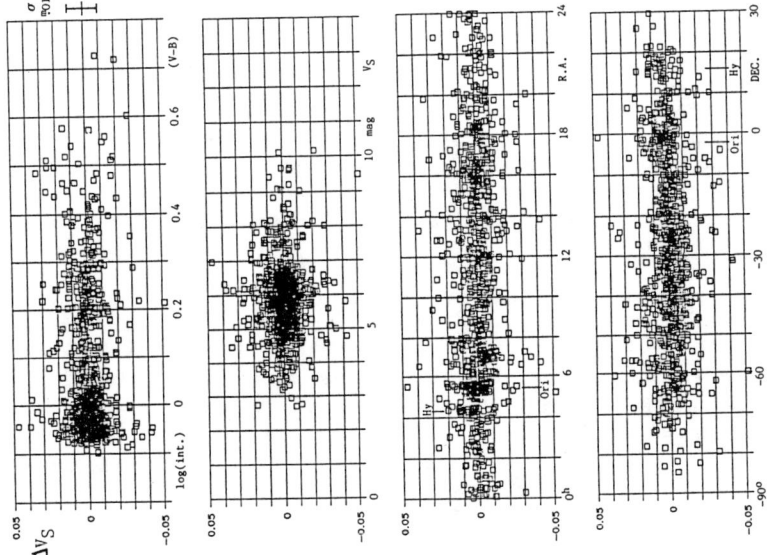

gravity, metallicity or reddening. The transformations from V_W to V_J and V_S are nearly identical but the relation between V_W and V_G is slightly different, indicating that the uvby V magnitudes are indeed in the V_J system, but that there is a small transformation between V_G and V_J.

The transformations from (V-B) to the corresponding colors are more complicated. For $(B-V)_J$ a third order equation was determined which is valid for all spectral types from O to K5, but this transformation depends slightly on reddening and strongly reddened stars had to be excluded from Fig. 1b. The transformations from (V-B) to (B_2-V_1) and (b-y) are even more complicated, as they contain significant dependences on reddening and, for the cooler stars, on surface gravity and metallicity. In both cases separate transformations (partly linear, partly second order) were fitted in three spectral intervals: OB-stars, A and early F stars, late F to early K. Stars with [Fe/H] < -0.50 were excluded, as well as supergiants for spectral types later than B5 and giants for types later than F5. For the transformation to (B_2-V_1) only stars with E(B-V) < 0.1 were selected; for (b-y) the reddening limit was even set at E(B-V) < 0.05 mag. These additional restrictions explain why the numbers of stars in Fig. 2b and 3b are significantly smaller than in Fig. 2a and 3a.

3. DISCUSSION

The results in Figs. 1 - 3 can be summarized as follows:

1) Comparison UBV - VBLUW (Fig. 1a, b). The scatter of the residuals is large, but there is no evidence of clear systematic effects as a function of R.A. or Dec. In most panels the concentrations due to Hyades, Pleiades and Ori-OB1 can be recognized easily. There are clearly systematic offsets for the Pleiades and, to a lesser extent for the Hyades, both in V and in (B-V). No such zeropoint problems are evident in the Ori-OB1 data. The V_J residuals are flat up to at least V = 2.

2) Comparison Geneva photometry - VBLUW (Fig. 2a, b). The scatter in the V residuals is larger than expected from the internal errors. There is a clear nonlinearity in the V residuals for stars brighter than about V_G = 4. More disturbing is the systematic dependence of ΔV_G on position: a nearly sinusoidal modulation of ΔV_G versus R.A. with full amplitude of about 0.02 mag, and a curvature with Dec. of about half that size. The (B_2-V_1) residuals have a much better scatter, consistent with the internal errors. The R.A. distribution is flat, but there is small curvature in the distribution versus Dec. Again there are small but significant zeropoint offsets, both in magnitude and color, for the Hyades and the Pleiades, but not for Ori-OB1.

3) Comparison uvby - VBLUW (Fig. 3a,b). The distribution in all panels looks excellent. The V_S residuals are flat to about $V = 3$. Unfortunately no uvby data were available for brighter stars. All other distributions are also flat to within a few millimagnitudes. The scatter is consistent with the internal accuracy claimed by both systems. Only the ΔV_S values for the Hyades may have a small positive offset. For the Pleiades no data in the "southern" uvby catalogs are available. The Kitt Peak data of Crawford and Perry (1976) were therefore reduced to Olsen's final standard uvby system by means of the small transformations given by Olsen (1983). The mean Pleiades (b-y) residuals thus obtained, indicated by the error bar in the lower panel of Fig. 3b, fit perfectly on the zero line. Since no V_S values have been published for the Pleiades, the corresponding comparison in Fig. 3a is not possible.

The overall conclusion from Figs. 1 - 3 is that, at least for the visual magnitudes and the blue-visual colors, the "southern" uvby and the VBLUW photometry have the most reliable standard systems. Considering that still some of the scatter in Fig. 3a, b will be due to shortcomings of the transformations, there is very little room left for systematic effects. This is particularly true for Fig. 3b, where position-dependent systematic effects larger than a few millimagnitudes can be excluded.

Comparing the average zeropoints for the Hyades, Pleiades and Ori-OBI in the various diagrams I conclude that the zeropoints of V_W and (V-B) for these stars are correct to better than 0.005 mag in V_W and 0.002 mag in (V-B). An explanation for the offsets of the Hyades and Pleiades data in some indices of the other systems can not yet be given, but they are most likely related to the local standards that were used for these clusters. The systematic effects encountered in the comparison with the Geneva photometry were quite unexpected. I can only conclude that the problem lies in the Geneva data, but an explanation for these effects requires further investigations in collaboration with my Geneva colleagues.

REFERENCES

Crawford, D. L. and Perry, C. L. 1976 AJ 81, 419
Grønbech, B. and Olsen, E. H. 1976 A&AS 25, 213
Nicolet, B. 1978 A&AS 34, 1
Olsen, E. H. 1983 A&AS 54, 55
Rufener, F. 1988 Catalogue of Stars Measured In The Geneva
    Observatory Photometric System, 4th ed., Geneva Observatory
Schuster, W. J. and Nissen, P.E. 1988 A&AS 73, 225

DISCUSSION

STERKEN: I understand that the VBLUW photometer at La Silla will be replaced by a CCD camera. Does that mean the end of VBLUW photometry?

PEL: No, but it means that we will soon discontinue the <u>permanent</u> use of the five-channel photometer on the 90-cm telescope (after more than 30 years). Like everyone else, we will try to switch to CCD photometry in the near future, to extend several of our VBLUW programs to much fainter magnitudes. The future philosophy will probably be to use the old photoelectric instrument exclusively for those programs that rely critically on the specific properties of the instrument, and use the CCD instrument most of the time.

BURKI: We just discovered these unfortunate differences. Of course, we don't know which system(s) are in error. We have to discuss this point during the next meeting on precision photometry!

PEL: I agree, these results were obtained very recently, and we now have to sit down and look for possible explanations of the systematic residuals.

THE EXTENSION OF THE GENEVA CATALOGUE

Noel Cramer

Geneva Observatory

ABSTRACT: During the two years following the publication of the fourth edition of the Geneva photometric catalogue (Rufener 1988), some 4200 additional stars have been measured in the system. We present here, in Section 1, a brief historical reminder of the genesis of the system and of its subsequent evolution. Some basic statistics regarding the present extension are discussed in Section 2 and a list of publications liable to assist a potential user of the catalogue with the interpretation of the data is given in Section 3, at the end of this paper.

1. HISTORICAL INTRODUCTION

The Geneva seven-color photoelectric photometric system was defined in the late fifties (~ 1958) by M. Golay and F. Rufener with the objects of: 1) taking advantage of the photometric accuracy of photomultiplier tubes; 2) reproducing the properties of the UBV system via an almost identical V-band but with slightly different U- and B-bands presenting less overlap at the Balmer jump; 3) approaching the classification properties of the BCD (Barbier-Chalonge-Divan) low resolution spectrophotometry, i.e. by the measurement of the Balmer jump and of gradients over the Paschen continuum by means of color indices obtained with four new filters (for the passbands, see Rufener and Nicolet 1988).

Together with this new system, Rufener introduced a novel measurement and reduction technique, the "M and D" method, which allows the accurate determination of atmospheric extinction and of its evolution during the night (see Rufener 1964, 1986). This technique is still systematically applied whenever the atmospheric conditions are judged to be good enough by the observer; otherwise, the program stars are measured at a constant air mass with a sufficient number of standard stars, and a mean extinction coefficient is then used.

All the measurements in the Geneva system have been obtained with a small number of photometers where great care was taken to accurately reproduce the passbands, and where the filters and photomultipliers are thermostatically controlled, the latter being also magnetically shielded. A Fabry lens makes the measurement insensitive to the location of the star in the diaphragm. The data that have been obtained up to the present can be roughly grouped into three periods in the following manner:

1960-1968: 40-cm telescope at the Sphinx observatory (3580 m) of the International Foundation Scientific Station Jungfraujoch, in the Swiss Alps. 1968: installation of a 76-cm telescope at Jungfraujoch and removal of the 40-cm instrument to Gornergrat (3130 m). Uninterrupted measurements with the 1-m telescope at the Swiss station at the Observatoire de Haute Provence (OHP), France.

1969-1977: Jungfraujoch or Gornergrat, or both together; 1-m at OHP. Two short campaigns at La Silla in 1971 and 1974 with ESO telescopes. 1975: removal of the 40-cm telescope to the ESO La Silla observatory. Installation at Gornergrat of a 1-m telescope (Observatoire de Lyon).

1977-1990: 1977: P7 differential photometer (Burnet 1976) operational at La Silla. 1980: installation of a 70-cm telescope at La Silla. 1985: 1-m telescope removed from Gornergrat; all subsequent northern photometry done at Jungfraujoch.

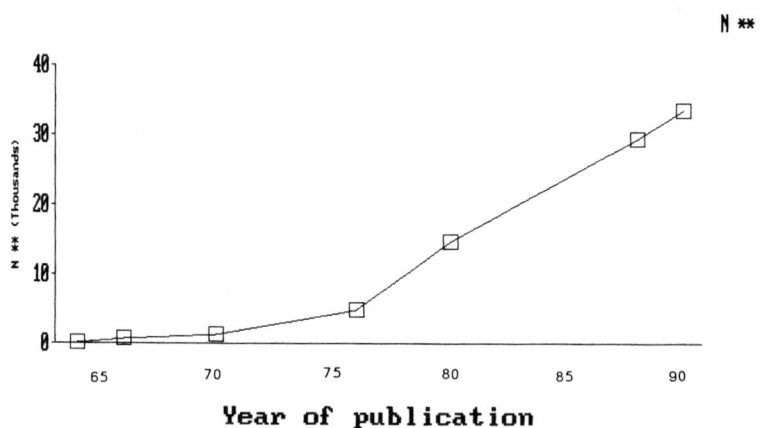

Fig. 1. Number of stars in each successive catalogue as a function of time. The last point represents the present extension.

All measurements have been successively compiled and published in a series of catalogues, the latest version of which appeared in 1988. The evolution of the acquisition of data is apparent in Fig. 1; the "takeoff" in 1976 - 1977 is due to the initiation of our systematic measurements in the superior atmospheric conditions offered by the La Silla site, and to the installation of the much faster P7 photometer

THE EXTENSION OF THE GENEVA CATALOGUE

in 1977.

A CCD photometer which reproduces the Geneva system has now been built (Blecha et al. 1990) and is being tested in the southern sky. This new technique should become operational towards the end of 1991.

2. THE EXTENSION

The 1988 version of the catalogue (Cat88) contains the mean colors and magnitudes for all the measurements made up to the end of 1987, i.e. over ~ 200,000 individual measurements of 29,400 stars of all spectral types. The following schematic representation, based as an example from the measurements made at La Silla, serves to illustrate the manner in which the successive editions of the catalogue are compiled:

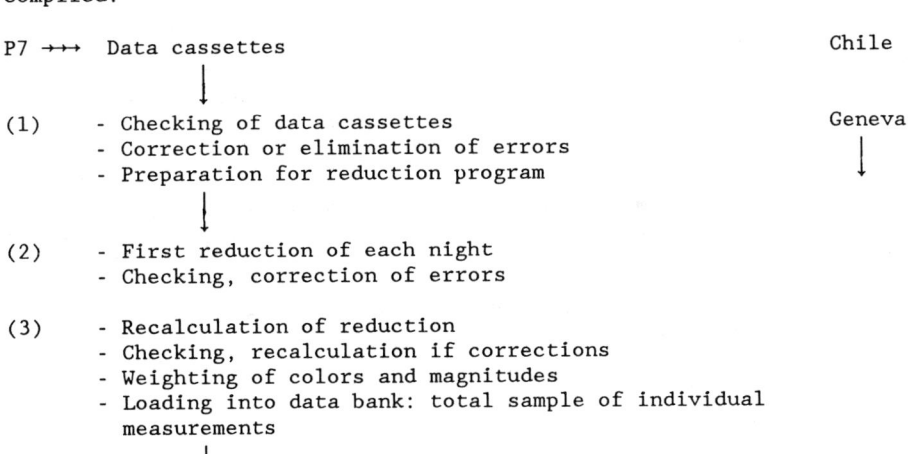

P7 →→→ Data cassettes   Chile
↓
(1) - Checking of data cassettes   Geneva
    - Correction or elimination of errors
    - Preparation for reduction program   ↓
↓
(2) - First reduction of each night
    - Checking, correction of errors

(3) - Recalculation of reduction
    - Checking, recalculation if corrections
    - Weighting of colors and magnitudes
    - Loading into data bank: total sample of individual measurements
↓
(4) - Calculation of means over new data (every year or two)
    - Critical evaluation of each mean and of whole data. Final corrections. Loading of these "provisional means" into data bank
↓
(5) - Every few years: final critical evaluation of provisional means and integration into catalogue. Tests of homogeneity and accuracy. Publication of new version of catalogue.

During the two years following the compilation of Cat88, a further ~ 43,000 individual measurements were obtained. These involve data for 4169 new stars as well as supplementary measurements of stars figuring in Cat88. The extension presented at this colloquium concerns only the sample of 4169 new stars. Furthermore, the time-consuming critical evaluation in (4) has not yet been completed. The extension distributed at the colloquium has therefore been subjected to a few simple selection criteria in view of reducing the possibility of gross errors occurring in the data:

1) All stars whose colors or magnitudes were, for various reasons, weighted 0 (see the definition of the weighting in Cat88) were discarded (most of these will not figure in the final catalogue anyway).

2) All stars with colors weighted 1 (weighting scale: P = 0 to 4) are presented in List #1 which contains 1560 stars. These are in the greatest majority stars with a single measurement (or in a few rare cases mean values over two poor measurements), and where the presence of an error is not excluded (misidentification, occultation by the dome slit, etc.).

3) All stars with more than one good measurement of the colors (P > 1) and visual magnitude weighted at least one (Q > 0) were then considered. As a supplementary measure to discriminate against misidentifications or other mishaps, the stars for which the standard deviations over the colors (see definition in Cat88), $\sigma(c1) > 0.01$ mag up to $m_v = 10$ and $\sigma(c1) > 0.02$ mag for $m_v > 10$, were eliminated from the sample. After removal of a few stars with obviously discrepant m_v deviations, the resulting selection is presented in List #2 which contains 1565 stars.

The following diagrams present some basic statistics regarding the new measurements:

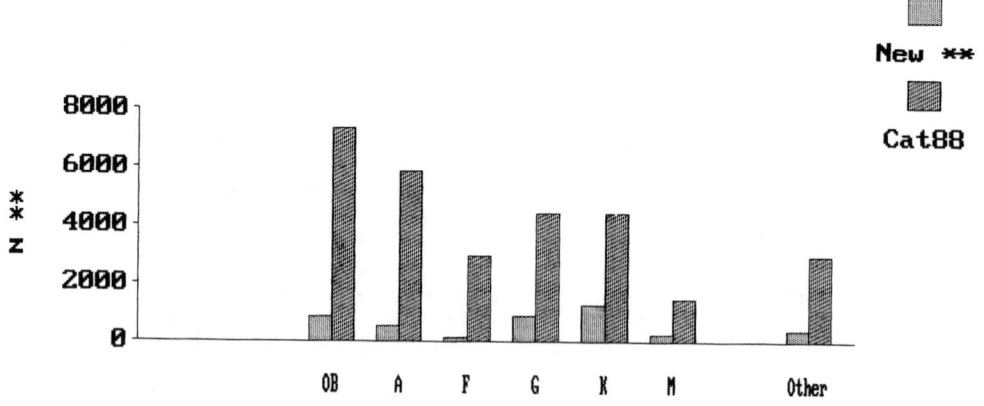

Fig. 2. Distribution of spectral types of the 4169 new stars, relative to those of the 29,400 stars in Cat88.

In Fig. 2, the comparatively strong increase of the B- and G-, K-spectral types is mainly due to two long term programs that are currently being carried out at La Silla: the systematic measurement of the B-type stars detected in the successive editions of the Michigan catalogue on the one hand, and the photometry of a selection of the Hipparcos astrometric satellite program stars on the other hand.

The contents of the two lists #1 and #2 are shown in Fig. 3 as a function of m_v. Fig. 4 compares the mean rms deviations of the colors and magnitudes (see definitions in Cat88) of the stars in List #2 with their respective values in Cat88.

The effects of the selections described in 3), above, are evident in Fig. 4a, and are the main cause of the generally lower deviations of these new measurements. The distribution of the stars removed from the candidates ("All **" in the diagrams below) for List #2 is shown in Fig. 5. The steady increase with m_v indicates that the selection criteria tend to be conservative, and reflects the naturally decreasing accuracy as the sources become fainter. Among the brighter stars in Fig. 5, there are a number of variable stars that have been monitored extensively. This is well shown in Fig. 6 for stars with more than four good measurements; the corresponding rms deviations leave no doubt concerning their variability.

3. SOME USEFUL REFERENCES RELATED TO THE CATALOGUE.

The following section contains a concise list of publications which also includes all the references mentioned in the text above, and can help the user of Geneva photometry to interpret the data contained in the various editions of the catalogue.

3.1 Concerning The Catalogue And Instrumentation

F. Rufener: Catalogue of stars measured in the Geneva Observatory photometric system (fourth edition).
A&AS 78, 469, 1989

F. Rufener,
B. Nicolet: A new determination of the Geneva photometric passbands and their absolute calibration.
A&A 206, 357, 1988

F. Rufener: Reduction to outside the atmosphere and statistical tests used in Geneva photometry. Proceedings of the workshop on Improvements to photometry, San Diego, June 1984, NASA Conference Publications 2350.

F. Rufener: The evolution of atmospheric extinction at La Silla.
A&A 165, 276, 1986

M. Burnet: Etude et réalisation d'un photomètre différentiel commandé par ordinateur.

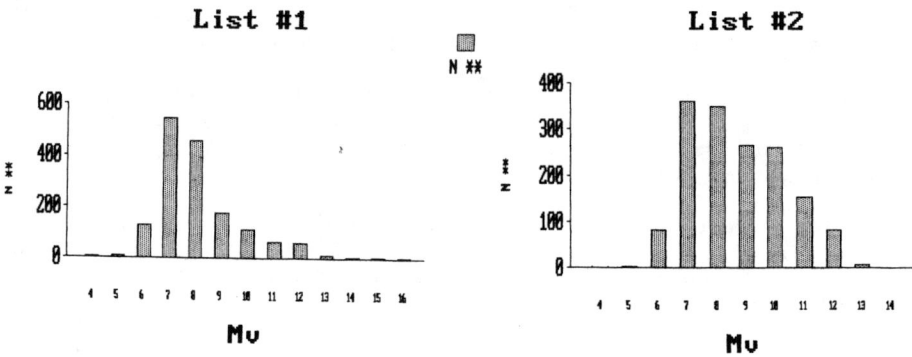

Fig. 3. (a and b): Histograms of the contents of lists #1 and #2 as a function of m_v.

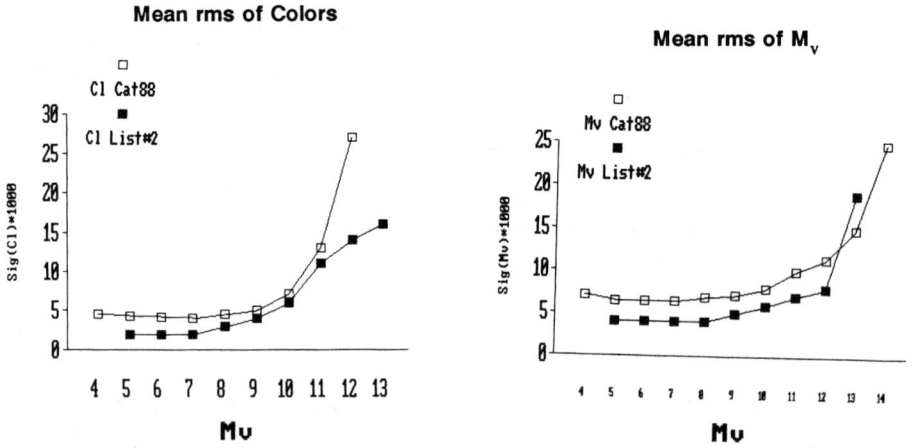

Fig. 4. (a and b): Mean rms deviations of colors and magnitudes for the stars of List #2, compared with those of Cat88.

THE EXTENSION OF THE GENEVA CATALOGUE

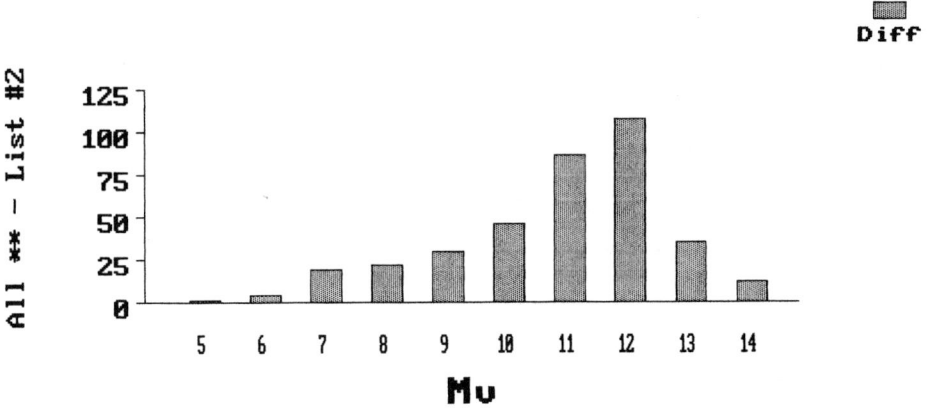

Fig. 5. Stars removed from List #2 candidates by criterion 3).

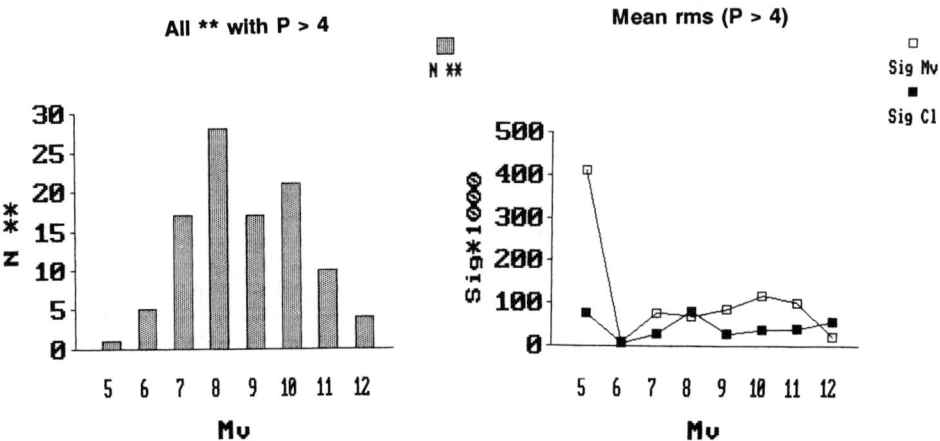

Fig. 6. (a and b). Stars with more than four good measurements. These are mostly variables, as shown by the rms deviations.

Thèse No232 (1976), Ecole Polytechnique Fédérale
de Lausanne

A. Blecha,
L. Weber,
G. Simond,
D. Queille: Scene dependent read-out of CCD.
In CCDs in Astronomy, G.H. Jacoby ed., Astronomical Society of the Pacific Conference Series, Vol. 8, 1990

3.2 Properties Of The Geneva Seven-Color Photometric System

M. Golay: The Geneva seven-color photometric system.
Vistas in Astronomy 24, 141, Pergamon Press, 1980

M. Golay,
N. Mandwewala,
P. Bartholdi: Spectral classification of stars with the same colors in intermediate multiband photometry. The concept of the photometric "star box".
A&A 60, 181, 1977

M. Golay: Analysis of the absolute magnitudes of stars in the same seven-color photometric "star box".
A&A 62, 189, 1978

B. Nicolet: Geneva photometric boxes: Distances and reddening for 43 open clusters.
A&A 104, 185, 1981

3.3 Calibrations

B. Nicolet,
N. Cramer: Ultraviolet and visible parameters for Am stars.
A&A 117, 248, 1983

N. Cramer,
A. Maeder: Luminosity and T_{eff} Determinations for B-type Stars.
A&A 78, 305, 1979

N. Cramer,
A. Maeder: Relation between Surface magnetic Field Intensities and Geneva Photometry.
A&A 88, 135, 1980

N. Cramer: Geneva [U,B,V] intrinsic colors of B-type stars.
A&A 112, 330, 1982

N. Cramer: Relations between U,B,V intrinsic colors and Geneva photometry for B-type stars. The effective temperature

scale.
A&A 132, 282, 1984

N. Cramer: Relations between the β-index and Geneva photometry. The B-type stars.
A&A 141, 215, 1984

M. Grenon: Propriétés photométriques des étoiles G, K, M en relation avec la structure et l'évolution galactique
Thesis, Publ Obs Genève, Série B, Fascicule 5, 1978

M. Grenon: Abundances, temperatures and reddenings of field and cluster population II giants.
In IAU Colloquium No 68, Astrophysical parameters for globular clusters, A. G. D. Philip and D. S. Hayes, L. Davis Press, Schenectady, p 393, 1981

G. Meylan,
B. Hauck: Relations between some photometric temperature parameters.
A&AS 46, 281, 1981

B. Hauck,
P. North: Photometric properties of Ap stars in the Geneva system.
A&A 114, 23, 1982

B. Hauck: Calibration in temperature of photometric parameters.
In IAU Symposium No 111, Calibrations of fundamental stellar quantities, D. Hayes, L. Passinetti and A. G. D. Philip, eds, p 271, 1985

G. Meynet,
B. Hauck: Geneva intrinsic colors of the supergiants of spectral types A and F.
A&A 150, 163, 1985

T. Lanz: Effective temperature of Ap stars: a comparison between several photometric estimators.
A&A 144, 191, 1985

B. Hauck: Metallicism among A and F giant stars.
A&A 155, 371, 1986

B. Hauck,
A. Slettebak: Effects of stellar rotation on the Geneva photometric system.
A&A 214, 153, 1989

S. Berthet: New calibrations of blanketing parameters Δm_2 and δm_1 in terms of [Fe/H].
A&A 236, 440, 1990

D. Kobi,
P. North: A new calibration of the Geneva photometry in terms of T_{eff}, log g, [Fe/H] and mass for main sequence A4 to G5 stars.
A&AS 85, 999, 1990

4. CONCLUSION

The data in the Geneva seven-color system have been increasing at an almost constant rate since 1976. This trend will be again modified in the near future by the application of CCD techniques which make use of the same passbands. We have briefly described here the new data acquired during the years 1988 and 1989. The final critical evaluation of these measurements will soon be completed. Meanwhile, the preliminary results described above are available upon request.

ACKNOWLEDGEMENT

I would like to thank Christian Richard and Jean Jacques Schwab who are in charge of the major part of the reduction work and the management of the related computer programs.

REFERENCES

Blecha, A, Weber, L., Simond, G. and Queille, D. 1990 in CCDs in
    Astronomy, K. Janes, ed, ASP Conf Series, in press
Burnet, M, 1976 Thesis, Ecole Polytechnique Fédérale de Lausanne
Rufener, F. 1964 Pub Obs Geneva, Ser A, 82
Rufener, F. 1986 A&A 165, 276
Rufener, F. and Nicolet, B. 1988 A&A 206, 357

DISCUSSION

PHILIP: The filters and calibrations for the Geneva System can be found in articles in Problems of Calibration of Multicolor Photometric Systems, the proceedings of an earlier meeting held in Schenectady.

HIGH PRECISION PHOTOMETRY IN THE STUDY OF VARIABLE STARS

G. Burki

Geneva Observatory

ABSTRACT: The precision of photometric data, the long-term homogeneity of the photometric system and the continuity of observations are important considerations in various fields of variable star studies. Three examples of long-term monitorings are given, based on Geneva 7-color photometric measurements obtained at La Silla Observatory:

• The RS CVn eclipsing system, RZ Eridani, showing various types of variability: eclipses, starspots, reflection effect, activity cycle.

• The mid-B-type variable HD 74560, in which three non-radial g-modes have been identified.

• Supernova 1987A, showing that photometric stations available for monitorings of first importance astrophysical objects are absolutely necessary.

1. INTRODUCTION

High-precision photometric data is not required for all studies of variable stars. In this context, a beautiful historical example is given by the first determination of the period of δ Cephei by John Goodricke (1786). Despite the poor photometric quality of these data (according to the actual criteria!), they are still used to study the variation of this period during the past two centuries (see Szabados, 1980).

However, in many circumstances, high-precision photometric data are absolutely necessary to describe the light variations of stars and to give good explanations for the observed complex phenomena. The following examples may be given:

• Studies in helio- and asteroseismology (e.g. Gilliland 1990).

• The determination of the physical parameters of eclipsing binaries from their light curves.

• The determination of the radius and distance of pulsating stars by applying the Baade-Wesselink method to the light, color and radial velocity curves.

• The detection and study of small-amplitude variables, such as supergiants, OB-type stars, Ap stars, white dwarfs, etc.

• Multicolor studies for the determination of temperature, gravity, chemical composition, turbulence, etc., based in particular on comparisons with stellar model atmospheres.

• The study of the exact shape of the light curves of variable stars, as example the intensity of the small hump at minimum radius in Cepheids and RR Lyrae (e.g. Gillet et al. 1989).

In the case of long-term studies of stellar variability, at least two conditions must be satisfied, in addition to the precision of the measurements:

1) The stability of the photometric system over the whole survey.

2) The continuity of the observations, depending on the "easy" access to a telescope, on the meteorological conditions and on the position of the star in the sky with respect to the telescope location.

At the ESO La Silla Observatory in Chile, the Swiss station is particularly well-adapted for long-term and for high-precision photometric surveys. We have a 70-cm telescope equipped with a photometer devoted to the Geneva 7-color system, continuously in operation since 1976. Among the long-term programs on variable objects, the survey of SN 1987A from February 1987 to August 1989 (Burki et al. 1991) and the monitoring of the quasar 3C 273 started in 1983 as part of a multi-wavelength survey (Courvoisier et al. 1990) can be mentioned.

2. LONG-TERM OBSERVATIONS OF VARIABLE STARS OF SMALL AMPLITUDE: THE EXAMPLES OF RZ ERIDANI (ECLIPSES, ACTIVITY, STARPOTS) AND OF HD 74560 (MULTI-PERIODIC SLOWLY PULSATING B-TYPE STAR)

During the past decades, many programs of long-term photometric surveys have given superb results. As examples, let us mention the observation of stellar activity cycles (Wilson 1975), the description of the supergiant variability by van Genderen and collaborators (e.g. van Genderen 1989) and the observations on many stars made by the group using the small telescopes at ESO, La Silla (e.g. Sterken 1986). With the Geneva system, several long term programs have been carried out. Among them, let us mention:

1. The survey of the Be stars in h and χ Persei during eight years by Waelkens et al. (1990). They show that at least half of the brighter stars in these clusters are variable.

2. The monitoring of supergiant stars (e.g. Burki 1987).

3. The study of the variability of Ap Stars (e.g. North 1987).

4. The analysis of the multi-periodicity of B-type stars by C. Waelkens and collaborators (e.g. Waelkens and Rufener 1985). An example is given below.

In the following, two examples of multi-periodic stars monitored during several years from the Geneva station in La Silla are presented.

2.1 RZ ERIDANI (HD 30050)

This is a summary of a paper by Kviz and Burki (1991).

This star is a member of the long period group of the RS Canum Venaticorum binaries (Hall 1976). They are systems with periods of between one day and two weeks, spectral class F or G for the hotter component and systems which display strong H and K emission outside eclipse. This classification has been extended to include longer and shorter periods and now includes about a hundred known systems. One of the characteristics common to many of these systems is a distortion of the light curve, which migrates in phase with respect to the eclipses. The most accepted theory for the explanation of the distortion wave is the starspot model (see Caton 1986, for some examples of light curves).

RZ Eri was monitored during 11 years and our results are the following:

1. The orbital period of the eclipsing system is 39.28254^d, the secondary eclipse is found at phase 0.67.

2. The analysis of the seven primary eclipses observed during the survey reveals that the primary component is an A-type star (possibly an Am). Its magnitude is constant, $V = 8.269 \pm 0.011$.

3. The variability of the secondary component outside eclipses was studied after removing the light from the primary. Fig. 1 shows this variability, outside eclipses. We see a very long-term variability which can be tentatively attributed to a stellar activity cycle. In particular, the V magnitude decreased by about 0.2 mag from 1978 to 1984.

4. The light from the secondary varies with a period of about 35^d (the migration wave), with changes in period, shape and amplitude from year to year. This is most certainly a starspot variation, and the period is that of the rotation of the secondary, at the latitude of the spot(s) location on the stellar surface. From $P_{rot} = 35^d$ and $P_{orb} = 39.3^d$, we deduce that the secondary is near the synchronization.

5. In 1988, a very complete set of observations was obtained, covering

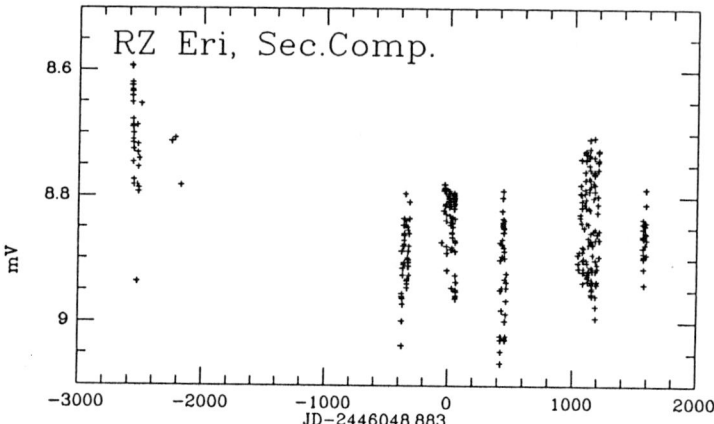

Fig. 1. Long-term variation of the secondary component of RZ Eri.

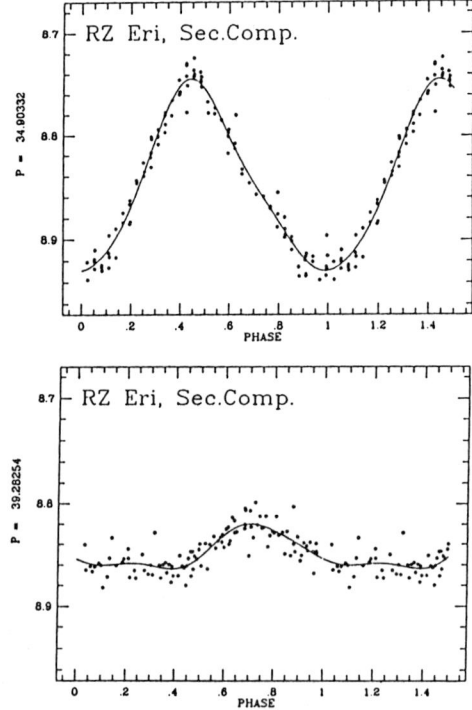

Fig. 2. Starspot and reflection effects in 1988 (secondary component of RZ Eri).

160^d. The Fourier analysis of these data (for the secondary component) reveals 3 periods:

a. 34.90^d with an amplitude (peak-to-peak) of 0.18. This is the starspot(s) variation.

b. 39.28^d with an amplitude of 0.05. This is the variation due to the reflection effect of the light from the primary. The period is the orbital one.

c. A long-term variation with an amplitude of 0.02. This effect could be due to stellar activity. These first two components of the variability of the secondary in 1988 are displayed in Fig. 2. Note the very good description of the reflection effect with a maximum occurring exactly at the phase of the secondary eclipse (0.67).

6. The variability due to the starspots changes from year to year. The periods were, in particular, 36.4^d in 1985 and 36.1^d in 1986.

7. After correction for the variability of the secondary, the light curve of the eclipsing system was analyzed by using the EBOP program (Etzel 1980). The light curve of the eclipsing system in 1988, is given in Fig. 3.

In summary, our survey of RZ Eri allows the study of various types of variability:

a. Eclipses $P = 39.28254^d$ (orbit)
b. Reflection effect $P = 39.28254^d$
c. Starspots $P = 35^d$ (rotation)
d. Stellar activity cycle (?)   Several years

2.2 HD 74560 (HR 3467, HY Vel)

These results have been obtained by Waelkens (1991).

HD 74560, a B3 IV star, is a member of a class of mid-B-type photometric variables as described by Waelkens and Rufener (1985). These stars can be identified with the 53-Persei stars and their variations are caused by stellar pulsations. This star was measured 413 times from La Silla between 1976 and 1989. The standard deviation in V is ± 0.014, thus HD 74560 appeared to be a small amplitude variable.

The period analysis reveals three distinct modes of periods; 1.5511^d, 1.6455^d and 1.7370^d. The residual standard deviation is ± 0.007 and the resulting light curves are presented in Fig. 4. It is important to note that these three pulsations modes have been stable during at least eight years.

In his paper, Waelkens (1991) also analyses six other mid-B-type

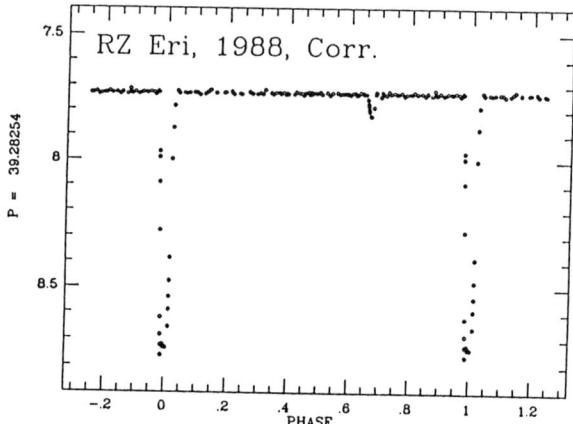

Fig. 3. Light curve of the eclipsing system RZ Eri in 1988, after corrections for the secondary variability.

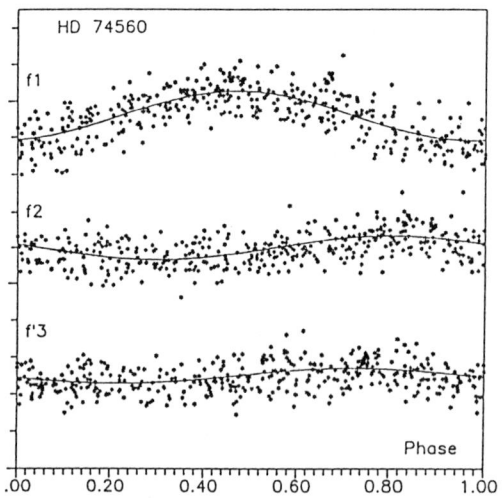

Fig. 4. The three non-radial modes of HD 74560 (Waelkens 1991).

stars which exhibit photometric variability similar to that of HD 74560. He interprets these variations in terms of non-radial g-modes of pulsation. This analysis clearly reveals the necessity of using photometric equipment stable over many years, and of organizing long-term monitorings to resolve the variability of such stars. Indeed, the periodograms calculated for each season of observation do not allow the detection of the complete set of frequencies.

Finally, no significant low frequency noise was detected in the case of the seven stars measured over several year by Waelkens (1991). This fact confirms the very good internal consistency of the Geneva photometric measurements carried out at La Silla.

3. THE MONITORING OF SN 1987A

It is quite logical to conclude this paper by a presentation of some results on SN 1987A, because this object is the best example showing the necessity of having photometric stations permanently in operation and available for long-term monitorings.

SN 1987A has been monitored in the seven colors of the Geneva photometric system from February 24, 1987 to August 28, 1989. The complete analysis of the data is published elsewhere (Burki et al. 1989, 1991), thus only an unusual light curve of SN 1987A is presented here: the track in a color-magnitude diagram. Fig. 5a gives the observational V vs. (U-G) diagram, and Fig. 5b summarizes the evolution in the absolute and dereddened M_{vo} vs. $(U-G)_o$ diagram. The dynamics of the variation is especially noteworthy: more than six mag in (U-G), about 10 mag in V (until Aug. 1989). The numbers one to five in Fig. 5b refers to the main phases of the supernova light curve and evolution (see Burki et al. 1991).

With the help of our photometric measurements it is possible to examine the variation of the dispersion of the data around the mean fitted light curve, in order to look for an eventual short-term variability, i.e. of the order of a few days.

In Fig. 5a, an increase of the dispersion is clearly visible from about December 1988, i.e. when the supernova became fainter than about V = 11. We have examined the magnitude-related dispersion of the light curve by calculating the residual standard deviation, $\sigma(V)$, around a polynomial fit on successive samples of the data with an air mass $F_z <$ 1.5. The result is compared in Fig. 6 with the $\sigma(V)$ versus V relation obtained for the well-measured southern field stars in the Geneva catalogue. Clearly variable stars were excluded from the latter sample and only the stars measured more than five times were taken into account. We see that:

1. Except for the first and the last two values, the dispersion of the visual magnitude measurements is extremely small and, on the whole,

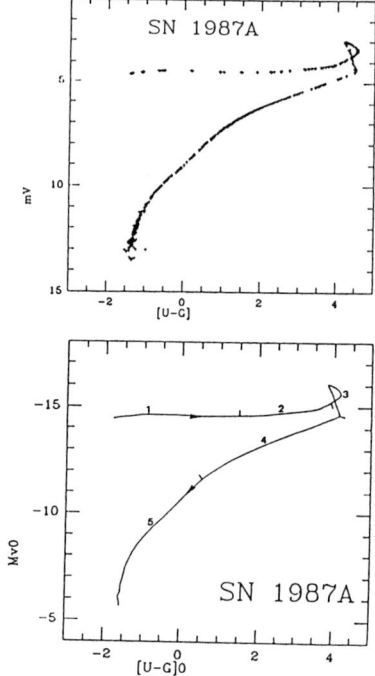

Fig. 5. The color-magnitude diagram of SN 1987A.

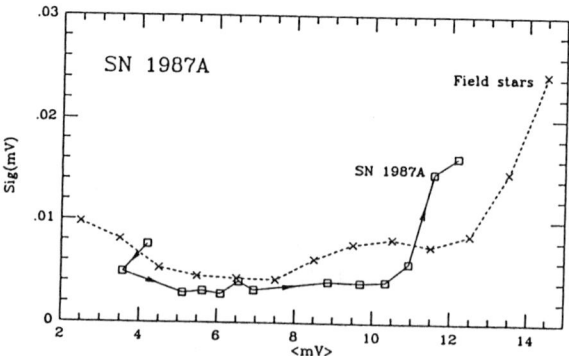

Fig. 6. Comparison of the variation of $\sigma(V)$ vs. V for SN 1987A and for the well-measured southern field stars in the Geneva catalogue (Rufener 1988).

lower than that of the field stars. This is probably due to the measurement of two nearby standard stars simultaneously to the measurement of the supernova.

2. The first point shows that the dispersion was slightly higher during the first days following the explosion. This is probably due to the fact that these first observations were carried out in spite of the poor photometric quality of some nights.

3. The last two points show a more rapid increase of $\sigma(V)$ with increasing V for the supernova than for the field stars. This is most probably due to the crowded field surrounding the supernova which makes precise aperture photometry difficult to achieve. Moonlight also became an increasingly important factor of dispersion at that time.

In conclusion, our analysis of the light curve of SN 1987A shows no significant short term variability in the visible part of the spectrum up to at least December 1988.

Finally, it is interesting to note that the use of the variable star exhibiting the highest possible amplitude was a very good test of the precision of our photometric equipment. A precision better than 0.004 in the V magnitude was obtained during the major part of the survey.

REFERENCES

Burki, G. 1987 in Instabilities in Luminous Early Type Stars, H. J. G. L. M. Lamers and C. W. H de Loore, eds., Reidel, Dordrecht, p. 23
Burki, G., Cramer, N. and Nicolet, B. 1991 A&AS, in press
Caton, D. B. 1986 AJ 91, 132
Courvoisier, T. J. L., Robson ,E. I., Blecha, A., Bouchet, P., Falomo, R., Maisack, M., Staubert, R., Terasranta, H., Turner, M. J. L., Valtaoja, E., Walter, R. and Wansteker, W. 1990 A&A, in press
Etzel, P. B. 1980 EBOP user's guide, 3rd edition, University of California, Los Angeles
Gillet, D., Burki, G. and Crowe, R. A. 1989 A&A 225, 445
Gilliland, R. L. 1990 in Precision Photometry, A. G. Davis Philip, D. S. Hayes and K. A. Janes, eds., L. Davis Press, Schenectady, p. 265
Goodricke, J. 1786 Phil Trans Roy Soc 76, 48
Hall, D. S. 1976 in IAU Colloquium No. 29, Multiple Periodic Variable Stars, W. S. Fitch, ed., Reidel, Dordrecht, p. 287
Kviz, Z. and Burki, G. 1991 A&A, in press
North, P. 1987 A&AS 69, 371
Rufener, F. 1988 Catalogue of Stars Measured in the Geneva Observatory Photometric System, Geneva Obs
Sterken, C. 1986 in IAU Symposium 118, Instrumentation and Research Programs for Small Telescopes, P. L Cottrell and J. B. Hearnshaw, eds., Reidel, Dordrecht, p. 255

Szabados, L. 1980 Mitteilungen der Stern, Ungarischen Akad.,
    Wissensch 76, 1
Van Genderen, A. M. 1989 A&A 208, 135
Waelkens, C. 1991 A&A, in press
Waelkens, C. and Rufener, F. 1985 A&A 152, 6
Waelkens, C., Lampens, P., Heynderickx, D., Cuypers, J., Degryse, K.,
    Poedts, S., Polfliet, R., Denoyelle, J., Van Den Abeele, K.,
    Rufener, F and Smeyers, P. 1990 A&AS 83, 11

DISCUSSION

DEMARQUE: I missed the name of the stars on which you detected non-radial g-mode oscillations.

BURKI: There are 7 mid-B-type stars in the paper by Waelkens (1990, in press). The results are shown in my paper concerning HD 74560.

WASHINGTON CCD PHOTOMETRY: ABUNDANCE DISTRIBUTIONS IN DIFFERENT
LATITUDES TOWARD THE GALACTIC BULGE

Neil D. Tyson

Columbia University

ABSTRACT: The Washington photometric system permits measurement of metal abundances for individual stars in the abundance range -3 < [Fe/H] < +0.5 . The broad-band passes of the system allow work on distant objects with small aperture telescopes. For objects such as resolvable stars in the Galactic bulge or nearby galaxies, the Washington system may prove superior even to multi-object spectroscopy. As a particular example, I present the first results of a survey of the stellar abundance distribution in different latitudes toward the Galactic bulge, undertaken in collaboration with R. M. Rich.

1. INTRODUCTION

There exists a variety of stellar systems that are distant yet resolvable from ground-based telescopes, such as the outer globular clusters, the Local Group dwarf spheroidal galaxies, the outer regions of M 31, and the Galactic bulge. The stars of these systems, however, are so faint that narrow and medium band photometry (e.g. Strömgren, Geneva and DDO) would require 4-m class telescopes or larger to obtain reliable color indices. It is from these indices that stellar physical parameters are deduced such as heavy element abundance, temperature and surface gravity. (See Philip 1979 for an extensive review of photometric systems). To render these distant, resolvable systems more accessible to photometric analysis one needs a broad band photometric system from which one can derive these same stellar physical parameters. Indeed, observations obtained with a suitably defined broad band system when compared with a narrow band system (or spectroscopy) allow smaller aperture telescopes to assume this task.

With these goals in mind, G. Wallerstein at the University of Washington designed, according to the guidelines put forth in Wallerstein and Helfer (1966), what now bears the name Washington System . It was developed with four broad ($\Delta\lambda \sim 1000$ Å) band passes with the intent to measure accurate abundances for G and K giants from photometric indices. As detailed further in Section 2, the Washington system is designed to remedy the problems incurred when one attempts to derive abundances from the UBV broad band system. A fifth, narrow-band

Mg "b" filter was later introduced by Geisler (1984) to help distinguish luminosity class.

I will briefly describe the history of the Washington system, its recent calibrations, and its new applications with CCD photometry. I conclude in Section 3 with how the Washington system has been used in the Galactic bulge.

2. WASHINGTON SYSTEM

The Washington system has been introduced and described elsewhere in the literature. Much of what follows is adapted from Canterna (1976), and Canterna and Harris (1979). Here, I simply outline the basic design.

2.1 General Features

The band passes of the Washington system are labeled C, M, T_1, T_2 and DDO 51. Their band centers and FWHM are listed in Table 1 with their response functions displayed in Fig. 1. These band passes were specially chosen to obtain the greatest amount of unambiguous information in a broad band system.

TABLE 1
The Washington Photometric System From Canterna 1976 and Geisler 1984.

Name	λ_{eff} Å	$\Delta\lambda$ Å (1/2 max)	
C	3910	1100	
M	5085	1050	
T_1	6330	800	
T_2	7885	1400	
DDO 51	5146	123	(DDO "Mg 51")

1) The C filter monitors the region near 3900 Å where we find primarily CH, metallic lines (inclusive of Ca II H + K), and CN blanketing. It is centered over 400 Å redder than the Johnson U filter which endows the C filter with two significant advantages: it is less affected by reddening, and the energy distribution of G and K giants favors detection through the Washington C filter when compared with the Johnson U filter. Consequently, exposure times in C are no more than one third of what.is required in U to observe G stars and later.

2) The M filter monitors integrated metallic line absorption from 4500 Å to 5500 Å, which is dominated by Fe I and MgH + Mg b. Unlike the

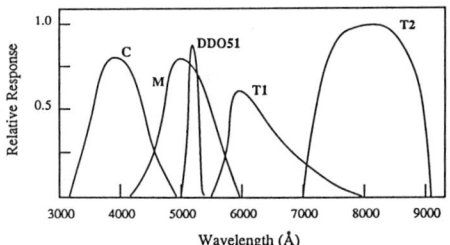

Fig. 1. Response functions for the Washington filter system. (From Canterna 1976, and Geisler 1984).

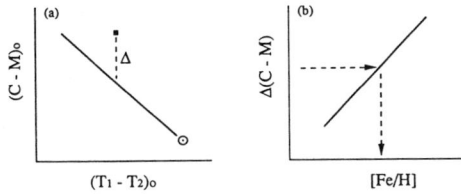

Fig. 2. To derive an abundance from Washington color indices, one first finds the difference in $(C-M)_o$ between the program star and the curve of solar abundance (Canterna et al. 1986). See panel (a). The abundance is then read directly from the empirical relation between $\Delta(C-M)$ and [Fe/H]. See panel (b).

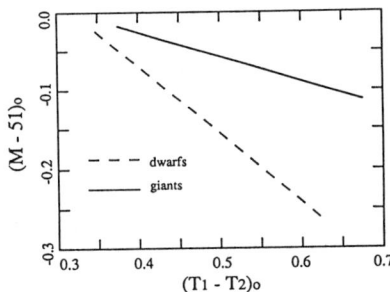

Fig. 3. Separation of dwarfs from giants is readily obtained with the (M-51) index. We illustrate the locus of giants (luminosity class III-IV; solid line) and the locus of dwarfs (luminosity class V-VI; dashed line), adapted from Geisler (1984). The luminosity sensitivity is preserved for all but the hottest stars.

strongly overlapping U and B band passes of the Johnson system, the M band pass was selected to minimize the contamination from the CN blanketing that the C filter measures. Hence, stars that are anomalously strong in CN absorption will not be confused (the way they are in the UBV system) with stars of genuinely high Fe-group abundance.

3) The T_1 and T_2 filters monitor color temperature through relatively clear red and infrared continuum regions. The two filters flank (and thereby avoid) the region of strong CN blanketing between 6800 Å and 7500 Å. These band passes are therefore insensitive to to variations in metal abundance and luminosity. They resemble the Cousins R and I filters.

4) The narrow band DDO 51 filter is obtained from the DDO photometric system to monitor exclusively the strong dependence of Mg (H+b) absorption on luminosity class. It was introduced by Geisler (1984) to discriminate dwarfs from giants.

The Washington system, by virtue of the blue-ultraviolet location of the C filter, is very sensitive to reddening (although much less sensitive than the Johnson U passband). An error in reddening of 0.1 in E(B-V) will result in an abundance error of 0.5 dex in [Fe/H] (using the reddening conversions in Harris and Canterna 1979). Note that the problem is of special concern in observations of the bulge where lines of sight pass within a few degrees of the Galactic plane; reddening corrections are not only large (e.g. E(B-V) = 0.41 at Baade's window; ℓ = 0.9°, b = -4°), but in many cases the reddening is variable (Blanco et al. 1984). These reddening corrections are a crucial issue, to which I will return later with a discussion of the reddening-independent Q-index.

2.2 Calibrations

As in most photometric systems, by convention all color indices are scaled to equal 0.0 for an A0 V star. The Washington band passes give rise to the following color indices from which physical parameters are derived.

(T_1-T_2) - Temperature-sensitive index. Other indices are typically plotted against this one. It is functionally equivalent to (R-I) in the Cousins system with a small linear transformation between them (Taylor 1986). The most recent conversions to T_{eff} are found in Bell and Gustafsson (1989) as a function of abundance and luminosity class. For middle to late M giants $(T_1-T_2 > 0.7)$, the continuum region monitored by T_1 becomes severely affected by TiO absorption which distorts the correspondence between (T_1-T_2) and temperature.

(C-M) and (M-T_1) - These indices have been calibrated to give abundances in [Fe/H]. The calibrating stars establish curves of constant abundance in the (C-M) or (M-T_1) versus (T_1-T_2) plane from which other abundances are interpolated. An equivalent method is to measure the Δ(C-M) (or Δ(M-T_1)) between the unknown star and the curve of solar abundance (Canterna et al. 1986). One then reads the abundance directly from the empirical relation between Δ(C-M) and [Fe/H]. (See Fig. 2.).

The earliest of these calibrations (Canterna 1976 and Canterna and Harris 1979) were based on the relatively small number of stars available with accurate spectroscopic abundances. This was particularly severe at the low abundance end ([Fe/H] < -0.5) where only 18 calibrators were available. Also, as emphasized in Canterna et al. (1986), theoretical colors and calibrations were strongly affected by uncertainties in the model atmospheres.

The calibrating set of Washington abundance standards was subsequently enlarged and tied to better high-dispersion spectroscopic abundances. The number of calibrators also grew to include large samples of field giants (Geisler 1984), metal-rich Pop I giants in open clusters (Canterna et al. 1986), and Pop II giants (Geisler 1986a). The calibration range now extends reliably from -3 < [Fe/H] < +0.5. Currently, Geisler and McWilliam (1991) are working to extend the calibration to [Fe/H] ~ 1.0. In all calibrations, the (C-M) color index has been shown to be more sensitive to abundance than (M-T_1). This is especially true at the low abundance end where (M-T_1) loses sensitivity for [Fe/H] < -1.0 (Geisler 1986a) while (C-M) may preserve its abundance sensitivity as low [Fe/H] ~ -4.0 (Twarog and Anthony-Twarog 1991).

In the Washington system, the (C-M) color was designed to flag the CN-rich stars because many giants, especially those that undergo thermal pulses on the asymptotic giant branch, have modified CN surface abundances that no longer correlate with Fe abundance. It was the Geisler (1986b) study that found the (C-M) index to be less sensitive to CN enrichment than previously suspected. In particular, Geisler found significant overlap in the (C-M) index among CN-rich and CN-normal stars for a given [Fe/H]. This may be due to the anti-correlation of CN and CH band strengths as found for giants in 47 Tuc by Norris, Freeman and DaCosta (1984).

For middle to late M giants, the metallic line region monitored by the M filter becomes severely affected by TiO absorption. For this reason, the Washington system is not (yet) calibrated redder than K giants.

(C-T_1) - Also calibrated and found to be a sensitive indicator of abundance. Friel and Geisler (1990) used it in their study of the metal-rich disk globular clusters NGC 5927 and NGC 6496, while Geisler and Forte (1990) used it to study the globular cluster system in the Fornax elliptical galaxy NGC 1399.

(M-51) - Readily distinguishes giants from dwarfs as was demonstrated in Geisler (1984). For hotter stars, however, the loci of dwarfs and giants converge. (See Fig. 3.)

2.3 Previous Work

Until very recently, most work in the Washington system has involved photoelectric photometry of selected stars in clusters to derive mean abundances. Notable among them are six globular clusters in Geisler (1986b), 13 metal-rich open clusters in Canterna et al. (1986), and seven metal-poor open clusters in Geisler (1987). In these studies, the mean deviation of abundances determined from Washington photometry and from other traditional methods (when available) such as intermediate and narrow band photometry, and spectroscopy is typically 0.1 dex in [Fe/H]. This is remarkable when we consider that this value is the same as the precision of the abundance calibration. Some earlier work (using the older, less accurate calibrations) determined the abundances of individual giants in the Local Group Draco and Ursa Minor dwarf spheroidal galaxies (Canterna and Schommer 1978) and used the integrated light of more distant systems such as globular clusters in Fornax and M 31 to establish mean abundances (Harris and Canterna 1977). To derive abundances from integrated light works the same way that one derives abundances from individual stars. For the Washington system, all that is required is that the integrated light resembles that of a G or K giant. This is typically what one finds for globular clusters since their light is dominated by stars on the first ascent of the giant branch.

2.4 Q-Index

A Q-index, akin to that derived for the UBV system (van den Bergh 1967), was developed for the Washington system in Harris and Canterna (1977). (See also Geisler and Forte 1990.) A Q-index can be formed from many combinations of indices. Its advantage (as is the advantage of most Q-indices) is that a reddening-independent estimate of the abundance can be obtained for fields where reddening is uncertain or possibly variable. The UBV reddening slope is the ratio of E(U-B) to E(B-V). It was found by van den Bergh (1967) (see also Hiltner and Johnson 1956) to be roughly constant at 0.72. One then forms the Q-index:

$$Q_{UBV} = (U-B) - [(E(U-B)/E(B-V))] (B-V) \quad \text{1a}$$

$$\text{where } [(E(U-B)/E(B-V))] = 0.72. \quad \text{1b}$$

For early-type stars, the Q-index is an excellent indicator of spectral class. For late-type stars, the reddening slope in the two color diagram is roughly parallel to the unreddened main sequence curve and thus the Q-index loses its utility as a classifier. Among these late-type stars however, the Q-index is found to correlate well with color, and therefore abundance, in the integrated light of globular clusters (Harris and Canterna 1977, Zinn 1980). It is important to recognize, however, that the reddening slope may not be constant for all clusters in all filter combinations. Recognizing this, Harris and Canterna (1977) calibrated Q_{CMT} vs [Fe/H] (updated in Geisler and Forte 1990) and parametrized the reddening slope $[E(C-M)/E(M-T_1)]$ as a quadratic function of [Fe/H] to give a Q-index for the Washington system:

$$Q_{CMT} = (C-M) - [(E(C-M)/E(M-T_1))] [(M-T_1) - 0.5] \qquad 2a$$

where $[(E(C-M)/E(M-T_1))] = 1.39 + 0.26 [Fe/H] + 0.08 [Fe/H]^2 \qquad 2b$

and $Q_{CMT} = 0.148 [Fe/H] + 0.655 \qquad 2c$

The Q-index, while reddening-free, is not as sensitive to abundance variation as the dereddened (C-M) index. For example, in the interval -2 < [Fe/H] < 0 the Q-index varies by $\Delta Q = 0.25$, while over the same interval in [Fe/H] we have $\Delta(C-M) = 0.7$. The Q-index is welcome, however, as a secondary abundance check for fields with questionable reddening.

2.5   CCD Photometry

Recently, Geisler (1987, 1990) established a set of Washington CCD photometric standards based on carefully selected 3' x 3' sub-regions of the Landolt (1973) SA fields and Eggen's (1969) photometry of the loose open cluster NGC 3680. Fields are selected based on the number of stars, their color range, and whether they are the appropriate brightness for high quantum efficiency CCD detectors. The field centers are listed in Table 2 along with the color and magnitude spread for the standards in each CCD frame. These new standards now permit CCD photometry that reliably transform to the photoelectric abundance calibrations. The accuracy of the Washington [Fe/H] calibrations is ± 0.1 dex, which corresponds to photometric precision of 0.035 mag in (C-M). In practice, however, uncertainties in CCD photometric offsets and zero-points in crowded fields will degrade the accuracy in [Fe/H] to typically ± 0.2 dex.

The first cluster to be published with Washington CCD photometry was the intermediate-age LMC cluster NGC 2213 as reported in Geisler (1987). Others have followed such as the metal-rich disk globular clusters NGC 5927 and NGC 6496 (Friel and Geisler 1990) which, as a class, have been given much less attention than the higher latitude, metal-poor halo clusters due to severe foreground contamination and reddening uncertainties near the Galactic Plane.

TABLE 2

Washington CCD Standard Fields (Geisler 1990)

Field ID	Field Center α (1988.5) δ		Number of Stars	Color Range $(T_1 - T_2)$	T_1 Mag Range (T_1)
SA 98	$6^h51^m32^s$	$-0°\ 18'\ 58"$	8	0.10 - 1.10	11.24 - 13.43
N 3680	11 25 00	-43 13 10	6	0.28 - 0.95	10.19 - 13.77
SA 110	18 42 37	+0 29 12	5	0.33 - 1.40	11.01 - 11.85
SA 114	22 41 02	+1 08 27	4	-0.03 - 0.54	11.89 - 14.01

In a recent extragalactic application, Geisler and Forte (1990) used integrated cluster light to derive the abundance distribution (from the 150 brightest clusters) and luminosity function (from 600 clusters) of the globular cluster system in the large elliptical galaxy NGC 1399. The clusters are found, on average, to be 0.75 dex poorer in [Fe/H] than those of the Galactic halo. The similarity of the T_1 filter with the Cousins R filter allowed Geisler and Forte (1990) to compare the turnover in the globular cluster luminosity function with the turnovers seen in the Galaxy and M 87. If the absolute magnitude of the turnover is a "standard candle" then the distance to NGC 1399 is readily derived to be 16 ±2 Mpc.

With CCD photometry, color-magnitude diagrams are readily constructed in the Washington system that convey all the features of stellar evolution that are found in other broad band systems. The T_1 vs (T_1-T_2) color is commonly used because of its resemblance with R vs (R-I) in the Cousins system. Both Geisler (1987) and Friel and Geisler (1990) present color-magnitude diagrams that nearly reach the cluster main sequence turnoffs. Color-magnitude diagrams are especially useful for the bulge population. They permit one to identify an unbiased subset of stars from which representative abundances are to be drawn. For example, M-giants signal a metal-rich population that, if isolated, would be a poor indicator in a mixed system of abundances.

3. THE GALACTIC BULGE

3.1 Stellar Abundances and Structure

Unlike individual star clusters, the Galactic bulge contains a rich mixture of stellar abundances that extends at least from -1.0 < [Fe/H] < 0.5. This was demonstrated by Rich (1988, 90) for one bulge field of low extinction (Baade's window; $\ell \sim 0.9°$, $b \sim -4°$) with his spectroscopic analysis of 88 K-giants. When there is a genuine abundance spread then one can seek more information than just the mean abundance. The detailed shape of the abundance distribution and the abundance dispersion each provide critical information about the history of heavy element enrichment. Clearly, a well-defined distribution requires dozens if not hundreds of stars. Fig. 4 presents three plausible abundance distributions that indicate three different histories of heavy element enrichment.

The Galactic bulge is a compact stellar system that contains approximately 10^{10} solar masses with over 95 percent within 2 kpc of the Galactic center (Blanco and Terndrup 1989). In the plane of the sky, assuming an 8 kpc solar circle, 2 kpc corresponds to 14°. There is mounting evidence for an abundance gradient within these two kiloparsecs of the bulge from M giant surveys (Blanco 1988, Blanco and Terndrup 1989, Frogel et al 1990, and Terndrup, Frogel and Whitford 1990), color-magnitude diagrams (Terndrup 1988), and carbon stars (Tyson and Rich 1991). An abundance gradient can signal dissipative collapse in the bulge's early history where chemical ejecta from already-formed stars descend to inner regions thereby enriching the center more than the outer zones. It can also signal systematic mass outflow with the outer region ejecting its enriched gas more effectively than the deep gravitational potential of the inner bulge.

As with elliptical galaxies, the bulge is a "spherical" gas-poor system. According to Terndrup, Frogel and Whitford (1990) there is also a remarkable resemblance between their luminosity functions and between their infrared properties. Spiral bulges also lie on the same radius-luminosity-surface brightness relation as elliptical galaxies (Kormendy 1987). These similarities suggest that if we understand the formation history of the bulge that some of this understanding may be transferable to elliptical galaxies.

3.2 Washington CCD Survey of the Bulge

In an attempt to characterize the star formation history and perhaps the collapse history of the Galactic bulge I have undertaken, in collaboration with R. M. Rich, an extensive survey of the entire bulge abundance distribution via Washington CCD photometry. The heavily obscurred Galactic bulge is actually accessible optically through a variety of low extinction "windows" through the disk. Seven such windows were targeted for this study along $\ell = 0°$ that range from -2.5° to -17° in Galactic latitude (0.3 to 2.3 kpc in projected distance from the Galactic center).

I present in Fig. 5 the dereddened $T_1 T_2$ color-magnitude diagram (similar to RI) for 400 stars in a 3' x 3' sub-area of the bulge window at b = -6°. Most of the scatter brighter than $T_1 = 16$ represents the genuine thickness of the red giant branch. In a population of mixed abundances this is precisely what is expected. There is some hint of a red horizontal branch near $T_1 = 14.2$ but it is poorly defined from only 400 stars. The foreground blue main sequence is also evident as well

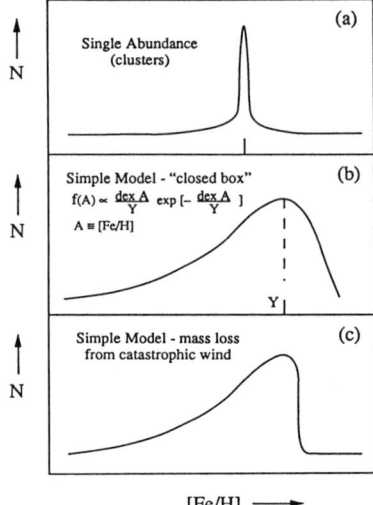

Fig. 4. Three plausible abundance distributions from very different histories of chemical enrichment. Panel (a) is the single burst, single abundance that is expected and found for nearly all star clusters. Panel (b) shows the expected distribution from the closed box simple model of chemical evolution (see text). The yield "Y" is the mass fraction of metals produced in a generation of stars. For the closed box model the yield equals the mean metallicity *and* the peak in the distribution of [Fe/H]. Panel (c) is the simple model where catastrophic mass loss has occurred from supernova-driven winds. The high abundance end displays the characteristic truncation from this process.

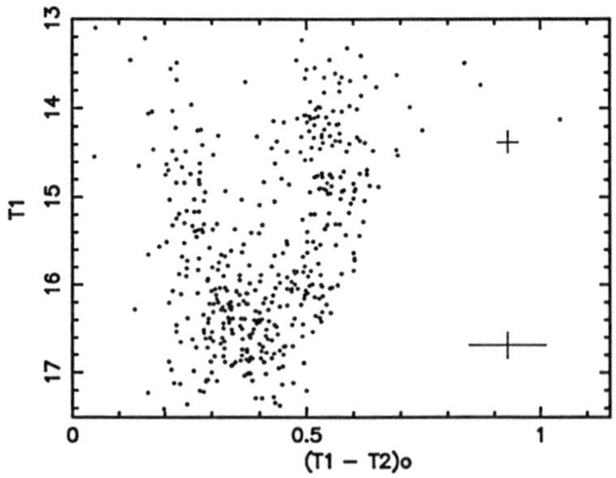

Fig. 5. Reddening-corrected T_1T_2 color-magnitude diagram of 400 stars derived from CCD Washington photometry of a 3' x 3' sub-area of the bulge window at $b = -6°$. Notice the broad giant branch which is a signature of a population with a significant abundance range. The width is much larger than photometric errors alone. Also evident is the extended blue foreground main sequence, the hint of a red horizontal branch at about $T_1 = 14.2$, and the very red M-giants (with $(T_1-T_2)_o > 0.7$) that are characteristic of a metal-rich population.

as several very red M-giants with $(T_1-T_2)_o > 0.7$ The current Washington calibration targets late G and K giants that fall within $0.4 < (T_1-T_2)_o < 0.7$.

As in any photometric system there should be agreement among the inferences drawn from various indices. For example, while the (M-51) index was introduced to identify dwarfs, one can also verify that the strong magnesium sensitivity of the narrow-band DDO 51 filter is in qualitative agreement with the [Fe/H] abundances derived from the (C-M) index. For the stars in field $b = -6°$ the 20 percent most metal rich and the 20 percent most metal poor with the best photometry (as derived from the (C-M) index) were selected for Fig. 6 where I present (M-51) versus (T_1-T_2). As anticipated, there is excellent separation between the metal rich subset (represented by "+") and the metal poor subset (represented by "o").

Thus far, we have derived a preliminary abundance distribution for two fields; one at $b = -8°$ and one at $b = -6°$. Given the early status of the photometric analysis, however, I present only the abundance distribution for the $b = -6°$ field (see Fig. 7a). The distribution shows an extended tail toward low abundances and an abrupt turnover at high abundances. In the space of [Fe/H], this is the characteristic signature of the closed box simple model of chemical evolution and the modified simple model with outflow (Rich 1991, see also Searle and Zinn 1978, and Mould 1984). In each scenario, a system processes its gas to completion as it recycles supernovae ejecta until all remaining gaseous mass is locked up in long-lived stars. In the simple model, however, only the version with outflow will reproduce an abundance gradient.

Our choice of fields includes the well-studied Baade's window for cross-checking with other studies. In particular, Geisler and Friel (1990) have used Washington photometry in Baade's window which we display in Fig. 7b superimposed on Rich's (1990) abundance distribution. Preliminary results support a small trend for the mean abundance to increase toward the inner latitudes and for the general shape to be preserved at both latitudes. The error in the mean abundance, however, is large enough so that the two distributions are consistent with no gradient at all. Distributions from other fields will be required to establish whether a systematic trend exists. It is also too early for us to decide whether the sharp, high-abundance turnovers are uniquely traceable to the abundance distribution expected from the simple model, or whether there was a catastrophic wind that halted star formation thereby preventing further enrichment. This is because the convolution of a small Gaussian error with the abundance distribution from a wind model will mimic the abundance distribution expected from the simple model. A forthcoming, detailed analysis of our observational errors will permit us to address this uncertainty.

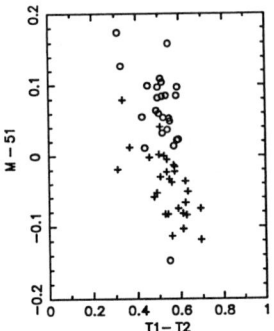

Fig. 6. A test of the consistency between the magnesium index, (M-51), and the [Fe/H] derived from the (C-M) index. Plotted are the stars in field $b = -6°$ that are the 20 percent most metal rich (noted by "+") and the 20 percent most metal poor (noted by "o") as derived from the (C-M) index. There is excellent separation between the populations as anticipated.

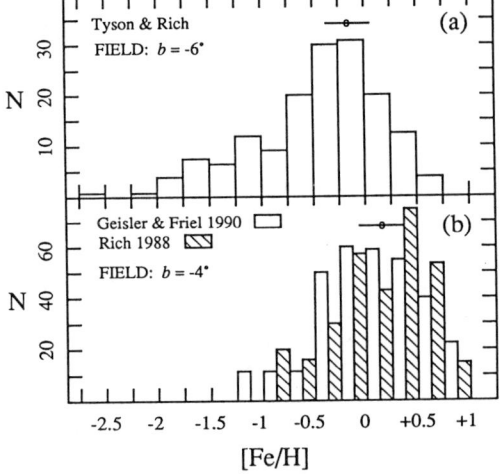

Fig. 7. Abundance distributions derived from Washington photometric indices of two latitudes toward the galactic bulge. Panel (a) contains a recent result from bulge field $b = -6°$ in the Tyson and Rich study (in progress). Panel (b) contains the abundance distribution at $b = -4°$ (Baade's window) from Rich's (1990) K giants, and from Geisler and Friel's (1990) Washington Photometry scaled to the same total N. For display purposes, the bins in panel (b) are presented at half their true width. There is good agreement between these two nested histograms. When comparing panels (a) and (b) notice the similar shape of their distributions and the small shift in mean abundance. Error flags indicate the abundance uncertainty inferred from the photometric uncertainty of 10.04 mag.

4. DISCUSSION

It is clear that Washington photometry sees its greatest utility with CCD photometry. Now that the reduction of crowded CCD fields is a tractable problem (with point spread function oriented data reduction packages such as Stetson's (1986) DAOPHOT, or M. Mateo's DoPHOT) severely crowded regions such as the inner Galactic bulge and inner regions of globular clusters can be the source of hundreds, if not thousands, of stellar abundances per CCD field. This will yield statistically significant mean abundances for clusters and reliable abundance distributions for other systems that are not represented by a single abundance. This includes systems of globular clusters such as those in NGC 1399 or in the Fornax dwarf elliptical galaxy. There is an additional utility to the Washington system that has yet to be exploited. Clearly, objects that compose the extreme sample in any distribution of properties tend to be the most interesting. Washington CCD surveys may prove to be the best (and fastest) means by which stars of very high or very low abundance can be discovered and flagged for further detailed spectroscopic analysis. Quite possibly, the most metal-rich (or metal-poor) star in the Galaxy may be found in the two kpc region that is the Galactic bulge.

ACKNOWLEDGEMENTS

I am grateful to S. Umar, who assisted in parts of the image reductions and computer programming. I enjoyed discussions with R. M. Rich, D. Geisler, and B. Twarog during the preparation of this manuscript. I also thank J. Patterson for his comments on a late draft.

REFERENCES

Bell, R. A. and Gustafsson, B. 1989 MNRAS 236, 653
Blanco, B. M. 1984 AJ 89, 1836
Blanco, V. M. 1988 AJ 95, 1400
Blanco, V. M., McCarthy, M. F. and Blanco, B. M. 1984 AJ 89, 636
Blanco, V. M. and Terndrup, D. M. 1989 AJ 98, 843
Canterna, R. 1975 ApJL 200, L 63.
Canterna, R. 1976 AJ 81, 228
Canterna, R., Geisler, D., Harris, H. C., Olszewski, E. and Schommer R.
    1986 AJ 92, 79
Canterna, R. and Harris, H. C. 1979 in Problems of Calibration of
    Multicolor Systems, A. G. D. Philip, ed., Dudley Obs.,
    Schenectady, p. 199
Canterna, R. and Schommer, R. A. 1978 ApJL 219, L 119
Eggen, O. J. 1969 ApJ 155, 439
Friel, E. D. and Geisler, D. 1990, preprint
Frogel, J. A., Terndrup, D. M., Blanco, V. M., & Whitford, A. E.
    ApJ 353, 494
Geisler, D. 1984 PASP 96, 723
Geisler, D. 1986a PASP 98, 762

Geisler, D.   1986b   PASP 98, 847
Geisler, D.   1987   AJ 93, 1081
Geisler, D.   1990, preprint
Geisler, D. and Claria, J.   1990, in preparation
Geisler, D. and Friel, E. D.   1991   in The Galactic Bulge, ESO/CTIO Workshop, D. Terndrup and V. Blanco, eds., La Silla, ESO
Geisler, D. and McWilliam, A.   1991, in progress
Harris, H. and Canterna, R.   1977   AJ 82, 798
Hiltner, W. A. and Johnson, H. L.   1956   ApJ 124, 367
Kormendy, J.   1987   in Structure and Dynamics of Elliptical Galaxies, T. de Zeew, ed., Reidel, Dordrecht, p. 17
Landolt, A. U.   1973   AJ 78, 959
Mould, J. R.   1984   PASP 96, 773
Norris, J., Freeman, K and DaCosta, G. S.   1984   ApJ 227, 615
Philip, A. G. D.   1979   Problems of Calibration of Multicolor Photometric Systems, Dudley Obs Rept No 14, Schenectady
Rich, R. M.   1988   AJ 95, 828
Rich, R. M.   1990   ApJ 362, 604
Rich, R. M.   1991   in The Galactic Bulge, ESO/CTIO Workshop, D. Terndrup and V. Blanco, eds., La Silla, ESO
Saha, A., Tyson, N. D., Gal, R. and Rich, R.M.   1991, in preparation
Searle, L. and Zinn R.   1978   ApJ 225, 357
Stetson, P.   1987   PASP 99, 191
Taylor, B. J.   1986   ApJS 42, 19
Terndrup, D. M.   1988   AJ 96, 884
Terndrup, D. M., Frogel, J. A. and Whitford, A. E.   1990   ApJ 357, 453
Twarog, B. A. and Anthony-Twarog, B. J.   1991   AJ, in press
Tyson, N. D. and Rich, R. M.   1991   ApJ, in press
van den Bergh, S.   1967   AJ 72, 70
Wallerstein, G. and Helfer, H. L.   1966   AJ 71, 350

DISCUSSION

RICHER: In your histograms of metallicity distributions in the different regions, you hinted that one could perhaps worry about them a bit because of the uncertainties of transforming the photometric indices into abundances, particularly at the metal-poor end. Why not just plot the indices and leave the metallicity transformations until later when it is better determined?

TYSON: My photometric precision is typically better than 0.06 mag in these crowded fields, and the precision for the selected stars used to determine the abundance distributions is better than 0.04 mag. The associated mean error in abundance (per star) converts to 0.15 dex in [Fe/H] (at [Fe/H] ~ -1.0). This error is better than what is necessary to 1) produce a reliable abundance distribution function and 2) to follow the dispersion and mean abundance trends from window to window in the bulge. As for the low abundance end, the precision in [Fe/H] degrades slightly for the same photometric precision. A plot of the distribution of abundance-sensitive indices will contain the same relative information as the distribution of abundances, but it removes the option to physically interpret the shape of the distribution in the context of star formation models. I am happy to simply report greater errors for the lower abundances.

DEMARQUE: How metal-rich are the RR Lyrae variables in the bulge?

TYSON: Using ΔS values, Blanco (1984) found $-2.0 < $ [Fe/H] $ < 0.5$ for 51 RR Lyrae stars in Baade's Window. These RR Lyraes show a strong maximum (FWHM ~ 0.5 dex) near [Fe/H] = -1.0. Rich (1988) demonstrated that the RR Lyrae distribution is significantly more metal poor than his entire sample of Baade's window K-giants. Although in a radial velocity study that is near completion, (Saha, Tyson, Gal and Rich 1991), we have found the Baade's window RR Lyraes to have a velocity dispersion (133 ± 24 km/sec that is consistent with the metal-poor subset of Rich's (1988) K-giants (126 ± 22 km/sec).

YOSS: Why do you call the Washington system broad band when you include the narrow band Mg b + MgH filter?

TYSON: Perhaps I should call it a narrow-band-assisted, broad band system. The Mg (DDO 51) filter was not in the original design of the system and therefore plays no role in establishing the photometric abundances. It is simply, and exclusively used as a dwarf discriminator-Q capitalizing on the luminosity sensitivity of the Mg complex near 5170 Å. In this regard, the Washington system can truly be considered broad band.

JANES: How did you decide on the reddening? Depending on what you choose, you could get very different answers.

TYSON: Reddening is a potential problem. There is no guarantee that the average reddening in a window (typically tens of square arc minutes) equals the reddening at the area of the CCD frame (9 square arc minutes). Clearly, all the data have been taken through these pre-determined windows of low extinction to minimize the reddening problems. Several of the windows have well-determined average reddenings. They range from E(B-V) = 0.025 at $b = -17°$ to E(B-V) = 0.6 at $b = -2.5°$. I have already formed an abundance distribution for field $b = -6°$ that is derived entirely from the reddening-independent Q-index. There is good agreement between the two distributions with the exception that the Q-index gives a slightly broader distribution than given by the dereddened $(C-M)_0$ index. This is purely a function of the greater errors incurred with the less sensitive Q-index. We plan to verify each window's abundance distribution with one derived from its Q-index. In this way we can properly characterize and estimate the reddening uncertainties.

ULTRAVIOLET EXCESS PROPERTIES OF THE THICK DISK AND HALO POPULATIONS

Steven R. Majewski

The Observatories of the Carnegie Institution of Washington

ABSTRACT: First results are presented from a new, complete survey of stellar multi-color (UBVI) photometry and proper motions to B ~ 22.5. Using high quality (σ_m < 0.02 mag) photometry, ultraviolet excesses and photometric parallaxes have been derived for a subsample of 250 stars with 0.3 < (B-V) < 1.10 in a field at the North Galactic Pole. Thus the chemical and kinematical distributions of late F, G and early K dwarfs have been probed to distances of up to 25 kpc above the Galactic plane, allowing for the first time an unbiased view of the properties of the thick disk and halo from unevolved stars *in situ*. An unexpected and dramatic change in the mean uv-excess is found at z ~ 5.5 kpc. This feature is tentatively associated with an "edge" to the thick disk. A similar discontinuity is also found in the kinematical distributions.

The data are consistent with no metallicity gradient in the halo and at most a negligible metallicity gradient in the thick disk. Previously found correlations between kinematics and abundance in surveys of high velocity stars are the result of distinct, yet broad and overlapping, metallicity dispersions for these two populations. The halo globular cluster system does not seem to share the same kinematical and chemical distributions as the field halo stars in this sample. The survey results are discussed in the context of collapse models for the Milky Way.

1. INTRODUCTION

With automated plate measuring and reduction techniques, it is now possible to produce catalogues of photographic photometry with moderate to excellent precision for thousands of stars (cf. Gilmore and Wyse (1987) for a compilation of such surveys). Such a large body of star count data, when combined with computer models of the Galaxy, has proven a boon to the analysis of stellar density distributions. Perhaps the most significant result of this work over the last decade is the repeatedly demonstrated need for an additional structural component to the Galaxy - the thick disk.

The *existence* of the thick disk can be shown from star counts in a single passband (cf. Gilmore and Reid 1983, Fenkart 1989). However, in order to gain insight into the chemical properties of the various stellar populations in the Galaxy, *multicolor* photometric data are required (the ability to obtain spectroscopic data for large samples of stars is only now beginning to be feasible to B ~ 21). In particular, substantial leverage in the determination of the gross properties of the stellar metallicity distribution may be obtained through inclusion of the U band, since the ultraviolet spectrum is most affected by line-blanketing for the most distant main sequence stars in such surveys - spectral types F and G. Many of the automated field star surveys have been conducted in only two bands, so that the distribution of only one stellar color may be explored.

To date, U band photometry has only been included in a few moderately deep surveys, namely those of the Basel group (Fenkart 1989, and references therein) and the work of Stobie and Ishida (1987). The latter compiled UBVI photometry derived from COSMOS measurements of over 18,000 stars on Kiso Schmidt plates taken of the North Polar Cap. The star count data and color apparent magnitude distributions were found to best fit Galactic structure models which included an intermediate component with an exponential scale height of 950 pc, a local normalization of 2%, and a mean metallicity of [Fe/H] = -0.9 (Yoshii, Ishida and Stobie 1987). Sensitivity to the thick disk component is clearly seen in the mean colors of the Stobie and Ishida sample (see their Fig. 7): in the range 11 < B < 18, the mean (B-V) color only slightly declines (by less than 0.1 mag), while the mean (U-B) color shows a significant drop from (U-B) = 0.13 for B < 15.5 to (U-B) = -0.03 for B > 17. This large shift in mean (U-B) is the result of a transfer in the dominance of the star counts at distances of about 1-2 kpc from the metal-rich disk population to a population with a more significant uv-excess. That the coincident shift in (B-V) is slight in comparison to (U-B) is a manifestation of the weaker effect of line blanketing on (B-V) colors.

Unfortunately, the depth of both the Stobie et al. and Basel surveys, V ~ 18 - 19, is insufficient to probe to distances (more than several kiloparsecs above the Galactic plane) where the density of halo stars is significant in comparison to the thick disk stars. Kron (1980), using much deeper plates from the KPNO 4-m, first showed clearly that the growing contribution of halo stars in faint magnitude-limited surveys is sizable at apparent magnitudes just beyond the reach of Schmidt plates: the color distribution becomes bimodal soon after his saturation limit on the 4-m plates, or at B ~ 19, with a blue cocentration of stars corresponding to halo subdwarfs and a red concentration corresponding to local M dwarfs.

Because of the saturation limit of B ~ 19, surveys with 4-m plates (e.g. Kron 1980, Jarvis and Tyson 1981) typically have minimal overlap with the Schmidt surveys. Unfortunately, the magnitude range where the Schmidt and 4-m surveys meet, V ~ 18 - 19, happens to be the point where one expects to see a change in F and G subdwarfs from thick disk to halo metallicities. Thus, a study of the thick disk/halo interface using these more luminous and accessible dwarf stars requires consistent photometric sampling throughout the important magnitude range $17 \leq V \leq 20$. To date, the only major survey encompassing this complete range has been that of Chiu (1980), but like Kron (1980), this survey was limited to one color. In addition, as shown by Majewski (1990), the Chiu survey suffered from some systematic photometric errors.

Analysis of stellar photometric surveys typically involves fitting a model with many free parameters for each of a given number of Galactic components: a density law, scale height and length, metallicity gradient, luminosity function, local density normalization, color-magnitude relation, etc. Deriving these quantities directly from the data is usually not possible because distances are not well determined for each star. Even when multicolor photometry is available, the precision has usually been such that metallicities and photometric parallaxes are poorly defined. For example, with 4% photometry, typical of the Basel and Stobie et al. surveys, the error in derived uv-excess is 0.06. For a star with (B-V) = 0.60 and [Fe/H] ~ -1.0, a metallicity typical of the thick disk population, this translates to an error in estimated [Fe/H] of more than 0.5 dex, and an error in estimated distance modulus of about 1.0 mag.

A new stellar survey has been undertaken (Majewski 1990) which fulfills the various photometric needs as described above: multicolor photometry including the U band, complete coverage in the range $15 \leq V \leq 21$, and moderately high ($\leq 2\%$) photometric precision.

2. PHOTOMETRY AND PROPER MOTIONS TO B = 22.5

Using Kitt Peak 4-m plates spanning a 17 year baseline, I have begun a deep survey of stellar proper motions and photometry to B = 22.5 (Majewski 1990). Every star in each of three 0.3 deg^2 fields (one each at the North Galactic Pole [NGP], anticenter and ℓ = 90 deg directions), is being measured on an extensive series of plates in the U, J (~ Johnson B), F (~ [V + R]/2) and N (~ Kron-Cousins I) photographic passbands with a PDS microdensitometer. Photometrically, the survey is an extension of the Kron (1980) work, with the addition of the U and N passbands, a substantial increase in the number of plates used in the photometry in all bands, and calibration of the saturated stellar images to increase the dynamic range. The latter is accomplished through use of the profile fitting code of Stetson (1979), which determines instrumental magnitudes by fitting only the unsaturated portions of stellar images.

Thus far, substantial progress has been made in the NGP field SA 57. All 1111 stars with either J < 22.5 or F < 21.5 in the Kron (1980) survey have been remeasured on 5 U plates, 20 J plates, 11 F plates, and 10 N plates. The large number of plates contributing to the photometry substantially reduces the random errors in the instrumental magnitudes so that they are less than about 2% to magnitude 20 in each band (Fig. 1). Calibration was performed using 72 photoelectric and CCD secondary standard stars taken from a variety of sources and spanning $13.5 < B < 22.6$ and $-0.1 < (B-V) < 1.6$.

In addition, proper motions have been derived for every star with a precision of better than 0.10 arcsec/cent to B ~ 21.5. Absolute proper motions are determined to better than 0.01 arcsec/cent (velocity errors less than 0.5 km/s/kpc) by tying the astrometry to the extragalactic reference frame established by 110 galaxies and 38 spectroscopically-confirmed QSOs in the field.

The photometric portion of the survey is presently limited by the errors in the U band photometry, which, as seen in Fig. 1, become large after U ~ 21.5. Discussion will therefore be limited to those stars with U < 21.5 and the additional color criterion $0.3 < (B-V) < 1.10$ will be imposed since the UBV photometric parallax technique (cf. Sandage and Kowal 1986) is not defined for redder stars where the UBV system is insensitive to metallicity. Thus, our sample is confined to 250 stars in the blue "Kron clump" (F, G and early K stars). Ultraviolet excesses are determined by converting the Hyades (U-B), (B-V) ridge line to the photographic UJF system. Photometric parallaxes are found by converting the Wildey et al. (1962) UBV blanketing vectors and the Hyades $[M_V, (B-V)]$ color-magnitude relation.

A few comments are in order concerning the derived photometric parallaxes. The choice has been made always to err on the side of *underestimating* distances (and hence, transverse velocities). First, it is assumed that all stars are unevolved. It might be expected that some of the stars in the blue sample are subgiants, but, based on simple density law arguments, such "contamination" should be small. Thick disk subgiants, even assuming a scale height as large as 1500 pc, would be expected to appear only brighter than V ~ 17, and thin disk (scale height 300 pc) subgiants would be even brighter (V < 14). Assuming an r^{-3} density law for the halo, it would be expected that there might be 15 - 20 halo subgiants in the present NGP sample (a 10 - 20% contamination). Second, I have assumed that no star is an unresolved binary. Only binaries with components of approximately equal mass are problematical. In this case, the distance error will be 40%, while for slightly unequal mass stars, there is a concern for color errors in the primary (a 10% error in color results in ~ 0.6 mag error in estimated absolute magnitude). The thick disk binary fraction is essentially unknown, but it is likely to be somewhere between that of the thin disk (55%) and the halo. If the halo binaries follow a van Rhijn distribution (Abt 1979) then the binary mass ratio peaks at about 0.25 ($\Delta M = 7.5$). The speckle work of Lu et al. (1987) indicates that only 16 - 28% of halo stars have $\Delta M < 2.5$). Finally, I assume that there is no reddening at the NGP.

3. PHOTOMETRIC PROPERTIES OF THE BRIGHT (U < 21.5), BLUE STELLAR SAMPLE

Fig. 2 shows the results of the photometric survey. The uv-excess parameter $\delta_{0.6}$, as defined by Sandage (1969), is plotted as a function of photometrically-determined distance. I would like to call attention to five major features of this figure:

1) There exists a substantial break in the mean uv-excess at a distance of about 5 kpc above the Galactic plane. Such a metallicity feature was hinted at in an earlier survey by Hartkopf and Yoss (1982) of polar giant stars, but the feature was not well defined since the depth of that survey was only about 5 kpc. More recently, Ratnatunga and Freeman (1989) have determined spectroscopic abundances for giant stars in several fields to greater depth. In Fig. 3 I have plotted their data with mine, with the horizontal axis giving the distance from the Galactic plane. For this comparison, I have converted my uv-excesses to [Fe/H] using the relation by Carney (1979). The combined data show the sudden decrease in the number of stars with relatively higher (> -1.0 dex) metallicity at z ~ 5 kpc.

Although the emphasis of this contribution is on the photometric aspects of my survey, at this point it is appropriate to mention a few of the results from the proper motion data. In particular, the kinematical distributions are also found to exhibit a feature at ~ 5 kpc above the Galactic plane. The V velocity distribution is clearly bimodal, with two overlapping populations: the first dominates for z < 5 kpc and shows a linearly growing (from ~ 0 km/sec at z ~ 0 to 120 km/sec at z ~ 5 kpc) asymmetric drift (-V velocity) and velocity dispersion (from ~ 30 to 65 km/sec) with distance. The halo appears at z ≥ 1 kpc and maintains a nearly constant reflex velocity (~ 275 km/sec; the implications of this large value, which indicate a halo in retrograde, are beyond the scope of the present discussion and addressed at some length in Majewski [1990]) and a large (~ 90 km/sec) velocity dispersion. The overlapping populations are difficult to separate in U velocity because they have the same (null) mean, but the range in U velocity for the 250 stars together is seen to steadily increase until about 5 kpc, where it reaches its maximum extent. Beyond 5 kpc, the range in U velocity remains relatively constant.

The distributions of uv-excess, V velocity, and U velocity with z-distance all appear to show two major populations, with an unexpected "edge" to the first population at z ~ 5 kpc (the two populations can be seen in Fig. 4d). For this discussion, I will identify this latter population as the thick disk, though in my survey its separation from the thin disk is not completely obvious. The F, G and K star sample here does not adequately sample nearby distances typical of thin disk stars: therefore, little may be said regarding the juncture of the thin and thick disk populations at z ~ 1 kpc. The remainder of this discussion concerns the nature of the thick disk/halo juncture. The

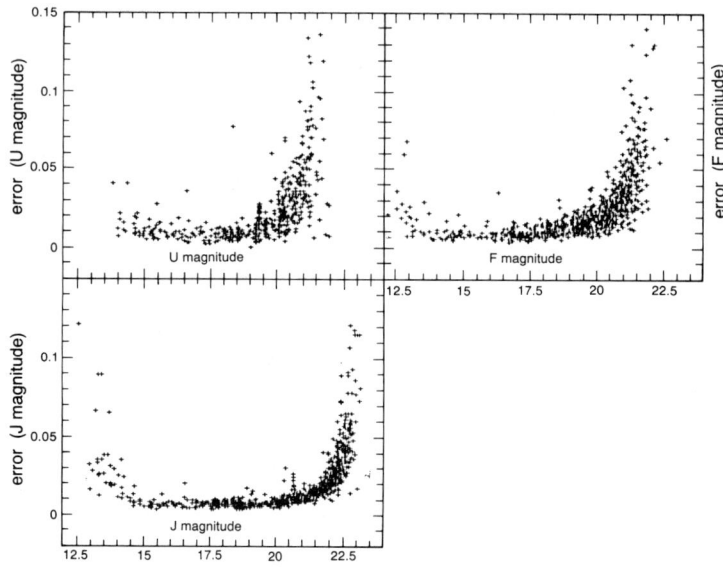

Fig. 1. Random photometric errors in the proper motion survey (Majewski 1990).

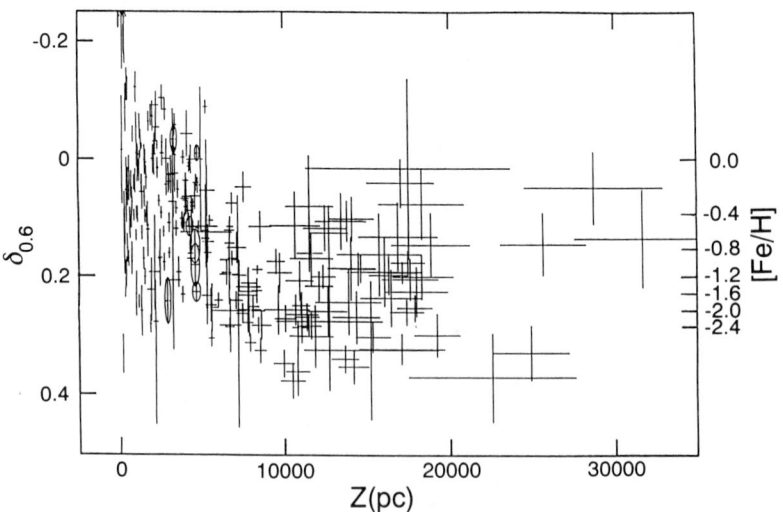

Fig. 2. Ultraviolet excess as a function of photometric parallax distance at the North Galactic Pole. The conversion of uv-excess to [Fe/H] is taken from Carney (1979).

term "edge" is used here with some reservation, since it refers to properties in the magnitude-limited distributions. The correspondence of these features to actual physical structure will not be clear until the data are compared to various Galactic structure models. However, it is tantalizing that the sharpness of the break in Fig. 2 is quite unlike the extended tail expected from an exponential density distribution.

2) There appears to be no halo star ultraviolet-excess gradient to $z \sim 20$ kpc. The mean metallicity, after converting the uv-excesses with the Carney (1979) relation is found to be $<[Fe/H]> \sim -1.4$. The lack of a gradient and the mean metallicity are more or less consistent with the results of surveys of the various halo tracer populations: giant stars ($<[Fe/H]> = -1.3$; Ratnatunga and Freeman 1989; see also Hartkopf and Yoss 1982), RR Lyrae stars in the range $5 < z < 15$ kpc (-1.5; Saha 1985), globular clusters (-1.6; Searle and Zinn 1978; Zinn 1985), and high velocity stars (-1.7; Carney et al. 1990; hereafter CALL90).

3) There is a large amount of scatter in halo uv-excess. Formally, the Gaussian dispersion is ~ 0.07 mag. Stars of extremely sub-solar metallicity are expected in the distant halo sample; however, as seen in Fig. 2, it appears that stars possessing relatively large (as high as $[Fe/H] \sim -0.5$) metal abundances may not be that rare at distances greater than $z \sim 5$ kpc. This would seem to agree with the wide range of metallicities found in Saha's halo RR Lyrae sample: $-0.5 \leq [Fe/H] \leq -2.2$. CALL90, for a sample of stars with $z_{max} > 4$ kpc, obtain typical Gaussian dispersions in their spectroscopically-determined metallicities which are large, ~ 0.55 dex, but find only a few stars with $[m/H] > -1.0$.

4) The thick disk data also are consistent with at most a slight (< 0.01 mag/kpc) uv-excess gradient. This translates to at most a gradient in $[Fe/H]$ of -0.05 dex/kpc. Hartkopf and Yoss (1982) also obtained a small (-0.14 dex/kpc) metallicity gradient in their sample of stars with $[Fe/H] > -0.5$. These stars would be associated with the thick disk. Carney, Latham and Laird (1989; CLL89 hereafter) found no metallicity gradient for the thick disk.

5) The thick disk also appears to have a large dispersion in uv-excess, although there is some contamination by near-solar metallicity thin disk stars for $z \leq 1.5$ kpc, and a few halo stars will also appear throughout. These contaminations will broaden the apparent distribution. However, in the range $1.25 < z < 2.5$ kpc, where both halo and thin disk contamination should be smallest, my sample shows the *broadest* distribution. Furthermore, Morrison, Flynn and Freeman (1990) have also shown that the thick disk contains stars as metal poor as $[Fe/H] = -1.6$.

4. COMPARISON WITH THE GLOBULAR CLUSTER DISTRIBUTION

It is interesting to compare our Fig. 3 with the similar figure for globular clusters as presented by Zinn (1985; cf. his Fig. 1). The globular cluster metallicity distribution as a function of galactoplanar distance shows a sharp feature, or "edge", similar to that seen in Figs. 2 and 3: no cluster with [Fe/H] $\geq$ -1.0 is seen at $|z| > 2 - 3$ kpc, but clusters with [Fe/H] < -1.2 are found at any distance. Thus, the *form* of the cluster and stellar distributions is similar; however, the cluster "edge" feature occurs at $z = 2 - 3$ kpc while the stellar "edge" feature occurs at $z \sim 5$ kpc.

Another difference between the distribution of field stars as presented in Figs. 2 and 3 and the globular clusters as presented in Zinn is that the latter shows a large number of low-metallicity objects at small z distances while the former shows a paucity. This difference is not likely to be physical but is instead related to selection effects: the cluster sample is more or less complete whereas the stellar samples are limited to pencil-beam slices through the Galaxy so that only a relatively small local volume is surveyed. Not many halo stars are expected at small z in Figs. 2 and 3 because of the tenuity of the halo density distribution.

Zinn has found the dispersion in halo cluster metallicities to be $\leq$ 0.4 dex. This is significantly less than the dispersion in halo stellar metalicity in my sample for $z > 5$ kpc, $\sigma_{[Fe/H]} = 0.73 \pm 0.08$. Norris and Ryan (1989b) have expressed concern over propagation of errors when UBV abundances are used in studies of metallicity distributions. Even if the error-broadening correction in my field star dispersion is somehow insufficient, it should be pointed out that CALL90 also obtain a larger field star dispersion ($\sigma_{[Fe/H]} \sim 0.55$) from their spectroscopic sample, and Laird et al. (1988) conclude that the field star metallicity distribution has broader wings than that of the clusters. The measured reflex velocity with respect to the LSR for the halo subdwarfs observed *in situ*, -275 km/s, is significantly different from the reflex velocity of the globular cluster system which is typically given as about -170 km/s (Frenk and White 1980, Zinn 1985). Thus, it appears that at present the globular cluster system does not accurately trace either the chemical or kinematical properties of the halo subdwarfs. This could imply that the two Galactic subsystems are of different origin, but certainly other scenarios are possible to account for the present dynamical and chemical arrangement. For example, tidal forces may act preferentially to disrupt clusters with certain kinematical characteristics, giving rise to a dynamical disparity with the field star distribution. The clusters may have also experienced some self-enrichment (Laird et al. 1988). An important endeavor will be to clarify the relationship of these two halo systems since it holds important clues to the order of events in the formation of the Galaxy. An additional issue to be addressed is the disparity between the reflex motion derived for the halo survey *in situ* and the smaller values derived in surveys of *local* stars selected to be halo

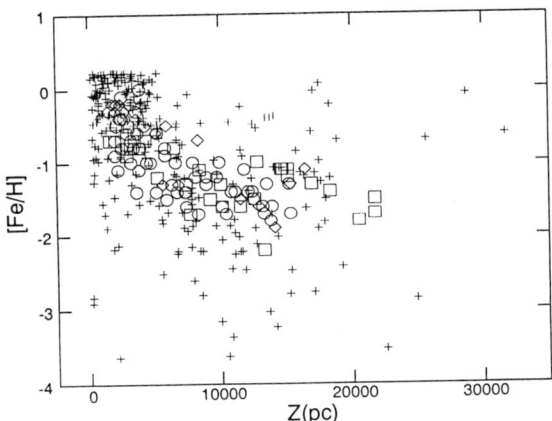

Fig. 3. Stellar metallicities as a function of distance from the Galactic plane. The plus symbols are for the Majewski (1990) survey. The remaining symbols are for giant stars in Ratnatunga and Freeman (1989): boxes are for the field SA 141 (South Galactic Pole), diamonds for SA 189, and circles for SA 127.

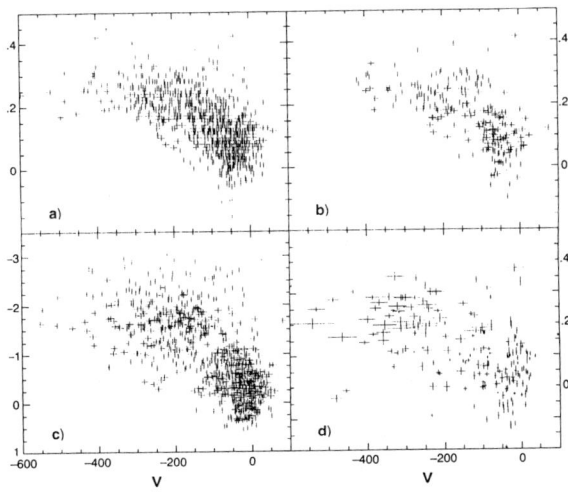

Fig. 4. Four similar figures relating stellar abundance and rotational velocity, V, in km/s from the literature: (a) Sandage and Fouts (1987), (b) Yoshii and Saio (1979), (c) CLL90 and (d) Majewski (1990). Panels (a), (b) and (d) show uv-excesses, $\delta_{0.6}$, while Carney et al. have derived spectroscopic abundances which are expressed as [m/H]. For (d), about 20% of the sample stars with the largest photometric errors have been excluded, and error bars of 0.5 σ are shown for clarity.

members; for example, Morrison, Flynn and Freeman (1990) obtain a reflex motion of -203 km/s for a sample of nearby ($|z| \leq 1$ kpc) stars with [Fe/H] $\leq$ -1.6.

5. IMPLICATIONS FOR COLLAPSE MODELS OF THE MILKY WAY

For several decades the standard picture for the formation of the Galaxy has been the homogeneous collapse model proposed by Eggen, Lynden-Bell and Sandage (ELS, 1962). Compiling data on orbits and metallicities for local high velocity stars, ELS found a number of correlations between kinematics and abundance which suggested that stellar enrichment proceeded simultaneously with collapse of an originally metal-poor, roughly spherical, primordial proto-Milky Way. As collapse continued, dissipation of energy through collisions of gas clouds increased until the gas settled into the metal-rich, circularly rotating disk.

The key tenet of the ELS model, as recently demonstrated with a much larger body of data by Sandage and Fouts (1987, SF hereafter), is that kinematics are correlated with abundance for halo stars (and, according to SF, for thick disk stars as well) as a result of this simultaneous collapse and enrichment of the Galaxy. As an example of such a correlation, Fig. 4a shows a plot of V velocities as a function of uv-excess as given by SF for their sample of proper motion stars. While the data in this figure seem to show a quite convincingly smooth correlation, similar figures as given by Isobe (1974), Yoshii and Saio (1979; Fig. 4b here), Carney, Latham and Laird (1990, hereafter CLL90; Fig. 4c), and Majewski (1990, Fig. 4d), are better described as having two separate loci of stars, one of low reflex velocity and high metallicity (the thick disk), and one of high reflex velocity and low metallicity (the halo). The apparent correlation of uv-excess to V velocity is a result of the fact that the two populations have broad and overlapping metallicity distributions. The discontinuities seen in Figs. 2 and 3 would not be expected in the ELS picture. This point has been emphasized by Freeman (1987) and CLL89 and further demonstrated by Norris and Ryan (1989b).

Two additional predictions that have been attributed to the ELS model are the existence of a halo metallicity gradient and a gradient in halo rotational velocity with z-distance. The latter is a manifestation of an expected "spin-up" as the collapse proceeded. It has now been shown rather convincingly by various surveys (cf. Section 3) that the halo does *not* exhibit a metallicity gradient. Nor does it apparently show a gradient in rotational velocity, or in the dispersions of U or V velocity for that matter, according to the deep proper motion survey. With no variation in either halo kinematics or metallicity, the idea of correlations in these properties becomes less meaningful. Another apparent contradiction to the simple ELS model is the large dispersion in halo metallicities, such that stars which are relatively rich ([Fe/H] ~ -0.5) are found high above (tens of kiloparsecs) the Galactic plane. Even SF found a "problem distribution" of stars with low (< 0.09 mag) uv-excesses on plunging orbits. CLL90 stress that the presence of these stars is a particularly thorny problem for ELS.

An alternative scenario for the formation of the halo component was proposed by Searle and Zinn (1978) as a result of their finding no metallicity gradient in the outer globular cluster system. In their picture, the halo formed through a continuous aggregation of transient protogalactic fragments. This picture for the formation of the halo seems to nestle much more naturally with the observations: with a drawn out, accretive formation, the halo is the product of a mixture of stars formed in different environments on different timescales, thus accounting for the lack of a metallicity gradient, a large abundance dispersion, and no gradient in kinematical properties. This model might even explain the surprising result of a possibly retrograde rotating halo (cf. Section 3), since there is no reason why the net angular momentum of accreted fragments might not randomly fall in one direction over another. Moreover, when the effects of dynamical friction are accounted for, it is found (Quinn and Goodman 1986) that retrograde bodies are much more stable to orbital decay, a fact which Norris and Ryan (1989a) have used to explain a significant asymmetry in halo field star orbits (a preponderance of retrograde over prograde motion) and to argue in favor of an accretive history for the halo. Given the greater compatibility with the data, does not the Searle and Zinn model present a more satisfactory scenario of halo formation than ELS?

In the evolution of a scientific theory, one starts with the simplest concepts and builds complexity as warranted by the observations. In this context, ELS might be considered the first order model of Galactic collapse, with the simplification of a protogalactic density fluctuation with a single constant density resulting in a "single, homogeneous, noiseless collapse" (Sandage 1990) as was appropriate to the contemporary data and Occam's razor. Sandage (1990), in an important discussion and reevaluation of the ELS picture, elaborates the ELS model in light of the much larger body of data now available. In his modified, "second order ELS" model, hereafter referred to as "ELS2", a Searle and Zinn-like spectrum of hierarchical, free-falling, proto-galactic shells each follows its own ELS-type collapse. In this picture, the halo is not expected to yield a metallicity gradient if one exists in the thick disk, which is proposed as the initial product of the onset of dissipation. Presumably, the "spin-up" would then be expected in the thick disk, since increasing circularization of orbits would be coincident with increasing dissipation and pressure support. In this case, the terms "thick" and "thin disk" might apply not so much to individual components as extremes of a kinematical and chemical continuum. The structure of Norris (1987), in which the thick disk is seen as an extreme tail in the properties of a thin disk population, might be seen as one example of this continuum. Thus, apparent differences of opinion on thin/thick disk discreteness are perhaps, at least between some authors, more a matter of semantics, with the real issues being the magnitude of the thick disk as a "tail" and the direction of evolution for the continuum (i.e. towards or away from the galactic plane).

Evidence for a thick disk "spin-up" is seen in the proper motion data in the form of the linearly increasing asymmetric drift with z-distance (Section 3). However, ELS2 falls short in explaining several of the abundance properties presented in Section 3. The first problem is the fact that the halo and the thick disk are not smoothly joined, either kinematically or chemically. The existence of discontinuities argues for a decoupling in star formation between the halo and the thick disk. This implies a respite between the final major phase of halo star formation and the initial major phase of (thick) disk star formation in order to allow time for enrichment and "spin-up" of the dissipative gas in the latter. In ELS2, which attempts to explain the formation of both the halo and disk in a unified, albeit noisy, sequence of events, would not the upper "layers" of the thick disk, which would have experienced the least amount of partial pressure support, be similar in character to the "lower layers" of the nearby halo? In other words, while the introduction of noise into ELS solves the problem of the lack of a metallicity gradient in the halo, as well as the problem of an age spread in the halo globular clusters, we are still left with the individual Searle and Zinn clumps each following an ELS-like smooth collapse. Even if there were some way to produce a discontinuity within each of the individual clumps, with time the net resultant galaxy would not be expected to exhibit a discontinuity because mixing of the incoherent properties of the spectrum of clumps should blur any such feature. A second problem is that the halo contains some rather metal-rich stars, the existence of which requires collapse-times for some halo proto-fragments which are long with respect to the recycle time. This implies either dissipation, extraordinarily low densities for long free-fall times, or very short recycle times in the halo.

A possible solution to both the discontinuity and the enriched halo star problems is to confine ELS-type models to the creation of disk components only, allowing an entirely different formation scenario and timeline for the halo. One possibility is that the halo formed through the capture and tidal disruption of smaller nearby galaxies (Rodgers and Paltoglou 1984). In this way, one could easily produce two populations with discontinuous but overlapping chemical and kinematical properties.

Finally and most importantly, recent observations do not appear to yield a sizeable metallicity gradient in the thick disk. The bearing of this observation on ELS2 is analogous to the original discussion of the implications of the lack of a *halo* metallicity gradient on ELS: the timescale of collapse for the thick disk must be short compared to the timescale of enrichment to produce a negligible gradient, obviating the notion of substantial dissipation in this phase. On the other hand, the added datum of a large dispersion in *thick disk* metallicity at all heights may argue for something like a "third-order" ELS, in which either the rate of dissipation within the collapsing thick disk has a spectrum of values or the age-metallicity relation is variable, or both. This would result in a range of stellar metallicities at all heights within the thick disk today as a result of rotational mixing. The variability of the age-metallicity relation is an observed feature in the Galaxy.

Is the thick disk an independent structure, formed in the first phases of pressure support in an ELS-type collapse, and therefore representative of an intermediate phase of Milky Way formation between the creation of the halo and the thin disk? Or is its origin unrelated to either or both of these other components? A key to answering this question will be understanding the relationship of the thin and thick disks. If the thick disk is found to be distinct in kinematics and chemistry from the thin disk, then, if an ELS2 scenario of formation is to be maintained, we are faced with analogous dilemmas to those discussed above for the thick disk/halo discontinuity: a mechanism must be found to account for a sudden change in the pressure support of the collapsing dissipative structure, and the persistence of a global discontinuity in stellar properties to the present demands that that mechanism operate in a fairly noiseless fashion. The latter problem may not be as severe for a more relaxed, mixed, and homogeneous disk structure as it would be for the freely-falling Searle and Zinn clumps. A possible mechanism is the sudden emergence of more efficient cooling through line emission from metals. Alternatively, models which create the thick disk through a transient dynamical phenomenon could easily produce discontinuities between properties of the thin and thick disks. Along these lines is the idea that the thick disk could have been created through the accretion of a Galactic satellite (Freeman 1987). The cannibalized satellite could either account for the thick disk population directly, or else act to dynamically heat thin disk stars into a thick disk tail. Perturbations of the thin disk by passing galaxies or transient bar potentials offer other possible heating mechanisms.

If, on the other hand, the thick disk is a less distinct entity, smoothly joined to the thin disk, then indeed we must proceed to a "third-order" ELS model, or provide an alternative scenario, to explain the apparent lack of a substantial metallicity gradient from $1 \leq z \leq 5$ kpc. Progress towards understanding the thin/thick disk transition is a priority because the character of this transition is a linchpin for various Galaxy evolution models. To date, the results of the present survey afford puzzling clues on the matter since the lack of a significant thick disk metallicity gradient is accompanied by a linear variation of the asymmetric drift with z-distance. A better understanding of the thin/thick disk relationship is one goal of future work in the deep proper motion survey, and one which will be better addressed with the red NGP sample. (It is a pleasure to thank Drs. Allan Sandage, Heather Morrison, and Bruce Carney for useful discussions and Elizabeth Doubleday for assistance with this manuscript.

REFERENCES

Abt, H. A.   1979   AJ 84, 1591
Carney, B. W.   1979   ApJ 233, 211
Carney, B. W., Aguilar, L., Latham, D. W. and Laird, J. B.   1990   AJ 99, 201 (CALL90)
Carney, B. W., Latham, D. W. and Laird, J. B.   1989   AJ 97, 423 (CLL89)
Carney, B. W., Latham, D. W. and Laird, J. B.   1990   AJ 99, 572 (CLL90).
Chiu, L-T. G.   1980   ApJS 44, 31
Eggen, O. J., Lynden-Bell, D. and Sandage, A. R.   1962   ApJ 36, 748 (ELS)
Fenkart, R.   1989   A&AS 81, 187
Freeman, K. C.   1987   ARAA 25, 603
Frenk, C. S. and White, S. D. M.   1980   MNRAS 193, 295
Gilmore, G. and Reid, I. N.   1983   MNRAS 202, 1025
Gilmore, G. and Wyse, R. F. G.   1987   in The Galaxy, G. Gilmore and B. Carswell, eds. Reidel, Dordrecht, p. 247
Hartkopf, W. I. and Yoss, K. M.   1982   AJ 87, 1679
Isobe, S.   1974   A&A 36, 333
Jarvis, J. F. and Tyson, J. A.   1981   AJ 86, 476
Kron, R. G.   1980   ApJS 43, 305
Laird, J. B., Rupen, M. P., Carney, B. W. and Latham, D. W.   1988   AJ 96, 1908
Lu, P. K., Demarque, P., van Altena, W., McAlister, H. and Hartkopf, W.   1987   AJ 94, 1318
Majewski, S. R.   1990, Ph.D. Dissertation, University of Chicago
Morrison, H. L., Flynn, C. and Freeman, K. C.   1990   AJ 100, 1191
Norris, J.   1987   ApJ Letters 314, L 39
Norris, J. E. and Ryan, S. G.   1989a   ApJ Letters 336, L 17
Norris, J. E. and Ryan, S. G.   1989b   ApJ 340, 739
Quinn, P. J. and Goodman, J.   1986   ApJ 309, 472
Ratnatunga, K. U. and Freeman, K. C.   1989   ApJ 339, 126
Rodgers, A. W. and Paltoglou, G.   1984   ApJL 283, L 5
Saha, A.   1985   ApJ 289, 310
Sandage, A.   1969   ApJ 158, 1115
Sandage, A.   1990   JRASC 84, 70
Sandage, A. and Fouts, G.   1987   AJ 92, 74 (SF)
Sandage, A. and Kowal, C.   1986   AJ 91, 1140
Searle, L. and Zinn, R.   1978   ApJ 225, 357
Stetson, P. B.   1979   AJ 84, 1056
Stobie, R. S. and Ishida, K.   1987   AJ 93, 624
Wildey, R. L., Burbidge, E. M., Sandage, A. R. and Burbidge, G. R.   1962   ApJ 135, 94
Yoshii, Y., Ishida, K. and Stobie, R. S.   1987   AJ 92, 323
Yoshii, Y. and Saio, H.   1979   PASJ, 31, 339
Zinn, R.   1985   ApJ 293, 424

DISCUSSION

YOSS: What will you do to confirm that the halo "normal abundance" stars are in fact what they appear to be? Classification spectra do not have the necessary resolution to derive reliable abundances for these spectral types.

MAJEWSKI: Two projects are planned to provide independent confirmation of the [Fe/H] ~ -0.5 halo stars, and indeed, all of the uv-excess data. First, I hope to obtain deep, higher precision (~ 1% at magnitude 22) CCD photometry in all three of my survey fields. With a TEK 2048 chip on the KPNO 4-m, each field can be covered in a 2 x 2 mosaic. The most laborious task is the U photometry, which will require several hours of integration to reach 2% at B = 22.5; however, such data will allow photometric parallaxes to be obtained to the depth of the proper motion data, B ~ 22.5. Thus, the survey will eventually probe to twice its current distances, or about 35 kpc and have higher precision at all magnitudes. The second project involves low resolution (1 Å/pixel, multislit spectroscopy using the Palomar 200" and the double spectrograph. The Ca II K line and the infrared calcium triplet will be measured for stars to B ~ 21. Both spectral features have been shown to be reliable metallicity indicators throughout the range 0 > [Fe/H] > -4. These spectra will yield radial velocities with better than 20 km/s accuracy and allow orbits to be determined for this subsample of the survey. Your concern regards the apparently high metallicity stars. While these must be confirmed, I also have concern for the low metallicity stars, since photometric abundances are unreliable below [Fe/H] ~ -2.5.

UPGREN: Why don't Sandage and Fouts (SF) show the break that you show for the other studies?

MAJEWSKI: It is not clear why that particular survey differs from the others in not clearly showing a break indicating two separate populations. To be fair, I should point out that SF mention the existence of an "abrupt break in the number of stars at δ = 0.15, with a much thinner distribution for δ > 0.15", and suggest that "a fundamental change occurs in the kinematical and chemical abundance distibutions...near δ = 0.15". But the apparently smooth transition toward more negative V velocities in their diagram is still an argument in favor of a correlation between abundance and kinematics. It has been suggested at this conference that perhaps, since the axes are not independent in the Sandage and Fouts diagram (because V velocities are a function of the photometric parallaxes), there is some smoothing across the break. In fact, the correlation of the errors runs perpendicular to the desired direction for this to be the explanation. Norris and Ryan (1989b) show how photometric abundance errors in SF would "linearize" decoupled populations. Curiously, in spite of similar photometric errors to SF and a comparable sample of pertinent stars, my own survey clearly shows the decoupling (Fig. 4d).

RATNATUNGA: Since errors in your metallicity estimate are larger at distances beyond five kpc, what confidence can you give that your high metallicity estimates are not caused by observational errors?

MAJEWSKI: Obviously, on a star-by-star basis little can be said beyond the error bars shown in Fig. 2. Independent confirmation of the abundances for these stars is desirable. Most of the stars shown in Fig. 3 with [Fe/H] > -1.0 and 5 < z < 10 kpc have 17 < U < 19, and so the photometry in all bands should be better than 2%. Some of these could be thick disk members, in which case their metallicities are not nearly as exciting, but then their z-distances are!

SARAJEDINI: Since you have so many stars and errors for your metallicities and distances, you can set a level of confidence on your claim that there is not a relation between distance and abundance.

MAJEWSKI: Simple weighted, polynomial fits to the data on either side of the uv-excess break yield no significant slope. For z > 7 kpc, the fit is statistically indistinguishable from a no correlation line, and in fact, the negligible slope is in the wrong sense. For z < 5 kpc, a slope of 0.005 mag/kpc is a five σ result.

PEL: You select your sample by a cut-off in (B-V), but your blanketing lines do not run at constant (B-V), which means that for very metal-poor stars your temperature cut-off is considerably cooler (and the absolute magnitude fainter) than for the solar-type stars in your sample. Is this selection effect a serious complication for your analysis?

MAJEWSKI: The blue cut-off in (B-V) eliminates only a half-dozen or so stars beyond the expected main sequence turnoff, so that they would be white dwarfs, horizontal-branch stars or blue stragglers anyway. At the red cut-off, the blanketing vectors are very small so that there is a negligible effect. In any case, the stars near the red cut-off are within several kpc and have little bearing on the issues discussed here. A more serious concern might be the cut-off in U magnitude, since this imposes a limit to the greatest volume surveyed at each (B-V) color for a given uv-excess. For stars redder than (B-V) ~ 0.8 the effect is small. For (B-V) ~ 0.5, stars with a uv-excess of 0.10 mag can be seen ~20% farther than stars of uv-excess 0.31 mag. Thus there is a slight selection for more metal-rich stars for $z \geq 8$ kpc, and essentially no effect closer than this. Therefore, only the conclusions on the halo metallicity gradient may be affected, but the results obtained are consistent with the other surveys (cf. Section 3).

uvby PHOTOMETRY OF A-TYPE STARS AT HIGH GALACTIC LATITUDE

John S. Drilling

Louisiana State University

A. G. Davis Philip

Union College and Van Vleck Observatory

Philip (1972) has presented a plan for a systematic study of stellar populations at high galactic latitudes. The original plan specified that 25 square-degree objective-prism plates be obtained at each 15 degrees of galactic latitude for the four galactic longitudes 0, 76, 180 and 290, and that the plates then be searched for stars of spectral class A7 and earlier down to as faint a limiting magnitude as possible. The fields for which finding lists have been published or for which uvby photometry has been obtained are listed in Table I. Successive columns give the name of the field, galactic longitude, galactic latitude, size of the field (square degrees), the name of the observatory at which the plates were taken, the spectroscopic dispersion of the plates (Ångstroms/mm), the average limiting B magnitude, the number of stars of spectral type A7 and earlier identified, and the number of stars for which uvby photometry has been obtained. In the notes to Table I, the latest published reference is given for each of the fields listed and for the North Galactic Pole, where the objective-prism survey of Slettebak and Stock (1959) was used.

To illustrate our methods, we have plotted the results for the 4 HLF 4 area in Figs. 1 - 4. These figures have been corrected for a constant reddening of E(b-y) = 0.054 magnitudes according to the standard extinction law (Strömgren 1966). In the case of 4 HLF 4, this correction has been justified by obtaining H-β photometry for the brighter stars, and then determining the run of E(b-y) with distance modulus for the stars with normal uvby-β indices using the calibrations given by Stömgren (1966) and Crawford (1970). In this case, and in general, one can assume that all but a few of the brightest A-type stars in these high-latitude fields lie beyond the reddening layer, and one can correct for reddening by simply fitting the zero-age main sequence relations in the c_1 - (b-y) and m_1 - (b-y) diagrams using the standard reddening law. The curves in Figs. 1 - 4 are the zero-age

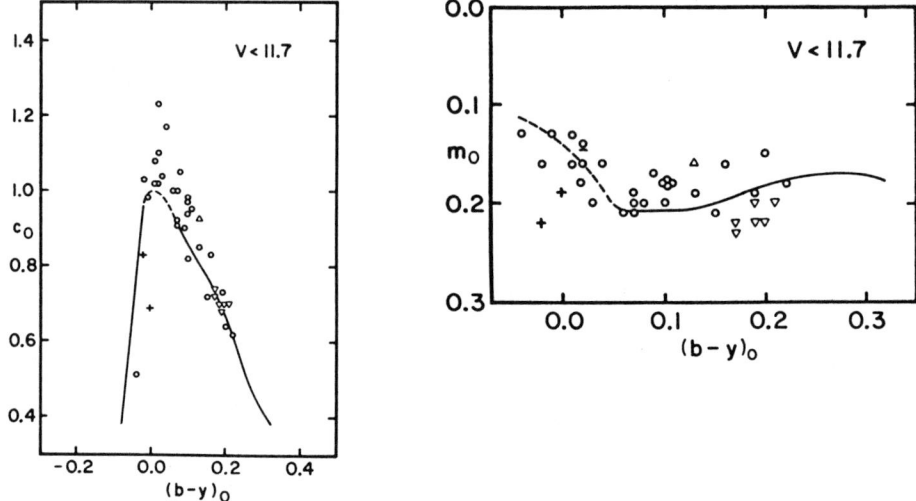

Fig. 1. c_0 versus $(b-y)_0$ for stars with $v < 11.7$ in 4 HLF 4.
Fig. 2. m_0 versus $(b-y)_0$ for stars with $v < 11.7$ in 4 HLF 4.
[Reproduced from the Astronomical Journal]

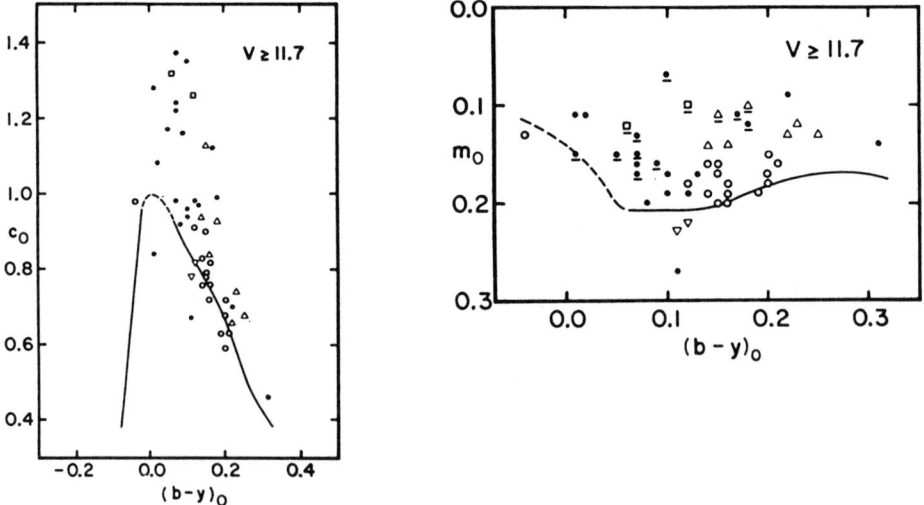

Fig. 3. c_0 versus $(b-y)_0$ for stars with $v > 11.7$ in 4 HLF 4.
Fig. 4. m_0 versus $(b-y)_0$ for stars with $v > 11.7$ in 4 HLF 4.
The underlined symbols represent stars which lie more than 0.2 magnitudes above the zero-age main sequence in Fig. 3.
[Reproduced from the Astronomical Journal]

main sequence relations determined by Crawford (1970) and from the data
given by Crawford et al. (1970).

TABLE 1

Fields for which a finding list has been published
or for which uvby photometry has been obtained.

Field	l	b	Size (sq.deg.)	Obs.	Disp. (Å/mm)	B limit	No. stars	No. uvby
SGP	--	-90	230	CTIO*	580	13.5	180	95
1 HLF 2	76	-30	33	Case	280	12.2	105	--
1 HLF 4	76	-45	25	Ton	280	12.5	34	--
3 HLF 4	0	-45	45	CTIO	280	13.5	146	55
4 HLF 4	180	-45	45	CTIO	280	13.5	88	85

* some plates taken at Tonantzintla.

References:

NGP   Philip and Tifft 1971 AJ 76, 567
SGP   Philip and Drilling 1990, submitted to ApJ
1 HLF 2 Philip 1966 ApJS 12, 391
1 HLF 4 Philip 1967 Bol Obs Ton Tac 4, 215
3 HLF 4 Drilling and Philip 1970 Bol Obs Ton Tac 5, 7
4 HLF 4 Drilling and Pesch 1973 AJ 78, 47

In Figs. 1 and 2, c_1 and m_1, respectively, are plotted against (b-y) for stars brighter than V = 11.7 in 4 HLF 4. If one ignores the highest point in Fig. 1 (which is a horizontal-branch star), the stars plotted as open circles are all seen to have normal uvby indices. The stars plotted as crosses and downward-pointing triangles are peculiar and metallic-line A stars, respectively. These stars are seen to lie below the other stars at a given (b-y) in the m_1 - (b-y) diagram. These classifications, which are based on the uvby photometry, have been confirmed by the MK spectral classification of Drilling and Pesch (1973). Radial velocities obtained from the same plates indicate that a large majority of the stars in Figs. 1 and 2 belong to Population I.

In Figs. 3 and 4, the same plots are shown for the stars with V > 11.7. The turnover to Population II is shown by the appearance of large numbers of horizontal-branch stars, which are those which lie more than 0.2 magnitudes above the zero-age main sequence relation in the c_1 - (b-y) diagram. It is seen that for (b-y) > 0.04, these stars lie above the other stars in the m_1 - (b-y) diagram. Philip (1990) has made extensive studies of these "field horizontal-branch" stars, and has shown that these stars are identical in all respects (including uvby indices, radial velocities, metallicity and continuous energy distribution) to the blue horizontal-branch stars in globular clusters.

No less interesting are the stars in Figs. 3 and 4 which have c_1 and m_1 indices which are normal for their (b-y) colors, because if these are normal A stars, one is faced with the problem of how they got to their present positions 1 kpc and more below the galactic plane. Rodgers (1971) isolated a similar group of apparently normal A stars at distances of up to 3 kpc or more from the galactic plane in the direction of the South Galactic Pole using UBV photometry and low-resolution spectra, and Philip and Drilling (1983) have shown that more than half of these stars have normal c_1 and m_1 indices for their (b-y) colors. A very extensive study of these stars has been made recently by Lance (1988), but the problem of their origin remains unsolved

Finally, one sees that six of the stars plotted in Figs. 3 and 4 (and one of the stars in Figs. 1 and 2) have normal c_1 indices for their b-y colors, but have ml indices which place them 0.04 magnitudes or more above the zero-age main sequence relation in the m_1 - (b-y) diagram. The possible explanations for these stars range from the field analogs of the blue stragglers in globular clusters (Bond and MacConnell 1971) to λ Bootis stars, which are metal-poor Population I stars (Gray 1988). Philip (1989) has shown that such stars in the direction of the South Galactic Pole have a space and velocity distribution which is intermediate to Populations I and II. Clearly, these stars deserve further study.

We would now like to present some preliminary results for the three fields where we have penetrated deepest into the halo: 3 HLF 4, which lies in a direction 45 degrees below the galactic center; the South Galactic Pole; and 4 HLF 4, which lies 45 degrees below the anticenter. The results are summarized in Table II, where successive rows give, for each of the three fields, the galactic longitude, galactic latitude, area (square degrees), E(b-y) as determined using the methods described above, and (for stars both brighter and fainter than V = 12) the total number of stars in the region, the number per square degree, the number for which uvby photometry has been obtained, and the percentages of normal Population I, intermediate, and horizontal-branch stars as determined from the uvby photometry according to the criteria described above. In the case of the South Galactic Pole, we have used the Slettebak and Brundage (1971) survey rather than the Philip and Sanduleak (1968) survey to calculate the numbers of stars per square degree due to the fact that some of the plates in the earlier survey were taken at Tonantzintla, as noted in Table 1, and hence have a limiting magnitude which is at least one magnitude brighter than the others.

uvby PHOTOMETRY OF A-TYPE STARS AT HIGH GALACTIC LATITUDE

It is seen that the numbers of stars with V < 12 per square degree are very similar in 3 HLF 4 and 4 HLF 4, and that the number per square degree in the South Galactic Pole does not differ by a ratio which is significantly different from the cosine of 45 degrees. Also, a large majority of these stars have uvby indices which indicate that they are normal Population I stars. Drilling (1972) has shown that the space distribution of normal A2 - A7 stars in 3 HLF 4 and 4 HLF 4 is similar to that given by Upgren (1963) for the North Galactic Pole. All of this is consistent with a plane-parallel distribution for the Population I stars with a scale height of around 100 parsecs.

The situation is quite different for the stars with V > 12. As we go from 4 HLF 4 to 3 HLF 4, the number of stars with V > 12 per square degree more than doubles. The reason for this can be seen in the percentage of the stars which are horizontal-branch stars, which rises rapidly as we go from 4 HLF 4 through the South Galactic Pole to 3 HLF 4. This would be consistent with a spheroidal distribution for the horizontal-branch stars with the space densities decreasing with increasing distance from the galactic center. It will be important to complete this work, and to try to model numerically the distribution of horizontal-branch stars in the galactic halo. Even more important will be the analysis of the thousands of faint A-type stars which are being identified in the extended surveys of Beers et al. (1988) in the southern hemisphere and of Sanduleak and Pesch (1988) in the north.

TABLE 2

Preliminary results for 3 HLF 4, 4 HLF 4 and the South Galactic Pole.

Region	3 HLF 4	SGP	4 HLF 4
Galactic Longitude	0	--	180
Galactic Latitude	-45	-90	-45
Area (square degrees)	45	230	45
E(b-y)	0.015	0.014	0.054
No. Stars w. V < 12	56	114	49
No. per square degree	1.24	0.50	1.09
No. w. uvby photometry	20	32	46
Population I	95%	84%	93%
Intermediate	0%	6%	4%
Horizontal Branch	5%	6%	2%
No. Stars w. V > 12	90	66	39
No. per square degree	2.00	0.29	0.87
No. w. uvby photometry	35	63	39
Population I	29%	38%	51%
Intermediate	11%	13%	13%
Horizontal Branch	57%	41%	33%

REFERENCES

Beers, T. C., Preston, G. W. and Schectman, S. A. 1988 ApJS 67, 461
Bond, H. E. and MacConnell, D. J. 1971 ApJ 165, 51
Crawford, D. L. 1970 in Stellar Rotation, A. Slettebak, ed., Gordon and Breach, New York, p. 204
Crawford, D. L., Barnes, J. V. and Golson, J. C. 1970 AJ 75, 624
Drilling, J. S. 1972 AJ 76, 1072
Drilling, J. S. and Pesch, P. 1973 AJ 78, 47
Gray, R. O. 1988 AJ 95, 220
Lance, C. M. 1988 ApJ 334, 927
Pesch, P. and Sanduleak, N. 1989 ApJS 71, 549.
Philip, A. G. D. 1972 in The Role of Schmidt Telescopes in Astronomy, U. Haug, ed., Hamburger Sternwarte, p. 117
Philip, A. G. D. 1989 in The Gravitational Force Perpendicular to the Galactic Plane, A. G. D. Philip and P. K. Lu, eds., L. Davis Press, Schenectady, p. 157
Philip, A. G. D. 1990 in Precision Photometry, A. G. D. Philip, A. R. Upgren and K. A. Janes, eds., L. Davis Press, Schenectady, p. 153
Philip, A. G. D. and Drilling, J. S. 1983 BAAS 15, 924
Philip, A. G. D. and Sanduleak, N. 1968 Bol. Obs. Tonantzintla y Tacubaya 4, 253
Rodgers, A. V. 1971 ApJ 165, 581
Slettebak, A. and Brundage, R. K. 1971 AJ 76, 338
Slettebak, A. and Stock, J. 1959 Astron Abh. Hamburg No. 5
Strömgren. B. 1966 ARAA 4, 433
Upgren, A. R. 1963 AJ 68, 475

DISCUSSION

RICHER: Are any white dwarfs turning up in these surveys? There have been some suggestions recently that these may be important contributors to the dark matter content of the halo.

DRILLING: Only one white dwarf has turned up. This is not surprising, as the limiting magnitude of these surveys is only B = 13 - 14 at best, and in many cases less than this. The one white dwarf discovered was found on of the UV prism plates taken of the South Galactic Pole, and these plates go a little bit deeper than the others.

WARREN: Is the reddening for the 4 HLF 4 stars mostly in the foreground?

DRILLING: Yes. A plot of E_{b-y} vs $V_o - M_v$ for the normal Pop I stars, obtained from uvbyβ photometry shows a steep rise to a constant reddening at $E_{b-y} = 0.054$, with only the three nearest stars lying within the dust layer.

THE LATE-B-TYPE STARS: REFINED CLASSIFICATION, CONFRONTATION WITH STROMGREN PHOTOMETRY, AND THE EFFECTS OF ROTATION

R. F. Garrison

David Dunlap Observatory, University of Toronto

R. O. GRAY

Appalachian State University

Two hundred late-B-type stars have been classified on the MK system as defined by Morgan (Morgan, Keenan 1973 and Morgan, Keenan, and Tapscott 1978). Nine carefully determined secondary standards have been proposed to complete the grid of rapidly and slowly rotating stars in the B7 - B9 region.

Rotation effects on classification indices and photometric indices have been analyzed. Reddening effects are more dominant for the late-B-type stars than for the early-A-type stars, so rotation effects are masked. The beta index is not observably affected by rapid rotation in the late-B-type stars.

Compared with previous types by Garrison for some of the stars, the temperature types are only 0.01 class later, with a scatter of 0.29. Luminosity classes are 0.18 class more luminous, with a scatter of 0.50.

DISCUSSION

STETSON: Could you please go into more detail on the criteria which allow you to distinguish a rapid rotator viewed pole-on from a non-rotator of slightly different temperature and gravity?

GRAY: The ability to discriminate between pole-on rapid rotators and intrinsically slow rotators is strongest in the late A-type stars. Between spectral types at about A5 and A8 I found a population of low-vsini dwarfs with large δm_1 indices (more typical of giants) and moderately large δm_1 indices. These stars have been mistaken by some authors as being slightly metal-weak. However, a close examination of the theory of Collins and co-workers shows that the effects of rapid rotation on δc_1 and δm_1 in nearly pole-on rotators (i < 30°) can account for this discrepancy very nicely. Thus I hypothesized in my thesis that these spectroscopically normal low-vsini dwarfs are actually pole-on rotators. In our paper on the late A-type stars we showed that this idea was actually consistent with the statistics. I have hypothesized that Vega is a pole-on rapid rotator basically on the basis of its abnormal absolute magnitude.

WARREN: I made a careful study of the B-type stars (see Warren 1975, MNRAS) (mostly earlier B stars) in the mid seventies and found an effect on the β index similar to what you have found for the A stars. That is, there is essentially no effect up to 250 km/sec, after which there is a noticeable effect described by a decreasing envelope in β. Of course, that study did not make use of MK type corrected for rotation effects so such a study with corrections might be worth repeating.

HAUCK: You claim your MK types are unaffected by rotation. We can confirm that by using in the Geneva system the d vs Δ diagram. In this diagram isotemperature lines are well defined and if we plot stars with the same MK type, taken from various sources, we observe a large scatter around the constant T_{eff} line. With A. Slettebak (Hauck and Slettebak 1990 A&A 214, 13) we have shown that by using your MK types no important scatter is detectable and we concluded that your types are really not affect by rotational effects!

GARRISON: I'm glad to see this independently confirmed. We tried very hard to eliminate as many effects as possible, other than T_{eff} and log g.

PROBING THE STELLAR POPULATION OF THE OUTER GALACTIC DISK: STAR CLUSTERS AND "WINDOWS" THROUGH THE GALACTIC PLANE

Kenneth A. Janes

Boston University

ABSTRACT: It is likely that the most recently-formed large-scale structure of the Galaxy is the outer part of the disk; its development may even be continuing at the present time. However, because of the difficulty of observing through the disk, we know little about the character of the stellar population of those distant regions. This review is a report on the initial progress in a study of the outer disk, based on observations of distant open clusters and a photometric survey through low-galactic latitude windows of low obscuration.

The cluster observations indicate a prolonged period of star formation in what can now be considered to be the outer disk. Nevertheless, the metallicities of clusters there are extremely low. At least one cluster, Tombaugh 2, has an age of little more than 1 Gyr, yet it has a metallicity of -0.68. To 2 also appears unique in that it may be associated with the warp seen in the galactic H I disk.

There are a small number of clear windows, exactly in the galactic plane, which permit one to study the stellar population along a line of sight right through to the edge of the Galaxy. This review will include a discussion of some initial results from a photometric survey of one of these windows located at $\ell = 243$ degrees. The obscuration is low enough in this window that several galaxies are visible within a few arcminutes of the galactic plane.

1. INTRODUCTION

1.1 Formation of the Galactic Disk

The rapid collapse concept for the early history of the Galaxy, originally formulated by Eggen, Lynden-Bell and Sandage (1962 "ELBS"), has long been used as the starting-point for understanding galactic evolution. The ELBS idea is that the first stars in the Galaxy, those comprising what are now the Population II stars, formed from the original gaseous proto-galaxy while it was in a state of nearly free-fall collapse. Thus the entire collapse phase of the Galaxy took place in a very short period of time, during which the remaining gas was greatly enriched in the heavy elements. The disk formed soon afterward from this remaining gas.

A. G. Davis Philip, A. R. Upgren and K. A. Janes (eds.)
PRECISION PHOTOMETRY
233 - 243 © 1991
L. Davis Press

Kenneth A. Janes
Dept. of Astronomy
Boston University
Boston, Mass.
02215

A considerably different picture of galactic evolution emerged in papers by Larson (1976) and Tinsley and Larson (1978). Although there were a number of ad-hoc assumptions in their work, the Larson-Tinsley theory does a remarkably good job of explaining many of the modern observations. A particularly important aspect of their theory is the concept that at the end of the rapid collapse phase which formed the halo, the disk began to form only gradually, and in fact, formed inside-out: as outlying material (with correspondingly larger angular momentum) fell into the developing galaxy, a disk began to form, growing radially but becoming thinner at fixed radius as time progressed.

Yet another view of the formation of the halo and the relationship between the halo and the disk has come from the work of Searle and Zinn (1978), Gilmore and Reid (1983) and Zinn (1985). Searle and Zinn proposed that the halo has formed from separate globular-cluster- or dwarf-spheroidal-sized pieces which merged to form the halo. As Zinn showed, the metal-rich globulars are distributed in a thick disk shape, with a considerable net rotation. Gilmore and Reid showed that there is in the solar vicinity a population of stars with kinematics intermediate between, and distinct from, the "thin disk" and the true halo. If these ideas are correct, then the formation of the halo and the formation of the disk are separate events in the life of the Galaxy. The question then becomes, does the thick disk represent the extreme high-velocity tail of the distribution of disk population stars, the remnants of the halo population or is it a distinct entity? The picture is further complicated by the possible existence of a "dark halo" to the Galaxy, that is, some form of unseen material which may or may not be baryonic. The much-debated dark matter is cosmologically speaking of profound importance, but within our Galaxy, the principal manifestation of the dark matter is in the rotation curve of the Galaxy - specifically in the rotation of the outer parts of the galactic disk. The best available evidence (see, for example, Chini 1985, or Hron 1987) indicates that the galactic rotation curve is flat or even rising beyond the Sun's position in the Galaxy. This is entirely consistent with observations of external galaxies, and implies large amounts of unseen matter.

The outermost region of the galactic disk is also seen in neutral hydrogen to be severely warped (see, e.g., Burton 1988 for a recent review). The character of the warp is strongly affected by any dark matter that may be present. It would be especially useful to learn whether there is a stellar component of the warped disk.

Unfortunately we have very little information about the dynamics of the outer parts of our own Galaxy, so the true character of the rotation curve and the dynamical interaction between stars and the gaseous warp are unknown.

1.2 Galactic Chemical Evolution

Whether the Galaxy formed in a rapid collapse, a slow collapse, or grew from a condensation of smaller pieces, an inevitable consequence seems to be that there should be a radial gradient in heavy element abundance among disk stars and a relation between the age and metallicity of stars. These conclusions are supported by most observations. Nevertheless, there are still significant questions which call into question some of the basic assumptions about the history of the galactic disk.

Twarog (1980) found a substantial increase in the metallicity of F-type stars in the disk from the earliest times to the present, with most of the chemical evolution occurring early in the life of the Galaxy. His data indicate an increase of about a factor of two in the past 8 - 10 Gyr, which is consistent with chemical evolution models incorporating substantial amounts of inflowing material. However, the two oldest known open clusters, NGC 188 and NGC 6791, which are about 10 Gyr in age (see VandenBerg 1985 and Janes 1988), are, if anything, more metal-rich than the youngest clusters. Neither cluster is consistent with standard pictures of galactic chemical evolution. Unfortunately, Twarog's analysis contains a bias against old, metal-rich stars. In spite of his attempts to correct the color distribution of stars in his sample for metallicity effects, the fact remains that the sample consists almost entirely of F stars. As his own analysis shows, a 10 billion year old star of solar composition at the main-sequence turnoff will probably not be found in his data. If such stars exist, Twarog would not have found them.

Numerous surveys *along* the galactic plane have established the existence of a distinct gradient of metallicity decreasing radially outward from the galactic center (Janes 1979, Shaver et al. 1983). Neese and Yoss (1988) have confirmed the existence of a composition gradient, although with a lower slope than most other investigations have found. In contrast, Lewis and Freeman (1989) found evidence that while the velocity dispersion decreases with increasing galactic radius, there is at best only a very slight composition gradient. Unfortunately there is also a clear systematic bias in the Lewis and Freeman data. They attempted to restrict their sample to old disk stars by considering only stars with (B-V) greater than 1.2. In fact the color of a red giant is almost completely independent of its age, but is, however, a strong function of its metallicity. Most old, metal-poor giants are substantially bluer than (B-V) = 1.2. Thus their sample is heavily biased toward metal-rich stars, many of which will also be rather young. If there are old, metal-poor stars in the outer disk, they will be completely missed by the Lewis and Freeman survey.

2. PROBING THE STELLAR POPULATION OF THE GALACTIC DISK

There evidently remains some considerable uncertainty about the evolution of the galactic disk, particularly the true character of the galactic composition gradient. Additional studies of the stellar population the plane are needed to determine the global dynamical and chemical properties of the galactic disk stellar population. Most galactic disk stellar surveys have been confined to objects within a few kpc of the Sun. Furthermore, in almost every case, an inherent bias has been introduced because of the difficulty of seeing distant objects that are truly in the plane of the Galaxy. Thus the more distant stars observed are always further from the galactic plane than the more nearby ones. At the largest galactic radii, we still know very little about stellar properties in the plane. How far does the stellar disk extend? What is the scale height (perpendicular to the galactic plane) at large galactic radii? Do the stars follow the same warp traced by the neutral hydrogen gas? What are the compositions and kinematics of the most distant disk stars?

2.1 Open Clusters in the Outer Disk

Two of the most distant known old open clusters from the galactic center are Tombaugh 2 and Berkeley 21. Previous photographic studies (Adler and Janes 1982, Christian and Janes 1979) gave indications of their distances and ages, but because of their extreme distances and the limitations of the photographic work, the results were somewhat uncertain.

Both clusters have now been re-done, using CCD images from the KPNO and CTIO 0.9-meter telescopes. The results will be reported on in greater detail elsewhere, but the basic conclusions are summarized in Table 1.

Table 1

The outer disk clusters Be 21 and To 2

	Be 21	To 2
ℓ	187	233
b	-2.5	-7.0
E(B-V)	0.7 $\pm$0.2	0.2 $\pm$0.2
$(m-M)_o$	13.9 $\pm$0.2	14.7 $\pm$0.2
R_{GC}	14.5 kpc	16.0 kpc
Z	250 pc	1000 pc
[Fe/H]*	-0.32	-0.68
Age	2-3 Gyr	2-3 Gyr

*Friel and Janes, 1990

Both clusters are rather metal poor compared to the solar neighborhood, but Tombaugh 2 with an age of little more than 2 Gyr has a metallicity of less than one-quarter that of the Sun. To 2 is also unique in that it could be associated with the warp seen in the galactic HI disk.

Most of the old open clusters (that is, clusters older than roughly one billion years), are located further from the galactic center than the Sun. Furthermore, the few globular clusters known to be younger than typical for their metallicity, are also located far from the galactic center. Although there does appear to be a substantial gap in age between the youngest outer halo globular clusters and the oldest outer disk open clusters, the cluster observations indicate a prolonged, not quite continuous, period of star formation in what can now be considered to be the outer disk. The compositions of open clusters also provide perhaps the most direct measure of the composition gradient in the disk (Janes 1979). In an update to that study, Friel and Janes (1990) have obtained spectra of red giants in a number of open clusters, including To 2 and Be 21; we find that the data are consistent with a gradient of -0.07 dex/kpc. There are however, indications that there may not be a smooth (linear) gradient, but rather there is a suggestion in the data of a break in the abundance distribution, with near solar composition to about 10 kpc from the galactic center (assuming the Sun is at 8.5 kpc) and a much lower average metallicity beyond that point. In any case, the metallicities of clusters in the outermost parts of the disk are extremely low, compared to typical values of disk stars near the Sun.

2.2 Surveys Along the Disk

Unfortunately, in most directions the Galaxy becomes completely opaque within just a few kiloparsecs from the Sun. However, surprisingly few people have noticed that through a few "windows" near the galactic plane, one can see to great distances. There are even a couple of lines of sight through which some galaxies are visible exactly in the galactic plane, which means that we can see right through to the outer edge of the Galaxy. Most notably, near galactic longitude 243 degrees, several galaxies are visible on both the Palomar Observatory Sky Survey blue and red prints (FitzGerald 1974; see also, Dodd and Brand 1976). FitzGerald estimates the total absorption in the blue to be not more than 4.5 magnitudes in that direction (or assuming a normal reddening law, about 3.5 magnitudes in the visual).

In the region which Fitzgerald identified, there is relatively little obscuration, not only just along the galactic plane, but also extending north of the plane and to some extent the south, right out to large galactic latitudes. I have located another direction, near galactic longitude 197 degrees along which galaxies are also visible. Thus in both the ℓ = 243 degrees and ℓ = 197 degrees directions, it is possible to examine a slice of the Galaxy from the Sun's position right to the edge of the galactic disk. The ℓ = 243 degree direction is of particular interest, since the effects of galactic rotation will show up strongly in radial velocities, permitting either a determination of distances to stars if the rotation curve is assumed, or determination of the rotation curve if the distances are known.

3. A PRELIMINARY SURVEY OF THE PUPPIS WINDOW

A red giant star at absolute magnitude near +1 with 3.5 magnitudes of (V) extinction (the amount FitzGerald estimated in the ℓ = 243 degree direction) will have an apparent magnitude of about 19.5 at a distance of 10 kpc. Thus, a survey reaching magnitude 20 should reveal red giants in the outermost reaches of the galactic disk. Assuming an exponential law of stellar density in the galactic plane, and a density of about 1000 tenth magnitude red giants per 100 square degrees toward the galactic anticenter (McCuskey 1967), then one would expect to find about 800 red giants per square degree between magnitude 19 and 20.

To investigate whether a survey along the galactic plane at the ℓ = 243 degree direction is feasible, I obtained a series of CTIO 0.9-meter CCD images in February 1989. At that telescope, a TI 395 X 381 pixel CCD covers a field of view of just over 3 arc-minutes on a side. A series of five partially overlapping fields was observed, yielding a total field of about 3 minutes EW by 10 minutes NS. The northern edge of the field crossed the galactic plane. In each of the fields, a series of long and short exposures was taken in each filter - the long exposures being typically 600 seconds in V, R and I, 900 seconds in B and 1800 seconds in U, with the short exposures being a factor of ten shorter. Finally, exposures were obtained through a 5100 A filter, part of the so-called "extended DDO system" (Clark and McClure 1979). On the final night of four, all five fields were re-observed with short exposures, along with a large number of Landolt (1973, 1983) and Graham (1982, E-region) standard stars. Standard stars were observed on the first three nights, although the skies were somewhat hazy, although presumably photometric, typical of the weather in Chile at that time of the year. The final night was distinctly less hazy.

So far, the results for night 2, including two of the five regions observed, have been reduced. The data were processed using standard IRAF routines, and the photometry was done with a PSF-fitting photometry program developed at Boston University. In this preliminary report, the transformation to the standard system is based on standard stars observed on that night, following approximately the procedures outlined by Harris, FitzGerald and Reed (1981). The rms residuals of the fits to the instrumental colors are 0.038, 0.023, 0.018, 0.023 and 0.015 to U, B, V, R and I respectively.

3.1 Reddening and Stellar Population Along the Line of Sight

The region observed includes galaxy number 7 of FitzGerald's (1974) list, and in addition, a number of fainter galaxies are visible. Their visibility indicates that the region is generally relatively transparent, and therefore there is an expectation that one can investigate the stellar population through to the edge of the galactic disk.

The line of sight in this direction passes through the stellar associations Puppis OB1 and Puppis OB2. According to Havlen (1972), the association OB1 is at a distance of approximately 2.5 kpc, whereas the OB2 group is at a distance of 4.2 kpc. The maximum reddening seen among the O and B stars in this direction is about 0.6 magnitudes, corresponding to stars in the latter association. Additional interstellar absorption is of course possible for stars beyond this distance.

Fig. 1 shows the (U-B) - (B-V) two-color diagram for stars in the two fields measured to date. Also shown in the diagram are the standard two-color lines (Johnson 1966) for zero reddening and for E(B-V) = 0.6 mag, corresponding to the likely reddening to Puppis OB2. As the diagram shows, virtually all of the stars in the figure can be accounted for with reddening values between zero and 0.6 magnitudes. Most of the stars in the sample are of spectral types A, F and G, as should be expected of a random field sample. There are no indications in the diagram of either Puppis OB1 or Puppis OB2; since the two CCD fields sample a tiny fraction of the spatial extent of the associations, this is not surprising. Thus the dominant stellar component along this line of sight consists of ordinary disk stars at various distances (and presumably of various ages). There is no evidence in the sample for stars with reddening larger than about 0.6 mag or so.

3.2 Reddening to the Galaxy FitzGerald 07

Among the galaxies visible in this preliminary survey is one of those discovered by Fitzgerald (1974). His galaxy number 7 provides a suitable standard candle for determining the maximum reddening along this line of sight. From its appearance, the galaxy appears to be an S0 galaxy, with a bright, semi-stellar nucleus. Assuming that is the case, it will have a (B-V) color between 0.7 and 1.0 (de Vaucouleurs 1961). Aperture photometry (8.9 arcsec diameter) on the galaxy yields the following values:

$$V = 16.65$$
$$(B-V) = 1.69$$

Although the galaxy clearly extends beyond this radius, the total flux outside the aperture is small compared to the bright central core of the galaxy; furthermore, in going from a aperture of 3 arcsec to 8.9, the color changes less than 0.05 mag - a characteristic common to E and S0 galaxies, but not spirals (de Vaucouleurs 1961). The implication is that the total reddening through the galactic plane is between 0.7 and 1.0 magnitudes. The total visual absorption is thus about three magnitudes.

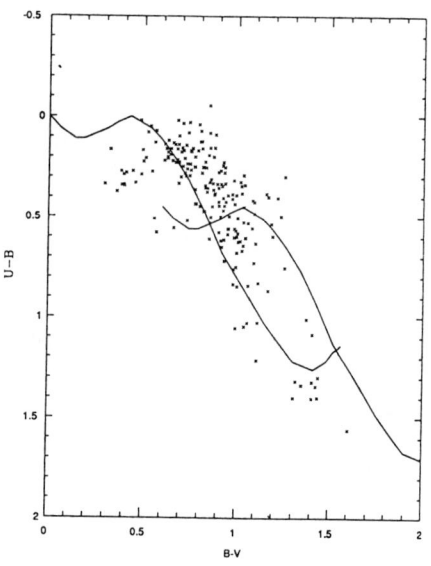

Fig. 1. Two-color, (U-B) - (B-V) diagram for stars in the Puppis Window. The standard main-sequence lines (Johnson, 1966) for zero reddening and for E(B-V) = 0.6 mag. are shown.

Fig. 2. Two-color diagram for stars in the Puppis Window. The index (51-V) is a measure of the MgH absorption in the 5100 - 5200 Å region. The approximate reddening vector for E(B-V) = 0.5 mag. is shown.

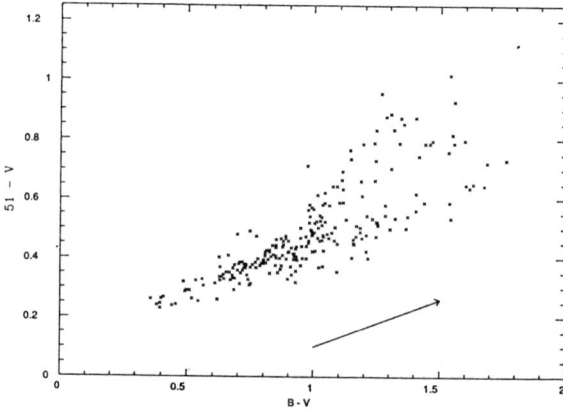

3.3 The 5100 Å MgH Band Index

From the standard (U-B) - (B-V) two-color diagram, it is not possible to separate distant red giants from more nearby dwarf stars. In order to identify red giant candidates, I have followed Clark and McClure (1979). They devised a color index based on a filter with a central wavelength of 5130 Å with a half-maximum bandpass of 154 Å to measure the MgH band and the Mg b triplet which appear prominently in the spectra of G and K stars. The band is very gravity dependent, being much stronger in dwarfs than giants; Clark and McClure made use of this dependence to set up a supplementary luminosity index to the DDO photometric system. Unfortunately, because their color index was defined by the standard star set of the DDO system, virtually all of the standard stars are too bright for a CCD even on a small telescope. For that reason, in this analysis a color index based on the 5100 Å instrumental magnitude minus the V magnitude will be used.

Fig. 2 shows a two-color diagram of (51-V) vs (B-V) for all the stars in the two subfields measured. Also shown in the figure is the approximate reddening line of E(51-V) = 0.32 E(B-V). Two distinct sequences are noticeable in the figure. The dwarf sequence deviates from a straight line at approximately (B-V) = 1, since the enhanced MgH band strength produces a redder (51-V) color at the same (B-V). The nearly straight portion of the diagram consists of main-sequence A and F stars, reddened by various amounts, and red giants, also reddened by various amounts. Thus with this diagram, a clear separation of G and K dwarfs from A and F dwarfs and G and K giants is possible. A metal-poor giant will fall even lower in the diagram. The only possible confusion would be with an extremely metal-poor dwarf star. Notice that the reddening has little effect on the separation of late-type dwarfs from giants.

4. FUTURE PLANS

Spectroscopy is needed to estimate reliable metallicities of individual stars and of course to determine their velocities. The rapidly-developing field of multi-fiber-optic spectroscopy is ideally suited to surveys such as this. Even though the most interesting stars are faint, the possibility of obtaining substantial numbers of moderate resolution spectra simultaneously makes this a viable project. At CTIO, the multi-fiber positioner, Argus, can feed light to a spectrograph at roughly one-quarter to one third the efficiency (*per star*) of a direct connection, but up to two dozen stars can be measured at once. At a spectral resolution of the order of 9 Å, spectra centered near 5100 Å can yield useful metallicity indices (Faber et al. 1985), and at this spectral resolution, exposures on the order of an hour or two will reach 19th magnitude. Thus, it will be possible to measure as many as 100 stars per night with multi-fiber systems such as Argus; the precise (relative) coordinates needed will be an automatic by-product of the photometry program.

REFERENCES

Adler, D.S. and Janes, K. A. 1982 PASP 94, 905
Burton, W. B. 1988 in Galactic and Extragalactic Radio Astronomy, 2nd Ed, G. L. Verschuur and K. I. Kellermann, eds., Springer-Verlag, New York, p. 295
Chini, R. 1985 in The Milky Way Galaxy, H. van Woerden, R. J. Allen and W. B. Burton, eds., Reidel, Dordrecht, p. 101
Christian, C. A. and Janes, K. A. 1979 AJ 84, 204
Clark, J. P. A. and McClure, R.D. 1979 PASP 91, 507
de Vaucouleurs, G. 1961 ApJS 5, 233
Dodd R. J. and Brand P. W. J. L. 1976 A&AS 25, 519
Eggen, O. J., Lynden-Bell, D. and Sandage, A. R. 1962 ApJ 136, 748
Faber, S. M., Friel, E. D., Burstein, D. and Gaskell, C. M. 1985 ApJS 57, 711
FitzGerald, M. P. 1974 A&A 31, 467
Friel, E. D. and Janes, K. A. 1991 in Formation and Evolution of Star Clusters, ASP Conf Ser, K. A. Janes, ed., p. 569
Gilmore, G. and Reid, N. 1983 MNRAS 202, 1025
Graham, J. A. 1982 PASP 94, 244
Harris, W. E., FitzGerald, M. P. and Reed, B. C. 1981 PASP 93, 507
Havlen, R. J. 1972 A&A 17, 413
Hron, J. 1987 A&A 176, 34
Janes, K. A. 1979 ApJS 39, 135
Janes, K. A. 1988 in Calibration of Stellar Ages, A. G. D. Philip, ed., L. Davis Press, Schenectady, p. 59
Johnson, H. L. 1966 Ann Rev A&A 4, 193
Landolt, A. U. 1973 AJ 78, 959
Landolt, A. U. 1983 AJ 88, 439
Larson, R. B. 1976 MNRAS 176, 31
Lewis, J. R. and Freeman, K. C. 1989 AJ 97, 139
McCuskey, S. W. 1967 AJ 72, 1199
Neese, C. and Yoss, K. 1988 AJ 95, 463
Searle, L. and Zinn, R. 1978 ApJ 225, 357
Shaver, P. A., McGee, R. X., Newton, L. M., Danks, A. C. and Pottasch, S.R. 1983 MNRAS 204, 53
Tinsley, B. M. and Larson, R. B. 1978 ApJ 221, 554
Twarog, B. A. 1980 ApJ 242, 242
VandenBerg, D. A. 1985 ApJS 58, 711
Zinn, R. 1985 ApJ 293, 424

DISCUSSION

STETSON: There was a paper in 1985 on remote OB Stars in Puppis which claimed, based on spectral classifications, that essentially all the OB stars were consistent with being at the same distance. As I recall the mean reddening was more like 0.7 than 1.0.

JANES: This is certainly consistent with the small survey I have here. I see no evidence for two distinct associations.

TYSON: Have you looked at the results of Henning and Kerr in your attempt to classify the mystery galaxy in your low extinction window? Their unbiased 21 cm search for galaxies in the galactic plane will help you decide if the galaxy is gas-rich (spiral) or gas-poor (S0).

JANES: I am not familiar with this work, but it is certainly worth checking into.

GARRISON: That region near 240° is very intriguing. I have also classified OB$^+$ stars in that region, with better quality spectra than used by FitzGerald and Reed. While I find a clump, I definitely find a string of OB$^+$ stars at different distances. These might be used to determine the reddening gradient.

JANES: Getting the run of reddening as defined by the OB stars is probably about the only way to do it.

SEITZER: Do you have a radial velocity for To 2? Does this agree for/against association with the warp?

JANES: Eileen Friel and I do have a velocity for the cluster. Assuming a flat rotation curve, the peculiar velocity of the cluster relative to its local standard of rest is 54 km/sec.

PRECISION PHYSICS FROM INFRARED PHOTOMETRY

R. Michael Rich

Columbia University

ABSTRACT: Infrared photometry in the 1 - 4 micron range permits us to measure effective temperature and bolometric magnitudes of late-type giants with unprecedented precision. Within the last decade, infrared photometry of galaxies, the Magellanic Clouds and the bulge of the Milky Way has revolutionized the field of stellar populations, and has challenged stellar evolution theory. Development of accurate model atmospheres for stars cooler than 3200 K promises to widen the use of narrow, as well as broad-band, photometry in the quest for accurate stellar parameters.

1. INTRODUCTION

The division between "infrared" and "optical" astronomers probably traces back to the days when infrared observations were made only by those groups who built their own equipment. Even today, infrared detectors often require cooling to liquid He temperatures, and are generally more difficult to use than the now familiar CCD detectors. Until recently, there were no infrared arrays, and observing in the IR was identical to traditional optical photometry.

Why observe with 58 X 62 pixel (or even 1 pixel) detectors, when CCDs of 1000 X 1000 pixel dimensions are available? Aside from the obvious capabilities of studying dust-shrouded sources, many fundamental stellar parameters - luminosity, effective temperature and gravity - are best measured using infrared photometry. By making good use of the J, H and K bandpasses (located at 1.2, 1.6 and 2.2 μm) observers have achieved significant advances in the study of the late type giants that represent the evolved stellar content of galaxies. Infrared observations sample the stellar energy distribution at its maximum, and color information is less sensitive to error from interstellar extinction. The dependence of colors and bolometric magnitudes on abundance also declines, because atomic lines account for very little blanketing beyond 1 micron. Table 1 gives the bandpasses for the Caltech infrared system, which derives originally from Johnson's filter system. All systems must cope with available windows of atmospheric transmission; variants of the Johnson system are used by active groups in Australia and South Africa.

Use of infrared photometry has conclusively demonstrated that the light of ellipticals and bulges is dominated by giants. The presence of luminous carbon stars has shown us that the Milky Way's retinue of dwarf spheroidal galaxies has a range in age. The evolved stellar population of the galactic bulge has been accurately described. We have found important disagreement with theory and observations of evolved stars. The last decade has seen much progress, but the field has suffered a serious blow with the loss of Marc Aaronson, whose energy gave the study of stellar populations a special sense of excitement.

TABLE 1

Bandpasses for the Caltech Infrared System

1 μm	FWHM μm	Band
1.25	0.24	J
1.65	0.30	H
2.20	0.40	K
2.20	0.11	CO-CONT
2.36	0.08	CO-FEAT

In this review, I will not consider observations of young stellar objects, although this is obviously an important and exciting area; some of the discussion is still of interest to workers in the area. I will also only touch briefly on those late-type giants that are so evolved as to have substantial circumstellar shells. These require observations longward of 5 μm (in the thermal infrared), and I consider that area to be also outside the scope of this review. Limitations of space and time prevent me from comprehensively reviewing the entire field, and I apologize in advance to those authors whose fine work is not referenced. Rather, I hope that the reader will conclude that although IR photometry has not yet attained optical precision, the resulting physical measurements are an impressive contribution to precision photometry in the Galaxy. In the next section, I consider how infrared photometry can be used to determine fundamental physical parameters for stars. Following, Section 3 considers case studies: application to the Magellanic Clouds, dwarf spheroidals, and the bulges of the Milky Way and M 31.

2. FUNDAMENTAL STELLAR PARAMETERS

2.1 Effective Temperatures

The first color-magnitude diagram of the galactic bulge (Baade's

Window) was measured by Arp (1965) from 5-m plates in the B and V bands. The CM-Diagram is difficult to interpret, primarily because the very cool M giants which dominate the AGB get no redder than (B-V) = 1.5. This can be contrasted with the very clear giant branch defined by the J, K photometry of Frogel and Whitford (1987). Blanketing by the TiO molecule corrupts the B and V bands; an infrared magnitude is needed if the color is to yield a temperature.

A color such as (V-K) is particularly useful because the baseline is long enough to minimize the effects of blanketing, and because the K band itself is virtually unblanketed by atomic lines or molecules. Ridgway et al. (1980) establish a temperature scale that is valid for photospheres as cool as 3200 K (or solar neighborhood spectra types as late as M6). The use of stellar diameter measurements is still preferred as yielding the most reliable effective temperatures. A photometric accuracy of 0.1 mag in the color can determine effective temperature to within ± 100 K. In the same paper, Ridgway et al. also show that temperature can be calibrated as a function of the 1.04 μm - L color. The use of a pure infrared color-index has potentially interesting applications in the study of OH/IR stars, which are not detectable at optical wavelengths, and in the study of highly reddened stars. Both M and C stars have a continuum point at 1.04 μm, as can be easily seen in the atlas of Gunn and Stryker (1983), as well as that of Smak and Wing (1979). Fig. 1 shows late M giants in the solar neighborhood, with strong TiO bands; the clean continuum near 1 micron is evident.

In the coolest stars, molecules and ultimately dust provide large sources of opacity. As infrared area detectors improve in size and performance, it becomes attractive to consider measuring effective temperatures of the coolest stars using the I(104) - L index of Ridgway et al. (1980). Bessell et al. (1989) use theoretical model atmospheres to calibrate a large number of narrow band indices to measure gravity, abundance and atmospheric extension as well as effective temperature. As the size and quantum efficiency of IR detectors increases, it becomes attractive to consider use of narrow-band infrared colors, which are potentially better because they measure effective temperature independent of other stellar parameters. Bessell et al. also find that (J-L) may be as effective a temperature measurement as the 1.04 μm - L index; the former is available in the standard InSb setup and will reach fainter stars. Both indices are potentially abundance sensitive, again emphasizing the need for the theoretical models and independent measurement of abundance. As a cautionary note, one cannot assume an abundance for a population such as the Magellanic Clouds or the galactic bulge. In the bulge, Rich (1988) finds a range from -1 to +0.7 dex.

The effective temperature of late-type stars has received much attention recently. Bell and Gustafsson (1989) also use model atmospheres and the infrared flux method to calibrate effective temperature and transformations between the many infrared color systems now in common use. McWilliam (1990) assembles recent observations (many of which are refinements of the Ridgway angular diameter technique) to construct a set of useful infrared color to effective temperature transformations.

Infrared and optical photometry for numerous field stars is reported by Johnson (1966) and Mendoza and Johnson (1965). Frogel et al. (1978) and Aaronson and Mould (1985) give empirical spectral type-color relations for the Caltech infrared system. It should be noted that thermally pulsing carbon stars are marked by their very red infrared colors (a property which confirms the carbon star classification). Few angular diameter measurements are available for carbon stars, so the effective temperature scale is not as well established. Aaronson and Mould (1982) find agreement between the angular diameter scale, an older empirical relationship from Mendoza and Johnson (1965), and Tsuji's infrared flux method, but remain concerned about differential blanketing as a function of metallicity.

The general correlation between infrared colors and effective temperature would suggest that giant branches of globular clusters should define a sequence which reddens monotonically with metallicity; Frogel, Persson and Cohen (1983) find this to be the case, but age and metallicity variations in complex populations such as the galactic bulge may erase this correlation (Frogel, Whitford and Rich 1984). The theoretical model atmospheres of Bessell et al. (1989) also indicate that conversion from broad-band color to effective temperature is more sensitive to metallicity than to gravity or extension.

2.2 Luminosities

Evolved giants emit most of their luminosity at 2 μm and redward. While attempts are often made to determine bolometric luminosities for such stars from V and I photometry in the optical, Frogel (1990) shows clearly that bolometric magnitude of late M giants from V, I photometry are often in error by > 1 mag. Such large errors are completely expected; photometry short of 1 micron samples stars short of the peak of their energy distribution, in a heavily blanketed spectral region.

The most luminous cool giants mark the intermediate age populations; IR photometry is essential to measure their bolometric luminosities. Here is a case where precision optical photometry is of potentially little value in measuring the important physical quantities; one must have infrared data. And we must also redefine what we mean by precision. If we neglect the infrared, we may be off by magnitudes; and comparisons with stellar evolution theory do not benefit from precision better than 0.1 mag. For populations older than a few Gyr, main sequence turnoff photometry is relatively insensitive to small age differences; the effects of Fe and O abundance variations, as well as fundamental theoretical problems (convective overshooting) pose difficulties for measurements of relative and absolute ages. The presence of an extended giant branch is a clear indication that the population in question is younger than the halo/old globular cluster age. The transition case of NGC 121 in the SMC, with both luminous carbon stars and RR Lyrae, illustrates the clarity with which an extended giant branch signals intermediate age.

Most late type giants have a peak in their energy distribution at the H band (1.65 μm) or redward. The tendency to radiate in the IR is not only due to the blackbody peak at T_{eff} = 3500 K, but also because there is much less opacity; the H⁻ bound-free and free-free opacity is minimum near 1.6 μm, and the TiO molecule contributes no substantial opacity beyond 1 μm. In cool, metal rich stars, atomic and molecular opacity is substantial in the blue (but CO overtone bands can be strong at 1.65 μm). It is often possible to integrate over the energy curve using just the J,H and K magnitudes, although the presence of dust will cause much flux to emerge at 10 μm. Unfortunately, emissivity of the atmosphere and the telescope itself makes ground-based measurements at 10 μm difficult; however, the measurement is required to determine the bolometric magnitude of OH/IR stars.

Bolometric corrections to IR magnitudes are usually determined by integrating over the energy distribution from U to L (or K, if L is not available). The methods of Cohen, Frogel and Persson (1983) continue to be used with little modification. In practice, the optical flux has little contribution in late type stars, and use of standard colors for the optical (cf. Mendoza and Johnson 1965) does not introduce unacceptable errors (Cohen, Frogel and Persson 1983).

It should soon be possible to use theoretical stellar atmospheres to derive bolometric corrections to the broad-band IR colors. Ideally, one would need only a color, luminosity measurement (or CO index) and an abundance estimate. The bins of luminosity functions are usually 0.5 mag wide, and stellar evolution theory rarely requires measurements exceeding 0.1 mag accuracy. It is clear, however, that the precision of the physics may be more affected by uncertainties about the emergent flux beyond 5 microns, and the atmospheric blanketing, than by instrumental precision. Blanketing and opacity changes do not pause in the bands blocked by the Earth's atmosphere; one hopes that the theoretical work now in progress will lead to bolometric corrections accurate for late giants with heavy molecular opacity, derived from integrating over the full emergent flux of a model atmosphere.

There are times when only one color is available, when one is pushing to the limits of detection. In this case, the measurements themselves will usually have 0.1 mag uncertainty. Sky emission is usually lower in the H band, and stars have the H⁻ opacity minimum at 1.6 microns. Wood and Bessell (1984) find that BC_H = 2.6 ±0.2 mag., *independent of spectral type or color*. Although the numerical precision might disqualify the method from inclusion in these proceedings, it is important to remember that the level of uncertainty is acceptable for comparisons with theory, and for identification of rare, luminous giants.

2.3 Surface Gravities and Abundances

We are fortunate that the strongest molecular absorber in the near-IR is the CO molecule, with vibrational bandheads (v = 2) starting at 2.29 μm. The spectral atlas of Kleinman and Hall (1986) illustrates the detailed spectrum of the molecule in a number of late-type giants of widely varying luminosity class. The CO band behaves in a counter-intuitive way in late-type giants, strengthening with *increasing* luminosity; Frogel (1971) proposes that this very saturated band strengthens due to the increase in microturbulent velocity in more luminous stars. The band is sufficiently strong that it is easily measured with narrow-band photometry from a [2.3] - [2.2] micron color. Frogel et al. (1978) presents the behavior of the band as a function of spectral type and luminosity for the solar neighborhood population. It is virtually undetectable in dwarfs, making it the best photometric dwarf/giant discriminator in late-type stars.

In late-type giants, there are additional factors other than the luminosity that affect the appearance of the spectrum. Bessell et al. (1989) find that TiO, VO and H_2O band strengths increase when the line forming region has a larger physical depth (they refer to this as the *extension*). The presence of this extra variable potentially complicates interpretation of JHK - CO photometry, but thankfully CO and CN are not sensitive to extension.

In populations with a narrow range in age, it may be possible to estimate abundances from varying CO band strengths at constant luminosity. This technique has been applied to giants in the globular cluster ω Cen (Persson et al. 1980). It has been argued by Frogel et al. (1990) that decline in the CO index in late bulge giants with galactocentric distance confirms the presence of an abundance gradient in the bulge. One must be cautious in making such inferences from the CO index because of its primary dependence on luminosity and temperature. Cross-comparison between stellar populations with a range in age and abundance could lead to spurious conclusions. One must also recall that the CO band strength depends on the C/O ratio, which can be affected by stellar evolution.

Finally, it should be noted that precision measurement of bolometric magnitudes (if one has accurate distances) and effective temperatures allow one to determine gravity directly (cf. McWilliam 1990). However, measurement of the CO index adds greatly to the information given by the broad-band colors. With care, it may ultimately prove to be useful in measurement of abundance as well as gravity.

The theoretical work of Bessell et al. (1989) shows that one may profitably exploit the 1.65 μm CO bands as a gravity indicator, especially in stars cooler than 3200 K. This theoretical work is now being extended to include cases in which the atmospheres are not static. Model atmospheres of late-type stars that incorporate the best molecular and continuous opacity sources promise a future classification system that can use precision narrow band photometry to sort out gravity, extension, temperature and abundance.

2.4 Physical Parameters for Variable Stars

Precision photometry of variable stars is a critical step in the cosmological distance scale. Considerable large telescope time has been devoted to the problem, and much effort will also be directed using the Hubble Space Telescope. There are many reasons to prefer the infrared for establishment of (P-L) relationships used in the distance scale. Total extinction in K is less than 1/10 that in the V band. In all variable stars - RR Lyrae, Cepheids and Miras, the total excursion in the K band is much smaller than in the visual. For RR Lyraes, Δm_K = 0.3, permitting random phase observations. Miras can vary by as much as 10 mags in the visible, but usually less than 1 mag at K. Changes in effective temperature and gravity have much less effect on the IR flux because of the decrease in atomic line blanketing beyond 1 micron. This decline in blanketing also reduces the color dependence of (P-L) relations. Because one is frequently comparing populations of different abundance, one wants minimize abundance effects.

Again, the decline in blanketing makes measurement of the effective temperature more accurate. For RR Lyraes, use of (V-K) colors can overcome the phase-dependent problems concerning use of <B-V> vs. <B> - <V>. For the RR Lyraes, Longmore, Fernley and Johnson (1986) established (P-L) relationships, and Liu and Janes (1990) have used the IR to improve Baade-Wesselink absolute luminosities for field RR Lyraes and those in the globular cluster M 4. Infrared photometry has also been profitably applied to Cepheid distances and two-dimensional detectors will further inspire this effort. An example (observationally) is the work of Welch et al. (1986). Hindsley and Bell (1990) investigate the theoretical and observational status of the Cepheid infrared (P-L) relation.

The Miras are late giants and most of the flux emerges in the infrared. The OH/IR stars are often invisible shortward of 2 μm. A (P-L-C) relationship for the Miras is measured by Feast et al. (1989); and infrared observations of the Miras have contributed to our understanding of the stellar populations of the galactic bulge and Magellanic clouds. Finally, calibration of the red supergiants may provide the ultimate distance indicator. The infrared study of the M 33 supergiants by Mould et al. (1990) is a pioneering effort in this direction.

3. CASE STUDIES IN THE APPLICATION OF INFRARED PHOTOMETRY

Since the 1970's, use of infrared photometry has given rise to a revolution in the study of stellar populations. We have been able to recognize and study in detail the class of *intermediate age* populations. These are populations of age in excess of a few Gyr; younger than the old globular clusters, but not young enough to have blue main sequence stars. We have also been able to demonstrate without doubt that giants dominate the red light of elliptical galaxies and bulges; this was at one time a subject of intense debate. Many fine investigators have contributed to the development of this field; the examples cited below are intended to illustrate the power of infrared photometry and do not do justice to the breadth of this subject.

3.1 The Integrated Light of Galaxies

Red galaxy populations have strong metal lines of CN, Mg and TiO. We now recognize these as being due to the presence of a population more metal rich than the solar neighborhood; this has been confirmed by study of the galactic bulge population (Rich, 1988). In the 1970's, the alternative hypothesis was advanced that the strong molecular bands might be due to a contribution by dwarfs to the integrated light. Measurement of the CO band in galaxies (Frogel et al. 1978) settled the issue unequivocally in favor of a giant-dominated population. The work of Aaronson, Frogel and Persson (1980) employs photometry in the 1.87 μm H_2O band to show that cool giants (M5 and later) contribute substantially to the integrated light of galaxies. Currently, there is a debate as to whether excess 10 μm emission in ellipticals is nonthermal or due to circumstellar dust (Impey, Wynne-Williams and Becklin 1986); more sensitive spectroscopic measurement of the H_2O band in galaxies may help to determine if enough cool stars are present to contribute the required dust.

3.2 The Magellanic Clouds

The complex formation history of the Magellanic Clouds render them an ideal laboratory for the study of stellar evolution in populations of widely varying ages. The red, luminous thermally pulsing carbon stars are easily detectable in the infrared, as they lie 0.5 - 3 magnitudes above the He-flash termination point of the first giant branch. Here, theory and observation are in gross disagreement. For the field population of the LMC, Cohen et al. (1981) find that carbon stars brighter than m_{bol} = -6 are absent, though they are predicted by the theory of Renzini and Voli (1983).

For the clusters, where age information is available from the turnoff, Mould and Aaronson (1986) show that the AGB tip stars are underluminous at all ages, particularly for those clusters younger than 1 Gyr. More recently, Frogel, Mould and Blanco (1990) show that the

AGB can contribute up to 40% of the bolometric luminosity of the intermediate age Magellanic clusters, and they demonstrate that the oxygen to carbon star transition luminosity differs between the SMC and LMC, presumably due to abundance. Some of these conclusions have been reached using optical I band photometry, but the most secure comparisons with theory require use of the infrared to determine accurate bolometric magnitudes for the most luminous (very red) AGB stars. As two-dimensional detectors become more widespread, we will begin to see more detailed comparisons of observations with theory.

3.3 The Dwarf Spheroidal Galaxies

Another chapter in the infrared revolution has been the discovery that five of the seven dwarf companions of the Milky Way have luminous AGB carbon stars (Aaronson and Mould 1985). Again, it is difficult to do justice to the many investigators whose surveys discovered the carbon stars and defined the AGB populations of these objects. The link between the luminous AGB carbon stars and a luminous turnoff was established for the Carina dwarf (Mould and Aaronson 1983). One is struck that the luminous AGB stars are a magnitude brighter than the tip of the giant branch, easily measurable with infrared photometry.

3.4 Bulges of the Galaxy and M 31

The red strong-lined stars comprising the bulges of galaxies bear more than a passing resemblance to the integrated light of ellipticals. For many years, astronomers focused on the presence of RR Lyrae stars in the bulge as an indication that the bulge was old and metal poor, like the halo. Blanco's grism surveys for late M giants (Blanco, McCarthy and Blanco 1984) ushered in a new era in the study of the galactic bulge. The bulge's remarkable AGB population had all but been ignored; no galactic globular cluster is known to have red giants as late as M9.

The identification of the bulge AGB stars came just as IR detectors at CTIO were becoming fully operational. The very complicated, difficult to interpret scatter plots of the optical CM-Diagram were transformed into a well defined giant branch and late-type variable star population (cf. Frogel and Whitford 1987).

The Blanco photometry yielded a sample of bulge K giants amenable to spectroscopy, which were studied by Whitford and Rich (1983) and Rich (1988). The problem of great interest was to determine the abundance distribution of the K giants. However, Baade's Window suffers from 1.5 mag of visual extinction. To meaningfully compare the bulge K giants to similar stars in the solar neighborhood and globular clusters, one must accurately deredden the colors. Fig. 2 illustrates the use of equivalent width as a function of (J-K), with the bulge giants appropriately dereddened. Because of the reduced reddening in the IR, it is unlikely that differential reddening, or even intrinsic

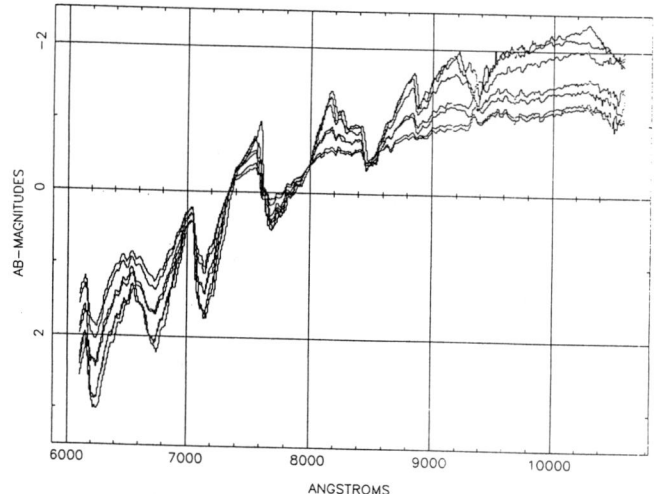

Fig. 1. CCD spectra of local M giants show that the 1.04 μm continuum point remains clean, even for very late (M8) giants. All of the absorption features are due to TiO and VO; atmospheric bands have been divided out of the data. Spectra were obtained by J. Gunn and J. Biretta on the Palomar 5-m telescope.

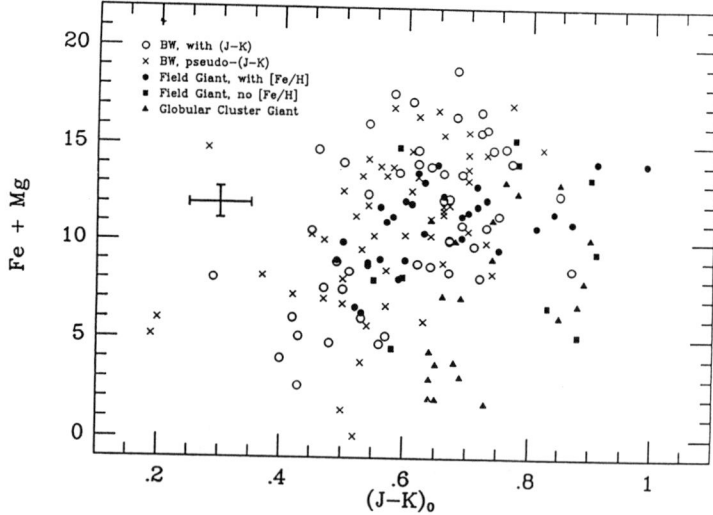

Fig. 2. Use of (J-K) infrared color to constrain effective temperature in bulge K giants (Rich 1988). J and K are both measurable in the standard InSb setup and $E(J-K) \approx 0.5E(B-V)$. Errors in effective temperature due to (1) uncertain reddening and (2) differential reddening are minimized. 1.5 mag of visual extinction obscure Baade's Window, at -4 deg galactic latitude.

blanketing differences (for the most metal rich giants) could cause serious error in the derived abundances.

Returning to the luminous M giants, Frogel and Whitford (1982) identify a problem which continues to be of great importance at the current time. They find that the Milky Way bulge contains giants as luminous as m_{bol} = -4.5. This is bright enough to cause one to think about the possibility that the progenitors of these stars are possibly younger than the globular clusters. More extensive photometry by Frogel and Whitford (1987) and Frogel et al. (1990) confirms the high luminosities. Again, the late M giants are brightest at K; the bolometrically brightest AGB stars are very faint in the optical. Frogel (1990) emphasizes that optical colors cannot be used to determine accurate bolometric magnitudes for the bulge M giants.

It is difficult to draw very accurate conclusions about the luminosities of Milky Way bulge giants, because the distance to the center of the Galaxy is uncertain. Some bulge giants may lie at smaller distances, causing them to appear brighter. These problems do not exist for the population of late M giants in M 31; even extinction is very small. The Blanco surveys showed us that such a population should be present in the M 31 bulge. J. Mould and I imaged the M 31 bulge in the J and K bands in an attempt to discover the Blanco M giant counterparts and determine their bolometric luminosities (Rich and Mould 1990). Figs. 3 and 4 illustrate the CM-Diagram and luminosity function for the stars in Fig. 5 (J band) and 6 (K band). Because the sky is not as bright at J, it is easier to image to fainter limiting magnitudes.

As the color-magnitude diagram and luminosity function show, M 31 contains many luminous giants, even up to m_{bol} = -5. The infrared (P-L) relationship suggests that these stars may be the counterparts of 800 day Miras found in the Milky Way bulge (Whitelock, Feast and Catchpole 1991). In the Milky Way, there is growing indication that the most luminous stars may not be so metal rich as to allow them to be very old, like the globular clusters. The infrared photometry of the most luminous bulge giants may be telling us that central bulges are younger than halos. Or we may be seeing very rare phases of stellar evolution. Again, it is interesting to note that the difference between the luminosity functions of the bulge of the Milky Way and M 31 is large. With proper theoretical models, one should be able to determine if these two bulges are indeed young. Again, infrared observations can be counted on to provide the necessary precision physics.

ACKNOWLEDGMENTS

The author acknowledges valuable discussions with J. Mould and K. Janes.

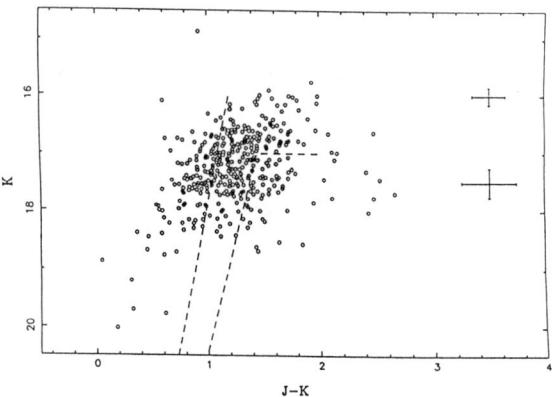

Fig. 3. CM-Diagram of luminous giants three arcmin SE of the nucleus of M 31, obtained with an infrared imaging array (Rich and Mould 1991). The photometric errors are larger than 0.15 mag in each color. Fiducial lines indicate the extent in color and luminosity of the giant branch in the bulge of the Milky Way. Notice that the high luminosity of the giants is seen clearly. Uncertainties in bolometric corrections are larger than those in photometric colors.

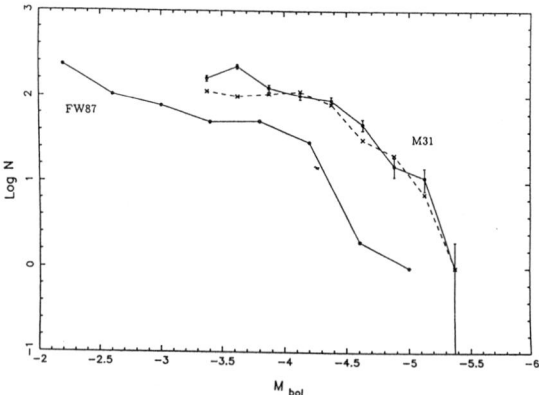

Fig. 4. Luminosity functions for the M 31 bulge, corrected for incompleteness due to crowding, compared with Frogel and Whitford's (1987) photometry in Baade's Window. Dashed connecting line is based on the CM-Diagram data, and the solid line is the full sample. The bulge luminosity function is below, lacking error flags. Milky Way stars were optically selected, and observed with a single-channel InSb photometer; the M 31 bulge sample is based on stars detected in the infrared K band.

PRECISION PHYSICS FROM INFRARED PHOTOMETRY

Fig. 5. Mosaic image of the M 31 bulge in the J band, from Rich and Mould (1991). North is to the left, East is at the bottom. The sub-frames are 58 X 62 0.3" pixels. The field is centered 3' SE of the nucleus. Each sub-frame required 240 sec of integration.

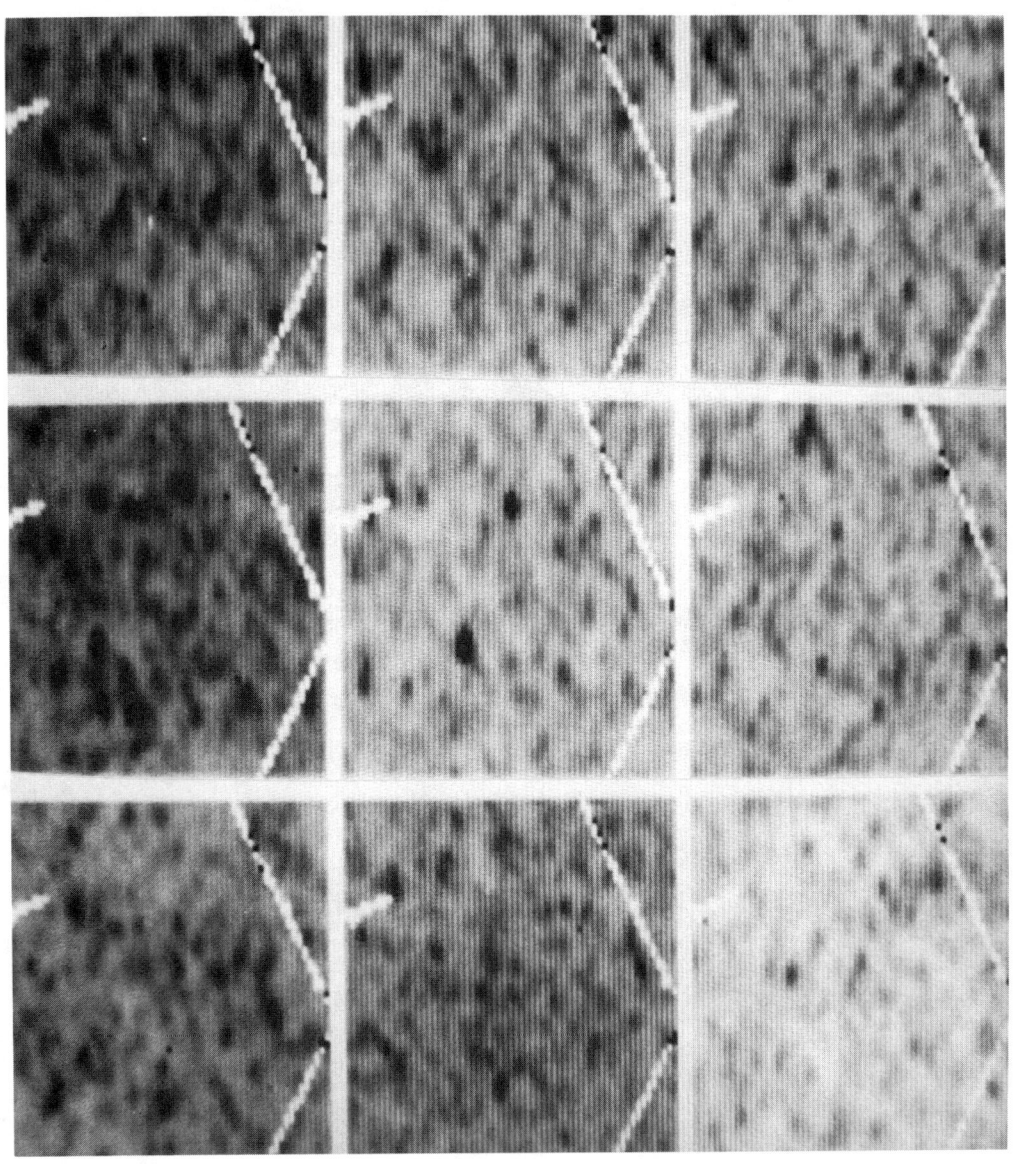

258   R. Michael Rich

Fig. 6. As in Fig. 5 for the K band. Each sub-frame represents 2 X 45 sec of integration, shorter because of the greater night sky emission at K. The star identifications correspond to stars in Figs. 3 and 4; Rich and Mould (1991) gives their magnitudes and colors.

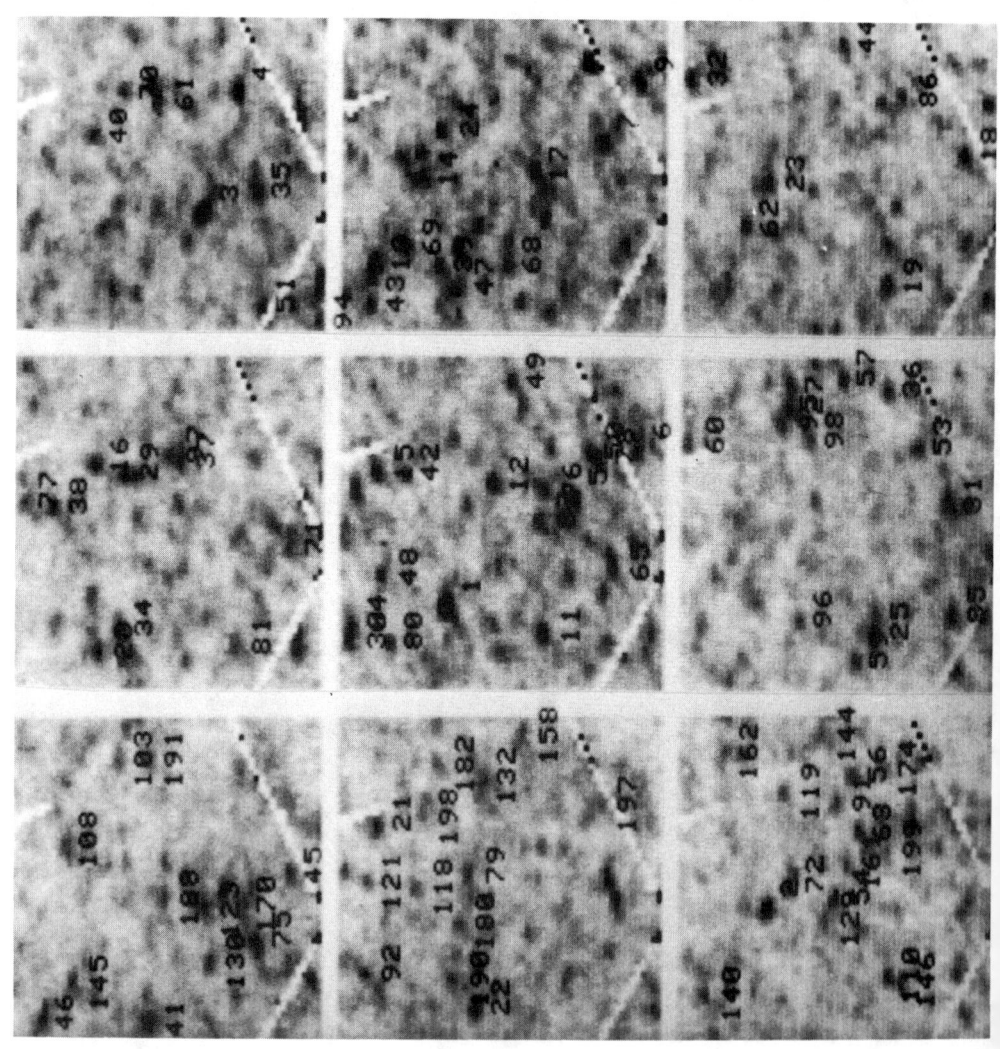

REFERENCES

Aaronson, M. and Mould, J. 1982 ApJS 48, 161
Aaronson, M. and Mould, J. 1985 ApJ 290, 191
Aaronson, M., Frogel, J. A. and Persson, S. E. 1978 ApJ 220, 442
Arp, H. 1965 ApJ 141, 43
Bell, R. A. and Gustafsson, B. 1989 MNRAS 236, 653
Bessell, M. S. and Wood, P. R. 1984 PASP 96, 247
Bessell, M. S., Brett, J. M., Scholz, M. and Wood, P. R. 1989 A&AS 77, 1
Blanco, V. M., McCarthy, S. J. and Blanco, B. M. 1984 AJ 89, 636
Cohen, J. G., Frogel, J. A. and Persson, S. E. 1978 ApJ 222, 165
Cohen, J. G., Frogel, J. A., Persson, S. E., and Elias, J. H. 1981 ApJ 249, 481
Feast, M. W., Glass, I. S., Whitelock, P. A. and Catchpole, R. M. 1989 MNRAS 241, 375
Frogel, J. A. and Whitford, A. E. 1987 ApJ 320, 199
Frogel, J. A., Mould, J. and Blanco, V. M. 1990 ApJ 352, 96
Frogel, J. A., Persson, S. E. and Cohen, J. G. 1983 ApJS 53, 713
Frogel, J. A., Persson, S. E. Aaronson, M. and Matthews, K. 1978 ApJ 220, 75
Frogel, J. A. 1971 Ph.D. Thesis, Caltech
Frogel, J. A., Terndrup, D. M., Blanco, V. M. and Whitford, A. E. 1909 ApJ 357, 453
Frogel, J. A., Whitford., A. E. and Rich, R. M. 1984 AJ 89, 1536
Frogel, J. A. and Whitford, A. E. 1982 ApJL 259, L 7
Frogel, J. A. 1990 in Bulges of Galaxies, B. J. Jarvis and D. M. Terndrup, eds. ESO, Garching, p 177
Gunn, J. E. and Stryker, L. L. 1983 ApJS 52, 121
Hindsley, R. B. and Bell, R. A. 1990 ApJ 348, 673
Impey, C. D., Wynn-Williams, C. G. and Becklin, E. E. 1986 ApJ 309, 572
Johnson, H. L. 1966 ARAA 4, 193
Kleinmann, S. G. and Hall, D. N. B. 1986 ApJS 62, 501
Liu, T. and Janes, K. A. 1989 ApJS 354, 273
Longmore, A. J., Jameson, R. F. and Fernley, J. A. 1986 MNRAS 236, 447
McWilliam, A. 1990 ApJS 74, 1075
Mendoza, E. E. and Johnson, H. L. 1965 ApJ 277, 149
Mould, J. and Aaronson, M. 1986 ApJ 303, 10
Mould, J. R. and Aaronson, M. 1983 ApJ 273, 530
Mould, J., Graham, J. R., Matthews, K., Neugebauer, G. and Elias, J. 1990 ApJ 349, 503
Persson, S. E., Frogel, J. A., Cohen, J. G., Aaronson, M. and Matthews, K. 1980 ApJ 235, 452
Renzini, A. and Voli, M. 1981 A&A 94, 175
Rich, R. M. 1988 AJ 95, 828

Rich, R. M. and Mould., J. R. 1991 AJ, in press
Ridgway, S. T., Joyce., R. R., White, N. M. and Wing., R. F. 1980 ApJ 235, 126
Smak, J. and Wing, R. F. 1979 Acta Astronomica 29, 187
Welch, D. 1., McAlary, C. W., McLaren, R. A. and Madore, B. F. 1986 ApJ 305, 583
Whitelock, P., Feast, M., and Catchpole, R. 1991 MNRAS 248, 276.
Whitford, A. E. and Rich, R. M. 1983 ApJ 274, 723

DISCUSSION

GARRISON: I always thought that the original CO work on galaxies was that of O'Connell, who predated Frogel by quite a bit.

RICH: O'Connell did much fine work on population synthesis, and was able to show from the TiO bands alone that giants dominate the light in E/S0 populations. However, Frogel and collaborators determined the standard CO vs. spectral type relationship for field stars, and also used the two micron H_2O absorption in galaxies to show that the infrared light is dominated by late-type giants.

STELLAR PHOTOMETRY WITH THE PC ON HST

Patrick Seitzer

Space Telescope Science Institute

There was tremendous promise offered by the Hubble Space Telescope and Wide Field/Planetary Camera in the field of accurate stellar photometry. Without the distortions and extinction caused by the atmosphere, the four Texas Instrument CCDs in the Planetary Camera (PC) could routinely provide images of 0.08" FWHM or better over a wide spectral range from 2000 Å to 1 micron. The most precise photometry was expected to be done with the PC rather than the Wide Field Camera (WFC) due to the much better sampling of the point spread function (PSF). Precision photometry of globular cluster main sequences was only one of the many projects that were to be undertaken.

1. "The Reduction of WF/PC Images" by Lauer (1989) describes in detail many of the characteristics of the CCDs in the WF/PC and the processing steps necessary to remove the instrumental signature. All of these steps are implemented in the STScI computers so that an observer will get both the unprocessed CCD images and processed data ready for photometric measurements.

2. "Theoretical Colours and Isochrones for some WF/PC HST Colour Systems" by Edvardsson and Bell (1989). Using the preflight measurements of the telescope and camera throughput together with theoretical calculations of spectral energy distributions for a wide range of stellar types, this paper presents the expected color dependencies for WF/PC observations.

3. "Photometric Calibration of the HST WF/PC: I. Groundbased Observations of Standard Stars" by Harris et. al. (1990). Using CCDs and filters similar to the flight ones, this paper presents the observational color dependencies. The filters and CCDs are not exactly the same as the ones flown, however, which will cause there to be small changes from these groundbased terms and those finally adopted after orbital data is obtained.

4. "Stellar Photometry with the WF/PC of HST" by Holtzman (1990) presents a series of experiments with and suggests modifications to DAOPHOT (Stetson 1987) for accurate PSF fitting. Although Holtzman deals explicitly with the WFC, many of the lessons

learned will be applicable to the PC as well.

Finally the camera and filter characteristics are summarized in the STScI's WF/PC Handbook (Griffiths 1990).

Tragically, as is well known, a significant amount of spherical aberration in the primary mirror was discovered shortly after the telescope was launched. Although stellar images still have a tight core of roughly 0.1" FWHM (roughly twice what was hoped for), the core has only about 15% of the total energy compared with some 80% that was expected if the mirror had been figured correctly.

The difference between PC images and very good groundbased images can be summarized by noting that observations with the ESO NTT can show images of 0.42" FWHM but with the 80% encircled energy radius being only 0.6". The PC still wins for some projects with its 0.1" FWHM core. However, the 80% encircled energy radius is now 1.75". The WF/PC functions as a very low contrast imager: the tight core (corresponding to the flux received from a 1 meter diameter telescope) surrounded by an extensive halo.

The impact on photometric programs is severe:

1. to obtain the same amount of energy within a 0.1" radius, exposure times must be increased by at least a factor of five.

2. most of the energy in a stellar image is now in the halo extending out to 4" radius or more. But the S/N of this is very low, and best photometry will come from just measuring the light in the core of the image. Furthermore, a small change in the background value can cause a significant change in the total measured energy since the stellar flux per pixel is so low. Determining the 'aperture correction' is now the most difficult part of measuring the total magnitude of stars on a PC image.

3. the PSF can vary significantly over the field of view of a single PC CCD and between CCDs, although for the telescope focus used for much of the Fall of 1990 the PSF was fairly constant over a significant amount of PC6.

4. the large amount of energy in the halo means that faint stars will be lost now at a much brighter level if they are close to brighter stars. The incompleteness as a function of magnitude will now be much worse than expected and will begin to be significant at a much brighter level. The effective dynamic range of the camera for luminosity functions is therefore reduced.

Clearly precision photometry will now be a tremendous technical and time consuming challenge.

According to current projections (February 1991), at the end of 1993 a replacement camera (WF/PC II) will be launched and installed in the telescope. Its biggest advantage is that the secondary mirrors on the reimaging cameras will be figured so as to remove most of the spherical aberration. Thus most of the science that is now impossible or extremely difficult should be possible. In addition, a number of improvements to the cameras and electronics will be made that should significantly enhance the capability of the camera for precision photometry. These include:

1. coating with liumogen instead of coronene. This will offer increased throughput in the UV.

2. all CCDs selected for flight will be checked for minimum quantum efficiency hysteresis (QEH). This should improve photometric capabilities particularly in the B passband and will largely remove the need for a UV flood.

3. a slightly different filter complement: four new broad band filters will be added in the UV and blue for increased sensitivity to faint sources. Deleted filters are mostly narrow and medium band.

4. a deep erase cycle on the CCDs to remove residual images from highly saturated stars in previous images.

5. an on board flat field source to provide accurate flat fields without the high frequency structure and large gradient across the field of view seen in the current WF/PC's internal light source.

6. fix the problem of missing values in the A/D converters. Unfortunately, the analog signal processing chain will still have only 12 bit A/D converters, and the maximum dynamic range will still be the same as the current camera.

With these changes and the successful installation of WF/PC II, the promise of HST for precision photometry in the UV and in crowded regions should be fulfilled.

REFERENCES:

Edvardsson, B. and Bell, R. A. 1989 MNRAS 238, 1121
Griffiths, R.G., editor 1990 Wide Field and Planetary Camera
 Instrument Handbook, STScI, Baltimore
Harris, H. C., Baum, W. A., Hunter, D. A. and Kreidl, T. J.
 (1990), AJ, in press
Holtzman, J. 1990 PASP 102, 806
Lauer, T. 1989 PASP 101, 445
Stetson, P. B. 1987 PASP 99, 191

DISCUSSION

JANES: What sort of photometric precision can you expect to get in the near future?

SEITZER: I expect that 5% or better relative photometry across a significant fraction of the field would be possible, since some people are already doing that well.

STETSON: Can you quantify the effective read noise after the A/D bits have been scrambled?

SEITZER: It effectively doubles the read noise - this is discussed in Lauer (1989).

MAJEWSKI: How stable is this spatially-dependent diffraction pattern you have shown? That is, will you be able to make a map of aperture corrections for photometry or WF/PC images?

SEITZER: If the system is stable, and once a focus is fixed, then such a map should be possible and will certainly be generated.

WARREN: What effect will the terminator instability (day/night transition jitter) have on all these observations?

SEITZER: My understanding is that a fix is underway for this problem. It will only affect exposures that continue across the day/night transition. Observations that are taken wholly on the night or day side will not show this problem.

RATNATUNGA: Testing of the HST is expected to take over six months. Do you see a similar period of testing after the instrument change to WF/PC?

SEITZER: Probably not, since there are five instruments that will be tested during this six month period. The amount of time dedicated to WF/PC is only a small fraction of this.

THE PROMISE AND CHALLENGE OF TIME-RESOLVED CCD ENSEMBLE PHOTOMETRY

Ronald L. Gilliland[1]

Space Telescope Science Institute[2]

ABSTRACT: CCD photometry allows precise monitoring of stellar intensities for many stars simultaneously at good temporal resolution. The photometric precision attainable from CCDs is as good as that with photoelectric photometry, but through the available spatial information, CCD photometry is much more robust. We review recent efforts to set empirical limits on the precision of time-differential CCD photometry of rich stellar ensembles. This observational approach holds the promise of allowing detection of stellar analogs to the solar 5-minute, p-mode oscillations. Such observations will provide additional new constraints on stellar evolution theory via indirect probing of stellar interior structure. An observing run with a network of 4-m telescopes would be both necessary and sufficient to define the limits of the technique, and to yield a good chance of success.

1. INTRODUCTION

The Sun shows a rich oscillation spectrum of p-modes with characteristic time scales of five minutes. These oscillatory modes may be characterized by three wave numbers, n the radial order, l the angular degree and m the azimuthal order, describing the number of nodes in various coordinates. On the Sun it is possible to obtain spatially resolved measurements, at high signal-to-noise, allowing the direct study of modes with many nodes around the surface. Such studies have revolutionized solar physics, for the first time providing an observational probe of interior solar structure. Major efforts are underway (e.g., the GONG project) to exploit solar oscillations to gain a detailed quantitative determination of solar interior properties. See Christensen-Dalsgaard (1988) for a general review.

[1]Guest Observer, Kitt Peak National Observatory, National Optical Astronomical Observatories, operated by the Association of Universities for Research in Astronomy, Inc., under contract with the National Science Foundation.
[2]STScI is operated by the Association of Universities for Research in Astronomy Inc., for the National Aeronautics and Space Administration.

The distant stars can be observed only in light integrated over the stellar disk, making the detection of oscillations more difficult. In particular the signal, from modes with many nodes over the surface, is averaged out. Only modes to an angular degree of l < 3 retain substantial amplitudes after averaging over the disk. Fortunately for the science of astroseismology, it is the modes of low l, and high (n = 15 to 30) radial order that carry information relevant to the global stellar structure. On the Sun these modes have (disk averaged) amplitudes up to 3 X 10^{-6} (intensity) and 20 cm/s (velocity). The solar modes, for different l and n, are separated by characteristic spacings of 68 and 9 MHz in a power spectrum. The larger separation responds sensitively to the stellar radius, while the smaller (and harder to detect) frequency spacing can be used to infer properties of the stellar core. Only rough theoretical estimates (Christensen-Dalsgaard and Frandsen 1983) exist for expected p-mode amplitudes on other stars. For stars somewhat more massive and evolved than the Sun, like the subgiants of M 67, the predicted amplitudes are a factor of about five times solar (15 μmag). These stars span a very interesting evolutionary stage, near main-sequence turnoff, for stellar structure theory. Furthermore, as the brightest, abundant stellar component of open clusters, are ideally suited for ensemble photometry observations. Detection of multiple oscillation mode frequencies, for the already well studied subgiants of M 67, should provide sufficient, new observational constraints to allow detailed confrontation with stellar structure and evolution theory. Astroseismology holds the promise of advancing stellar astrophysics to a new level of quantitative analysis; the nature of the oscillations makes this a challenging observational problem.

Several groups have claimed detection of solar-like oscillations on other stars, using either Doppler or photometric observations, sometimes at amplitudes well above solar, but none of these results are sufficiently robust to offer challenges to theory. It is now possible to measure stellar Doppler shifts to about 1 m/s per minute (Brown et al. 1991) using bench-mounted fiber-fed echelle spectrographs on 2-m class telescopes. Observations like these should allow study of oscillations on the brightest solar-like stars (e.g., Procyon, α Cen A). The current limitation for Doppler techniques is lack of light (and available telescope time), restricting this approach to only the brightest stars. For the case of direct photometric detection there are plenty of photons, but the atmosphere introduces various forms of noise. Harvey (1988) has provided an excellent review of observational approaches to detection of stellar oscillations. The most difficult noise term to deal with is atmospheric scintillation, which introduces a white noise source of amplitude 650 μmag per minute for a 0.9-m telescope and an 8 hour observing window (Young 1967). Atmospheric scintillation variations are uncorrelated for spatial scales greater than a few arcsec on the sky, and thus cannot be removed by the ensemble reference calibration. Since this term decreases only as the 2/3 power of telescope aperture, it remains as an irreducible noise of 230 μmag per minute on a 4-m telescope. (Our results, to be discussed in Section 4, suggest that this estimate may be slightly too pessimistic.) The photon count rate, 2×10^7 per minute, to keep photon shot noise to 230 μmag can be obtained to 13th magnitude on 4-m telescopes for broadband photometry.

The remainder of this paper will be devoted to reviewing the technique of CCD ensemble photometry in Section 2, discussion of applications in Sections 3 and 4, and a review of what could be done with current state of the art by using a network of 4-m telescopes in Section 5.

2. TECHNIQUE OF CCD ENSEMBLE PHOTOMETRY

The practice of CCD ensemble photometry is trivial in concept. Simply observe one region of the sky, containing an ensemble of stars, at the desired cadence. Perform all of the standard CCD reductions as well as possible. Determine the sky brightness with a nonlocal algorithm over the full frame and subtract for all pixels. Using appropriate algorithms (e.g., point-spread-function (PSF) fitting, or simply digital aperture summation) extract intensities for all stars on all frames. With raw intensities normalize on a frame-by-frame basis to the ensemble average of several nonvariable stars. In all reduction steps maintain a time history of "external parameters" (e.g., seeing, sky brightness, mean positions of ensemble relative to detector coordinates) that might be expected to introduce errors in the time series, but could not be causally related to real stellar variations. For each stellar intensity time series remove correlations against external parameters. (If interested in variations only over a restricted domain of frequency space, apply appropriate band-pass filter before the last step.) If all has gone well, time series limited only by photon shot noise and atmospheric scintillation should result. Detailed discussion of the above steps may be found in Gilliland and Brown (1988) and Gilliland (1990).

A point that has become evident with the S/N = 2500 observations discussed in Section 4 below is worth pointing out here. PSF fitting allows optimally weighted extractions of image intensity. PSF fitting will fail to provide good results if the PSF is not known to higher precision than the individual images being fit. If the PSF is spatially variant over individual frames, it becomes difficult to determine the correct PSF to use for individual stellar images (especially if the number of isolated candidate PSF stars is small to begin with). We found that simple digital aperture photometry out performed PSF fitting for the highest S/N cases discussed in section 4. The PSF fitting is still useful as a means of determining the various external parameters for each data frame.

CCD ensemble photometry has a number of advantages versus classical photoelectric photometry. For the sake of discussion let us compare the merits of CCDs versus use of a three channel (object star, comparison star, sky) photoelectric photometer (see Belmonte et al.

1990 for a discussion of excellent photoelectric results).

2.1 Advantages of a CCD

1. One may observe several candidate stars simultaneously, increasing the chance of success.

2. Atmospheric scintillation is coherent only over arcsec spatial scales. The multichannel photometer will incoherently add the scintillation from reference star to the object star. Using the mean of a large ensemble as a reference can result in an essentially noiseless comparison. There is a potential gain of sqrt(2) for CCD.

3. The CCD will provide a measure of the sky which is more robust than possible with a single photometer aperture.

4. Optimally weighted intensity estimates are possible with CCD data, or one can use digital apertures appropriate to conditions; with a photometer there is very little flexibility.

5. With a CCD one can easily measure and remove noise induced by seeing, position of images on detector, etc. One cannot even measure the external parameters with a photometer.

6. CCDs have better quantum efficiency.

2.2 Potential Disadvantages of CCD

1. The amount of data to be reduced is greater by 10^5 - to 10^6. It places greater hardware and software demands on the institution.

For the particular problem of detecting oscillations with characteristic time scales of 5 to 20 minutes for open cluster stars, CCD ensemble photometry is superior to photoelectric photometry approaches.

3. RESULTS FROM KPNO 0.9-m.

Over the period 30 January - 12 February 1988 the KPNO 0.9-m telescope was used to monitor the central region of M 67 (see full description in Gilliland et al. 1990). Since results have been described at other meetings and submitted for publication, we only summarize the general conclusions here:

1. The CCD observations resulted in the serendipitous discovery of two new U UMa systems, two blue stragglers with multi-mode oscillations at low amplitude (3 mmag), and one likely AM Her CV with a 91 hour period. Useful photometry over a full 10 magnitudes of dynamic range was extracted from one data set.

2. Demonstrated noise levels to less than 1 mmag for 13th magnitude stars over one minute integrations. There was no evidence of noise contributions (except in obvious cases with saturated image cores) from the CCD.

3. Full time series (~3000 individual data frames) could place upper limits to coherent oscillations of ~150 μmag (50 times solar) per mode.

4. RESULTS FROM KPNO 2.1-m REFLECTOR

To further test the limits to CCD ensemble photometry, and to obtain further data on the oscillating M 67 blue stragglers, we used the 2.1-m and TEK 512 x 512 CCDs with a B filter over 5 - 10 January 1990. These were the first observations acquired with the refurbished 2.1-m. At f/8 on the 2.1-m an image scale of 0.34 arcsec per pixel resulted in a 3 x 3 arcmin field of view. The region of M 67 covered included two oscillating blue stragglers, two W UMa systems, three nonvariable stars at B ~ 11 and five subgiants at B ~ 13. The greater image scale relative to the 0.9-m (which has a scale of 0.77 arcsec/pixel) allowed a factor of five more counts to be accumulated per image in a single CCD exposure. Coupled with using a narrower filter, and taking shorter exposures, the brightest stars were always kept from saturation. (In the 0.9-m experiment the brightest stars were allowed to saturate.) The greater number of counts obtained with the 2.1-m, coupled with a lower scintillation contribution, should allow time series precisions, limited again only by photon shot noise and atmospheric scintillation, that are much better than previous results.

The relatively small (three bright stars dominate) ensemble of nonvariable stars in the 2.1-m field of view led to problems with our standard (Gilliland and Brown 1988) ensemble normalization, which included color and x,y position terms in addition to the zero point. With the number of free parameters comparable to number of data points in each frame the standard ensemble normalization could produce artificially excellent results. To avoid this problem a frame-by-frame normalization only to the mean of the ensemble was adopted. This simplification will tend to bias results to larger errors than might be possible with a more robust ensemble averaging algorithm.

Using the scaling laws from Young (1967) for atmospheric scintillation, the contribution for 24 s integrations on the 2.1-m, averaged over an 8 hour observing window, is 560 μmag. The best time series star, with B = 11.1, had 7.2 x 10^6 counts per image for a further photon counting contribution of 370 μmag to the noise budget. Combining the photon shot noise and predicted scintillation noise incoherently leads to a predicted noise level of 720 μmag. For the night of 10 January 1990 a best result of 590 μmag was obtained for 20 - 24s integrations. The results are therefore better than expected, and allow an upper limit of 460 μmag (per 24 s) to be placed on the scintillation contribution. This result, 18% below nominal expectations, is well within the allowed range for scintillation (Young 1967). Such a favorable result demonstrates that no limitations due to the CCD detector are entering at the level of a few hundred μmag per readout. Scaled to the same cadence (84 s) used for the 0.9-m observations, the best time series on the 2.1m- has a 410 μmag rms. It is worth noting that this was obtained under less than ideal observing conditions (not photometric, and only one day from full moon yielding a bright sky), and with a less than optimal ensemble for the reductions. For the interesting subgiants of M 67 at B ~ 13, the precisions are shot noise limited to 800 μmag (per 84 s), with less data from the 2.1 m run this will allow results comparable to the 150 μmag per mode detection limit obtained on the 0.9-m. These limits are still a factor of 10 higher than theoretically expected amplitudes.

Fig. 1. shows time series results for the same star (B = 11.3) on a good night (10 January with 20 - 24 s integrations, top panels) and a poor night (9 January with 30 s integrations, bottom) with variable bright cirrus. The upper panel of each pair shows raw time series data in digital numbers (17.6 electrons/ADU) versus time in seconds after start of time series. The lower panel shows result (magnitudes) after normalization to ensemble mean, and high pass filtering. Note the change of exposure time with 10 January data demonstrates extreme linearity of CCD at high count level. The robust nature of CCD ensemble photometry is amply demonstrated by the data from 9 January. Although acquired in abysmal conditions the error is worse by only 33% relative to the best night (530 versus 410 μmag per 84 s cycle).

5. WHAT A NETWORK OF 4-m TELESCOPES COULD PROVIDE

With predictions of the p-mode oscillations of interest for the M 67 subgiants at about five times (but quite uncertain) solar amplitudes, even the 2.1-m results would not yield a detection with a reasonable amount of observing. (A convenient definition of reasonable may be taken as 4 - 6 days, corresponding to nominally expected oscillation coherence times.) Adopt $x = 2\sigma(SN/NO)^{1/2}$ (Scargle 1982) as the signal level, X, for a coherent oscillation which can be detected at a signal-to-noise level of SN, with NO observations, against a noise level σ. A network of 2.1-m telescopes could reach X = 20 μmags at B = 11.0. This is a scintillation limited result. For the M 67 stars of primary interest at B = 13 mag, the results would be limited by photon shot noise to X = 28 μmags, even with use of a filter with twice the throughput of the B-band. (Use of a still broader filter does not seem advisable. It would lead to too much color sensitivity to atmospheric variations over the ensemble.) These limits are not sufficient to predict success for detection of p-modes on the M 67 subgiants.

The CCD introduces basic constraints due to a finite well depth. On the 2.1-m, with 0.34 arcsec per pixel, 7×10^6 counts in one arcsec seeing could be accumulated per image before nonlinearity/saturation

THE PROMISE OF TIME-RESOLVED CCD ENSEMBLE PHOTOMETRY 271

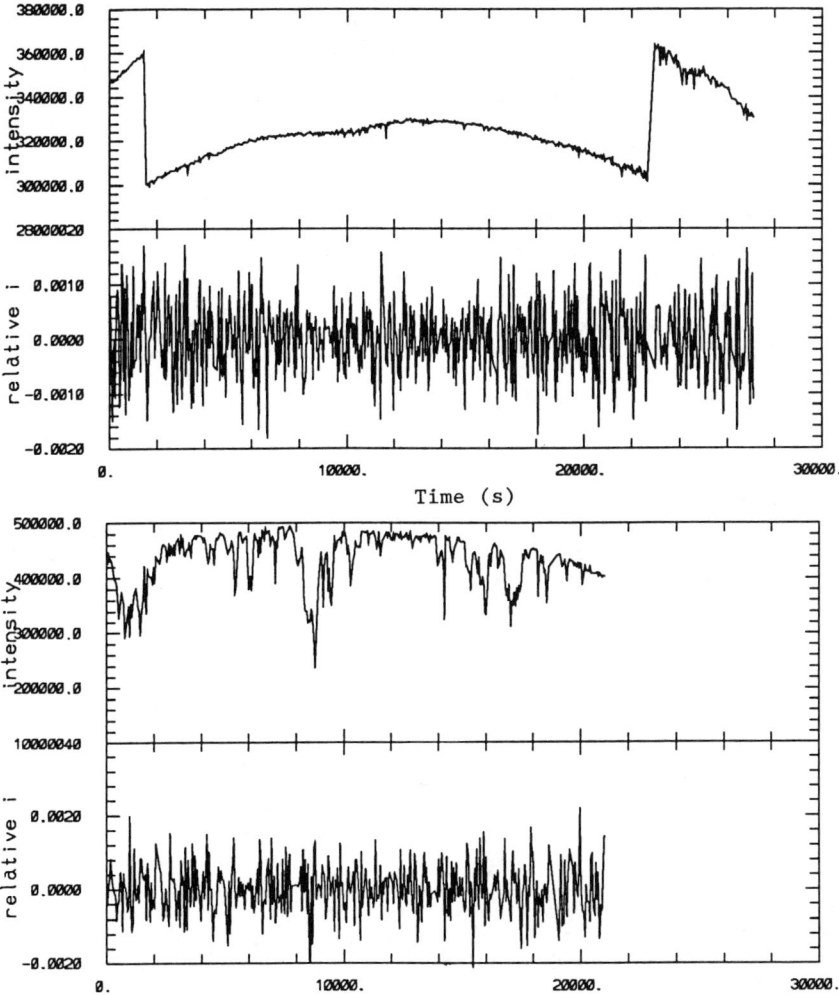

Fig. 1. Time series results for a star on a good night (top panels - 10 January with 20 - 24 s integrations) and a poor night (bottom panels - 9 January with 30 s integrations). The upper panel of each pair is a raw time series data sequence in digital numbers (17.6 electrons/ADU) versus time in seconds after the start of the time series. The lower panel shows the result (in magnitudes) after the normalization to the ensemble mean and high pass filtering. Note the change of exposure time with the 10 January data demonstrates the extreme linearity of the CCD at a high count level. The robust nature of CCD ensemble photometry is amply demonstrated by the data from 9 January. Although acquired in abysmal conditions, the error is worse by only 33% relative to the best night (530 vs 410 μmag per 84 s cycle).

effects became important. If significantly higher precisions are desired, more counts per unit time must be allowed for (and on a larger telescope for which scintillation is smaller).

More counts per unit time can be acquired (assuming the requisite photon flux is available) by: 1) using a larger image scale yielding more pixels per stellar image (but lose field of view), 2) using a detector with deeper wells (the TEK CCDs with 500,000 electrons are best known to author), 3) cycling the CCD faster (with significant time overhead per readout this lowers the duty cycle, thus increasing scintillation), and 4) defocusing the images (brings in more sky, increases confusion with close neighbors).

With a 4-m telescope the (8 hour averaged) scintillation limiting precision is 300 μmag per 24 s of integration scaled from the realized 2.1-m results. Scaled to the standard case of 60 s integration this is a limit of 190 μmag per cycle (typically additional 24 s of overhead is expected).

Given our primary interest in detecting oscillations on the subgiants of M 67 (B = 12.5 to 13.8) can we identify a field of view and observing strategy allowing nearly scintillation limited results? At f/8 on a 4-m the image scale at 0.18 arcsec/pixel would safely allow 3×10^7 counts per readout for the brightest image (assumes TEK 512 x 512 CCD and one arcsec seeing). At this scale the field of view is only 1.5 x 1.5 arcmin. There is a field in M 67 (centered on the dipper asterism) with 11 subgiants over B = 12.5 to 13.8, these would provide an excellent ensemble. With a B filter 2×10^7-counts could be acquired in 60 s; this would result in photon shot noise as largest term, even on the brightest star (total noise = 275 μmag per 84 s cycle). With a broader filter, passing twice the light of B we would reach 3×10^{17}-counts in 45 s, here the scintillation and shot noise contributions are 220 and 183 respectively for a total of 286 μmag. Adjusted to the standard 84 s cycle length this yields 250 μmag/cycle. *Without major changes in more than one of the components (stellar ensemble brightness, CCD well depth, and telescope aperture) it is not possible to improve on this result substantially.* Brighter stars (or broader filter) wouldn't help, since to count more photons would require more readouts, a lower duty cycle and increased scintillation. A CCD with deeper wells (or quicker readout) would help only to the extent that an improvement in duty cycle would lower scintillation noise, only 15% improvement is possible (although with more counts per pixel allowed, a larger area of sky, and thus larger ensemble, could be followed). A larger telescope wouldn't help substantially without a correspondingly improved CCD to handle the better photon flux (8-m at f/4 with same CCD would allow 25% improvement). The above argues that a 4-m network, using current CCDs, observing an identified field of M 67, would be able to do essentially as well as can be envisioned (from the ground). If the marginal gain with a larger telescope, or better CCD, or better ensemble of stars is small, is it equally true that compromising any single component is a small loss? No! For the stars in question (B ~ 13) a smaller aperture would directly translate to a nearly linear loss from increased photon shot noise. A network of 2-m telescopes would be a factor of two poorer (an important factor) than 4-ms. Similar arguments would follow for less capable CCDs, or less favorable ensembles.

With observations from space, scintillation noise would not be a problem. In principle a 1-m class telescope in space would be able to vastly outperform any ground based network (assuming the selection of brighter stars). Given the cost, time, and effort involved in developing a dedicated space mission, it would be prudent to fully test the limits for ground based work. A single networked observing run consisting of four, 4-m telescopes over five nights would provide a combined sensitivity (per oscillation mode) at 4 σ level of 14 to 19 μmags over the 11 star ensemble proposed for detection of coherent oscillations. This is a necessary, and based on current expectations, sufficient level of precision to allow scientific success for study of stellar oscillations. Such a network experiment would also yield a definitive measure (to 10's of percent) of limitations to ground based photometric precision for many years to come.

ACKNOWLEDGEMENTS

I thank Drs. Tim Brown, George Jacoby, Larry Petro and Peter Stetson for discussions concerning various aspects of CCD ensemble photometry practice, and Dr. Ronald Probst and the Kitt Peak staff for the time-shared access to the 2.1-m in its recommissioning in January 1990.

REFERENCES

Belmonte, J. A., Perez Hernandez, F. and Roca Cortes, T. 1990 A&A 231,383
Brown, T. M., Gilliland, R. L., Noyes, R. W. and Ramsey, L. W. 1991 ApJ 368, 599
Christensen-Dalsgaard, J. 1988 in IAU Symposium No. 123, Advances in Helio- and Astroseismology, J. Christensen-Dalsgaard and S. Frandsen, eds., Reidel, Dordrecht, p. 3
Christensen-Dalsgaard, J. and Frandsen, S. 1983 Solar Phys 82, 469
Gilliland, R. L. 1990 in CCDs in Astronomy, G. Jacoby, ed., PASP Conf Ser p. 281
Gilliland, R. L. and Brown, T. M. 1988 PASP 100, 754
Gilliland, R. L., Brown, T. M., Duncan, D. K., Suntzeff, N. B., Lockwood, G. W., Thompson, D. T., Schild, R. E., Jeffrey, W. A., Penprase, B. E. 1991 AJ, to appear Feb, 1991
Harvey, J. W. 1988 in IAU Symposium No 123, Advances in Helio- and Astroseismology, J. Christensen-Dalsgaard and S. Frandsen, eds., Reidel, Dordrecht, p. 497
Scargle, J. D. 1982 ApJ 263, 835
Young, A. T. 1967 AJ 72, 747

DISCUSSION

STETSON: On the question of the origin of the blue stragglers: It seems to me that an important criterion for distinguishing mixing models from mass-exchange models from coalescence models would be the envelope helium abundance. Is the period splitting of the multiple modes sensitive to this?

DEMARQUE: I do not know but I suspect that it would be quite difficult. In fact, Guenther and Sarajedini (ApJ 1988) show that for the Sun, the oscillation spectrum is highly insensitive to the helium abundance.

JANES: You refer to limitations in the current generations of CCDs. What is it about them that is deficient?

GILLILAND: For this project the primary limitation is CCD well depth. To reach the atmospheric scintillation limit on a large telescope requires detection of a very large number of photons; increaased well depth would help considerably.

THE ACTIVITIES OF THE NSSDC AND ADC IN THE AREA OF ARCHIVING PHOTOMETRIC DATA

Wayne H. Warren Jr.

Astronomical Data Center (ADC)
National Space Science Data Center (NSSDC)
NASA Goddard Space Flight Center

ABSTRACT: Participating members of the international network of astronomical data centers have been archiving and distributing photometric data for many years. In the past, these data have been almost exclusively in the form of compilation catalogs of stars observed on particular photometric systems, many of which are assembled at the photometric data center of the Geneva Observatory. The recent development of automatic photoelectric telescopes (APTs) nd a corresponding explosion in the volume of data to be archiv l and disseminated will require the development of new procedures for ecord keeping, data storage, and distribution. This paper discusses the problems involved and some possible solutions.

1. INTRODUCTION

The astronomical data centers have been archiving and disseminating photometric data since their inception in the early 1970s, and the collection of catalogs has grown continuously over the last 15 - 20 years. A special category (2) in the lists of catalogs available contains all catalogs and data compilations in which the primary data consist of photoelectric or photographic photometry. While many of the data sets contain photometry in special regions or of particular collections of objects, and even some theoretical or computed data, the most often requested ones are the generalized compilations of photometric data in particular systems. Although individual photoelectric observations have generally not been archived in the past, recent developments in the area of automatic observing and APTs have raised questions about how to archive individual observations and how to make them available to other researchers.

This paper briefly discusses photometric catalogs in general and provides a list of currently available compilation catalogs in existing standard photometric systems. Considerations for the storage and retrieval of individual photoelectric observations for large numbers of objects are then discussed.

2. PHOTOMETRIC CATALOGS

The category of photometric data contains the largest number of active catalogs (presently 122) in the astronomical data center archives. This is due primarily to the fact that there are so many different photometric systems in which the same objects have been measured and because catalogs are compiled separately for each system. Although there are many small collections of photometric data for particular objects, the most widely used photometric catalogs are the compilations of all published observations in a particular photometric system. These compilations have been done mostly at the photometric data center located at the Institut d'Astronomie de l'Université de Lausanne by Mermilliod, Hauck and their collaborators. These photometric compilations generally consist of a catalog of individual measurements, as published in the literature, and a catalog of homogeneous means determined via the rather complex procedure of referring each published paper back to a standard system to assign weights, then computing weighted means for each multiply observed object. Although this procedure may not always yield the best mean for every object observed, in principle, it is the optimal overall method for determining mean values for large numbers of objects.

The photometric compilation catalogs presently available in machine-readable form from the data centers are summarized in Table 1. All of the most frequently used photometric systems are represented. Special systems used primarily in non-visible (ultraviolet and infrared) regions of the spectrum are not included.

As can be seen from the references in the table, many of the photometric compilations date back 10 - 15 years and need to be updated. It is clearly difficult to keep such compilations up to date when so many systems are involved and a complete literature search must be maintained for each system. The archiving of machine-readable data from individual papers and subsequent submission of data to the photometric data center in machine-readable form would be of great help toward facilitating the incorporation of new data into compilation catalogs, but that would be a mammoth undertaking for the data centers and is probably not feasible at this time. Electronic scanning and digitization of published tables should assist with this work in the interim.

In addition to the photometric catalogs available in machine-readable form, it is desirable to prepare small computerized files of standard stars for all of the currently used photometric systems. (This should also be done for other observational systems using standard stars that must be compared frequently or used to reduce observations, such as radial and rotational velocities, MK spectral types, etc.) Although tables of standard stars have been compiled by Philip and Egret (1985), a collection of such tables has not yet been

TABLE 1

List of Photometric Compilation Catalogs

System	Catalog Number	Number of Stars	Reference
DDO	2017A	915	Magnenat 1974
	2080	2196	McClure and Forrester 1980
Balmer lines	2063	18000	Mermilliod & Mermilliod 1980
Geneva	2072	14633	Rufener 1981
Hα	2066	2299	Ducati 1981
Kron	2075	11738	Jasniewicz 1982
UBV	2051	53839	Nicolet 1978
	2122A	87267	Mermilliod 1987
UcBV	2027	7141	Nicolet 1975
UBVr 20	2015	368	Magnenat 1973
UBVRI	2116	6849	Lanz 1986
UBVRIJKLMNH	2007A	4500	Morel & Magnenat 1978
UVBGRI	2008	1297	Nicollier & Hauck 1978
uvbyβ	2107A	40482	Hauck & Mermilliod 1985
VBLUW	2014	2687	Python 1979
Vilnius	2157	4849	Straizys et al. 1989

made available in a machine-readable form useful for observations at the telescope. Each file of such a compilation would require star identifications, equatorial coordinates, and the standard data for the relevant system. Appropriate small files could then be uploaded to each observatory's computer and used interactively for observations and possibly even for data reduction.

3. PHOTOMETRIC OBSERVATIONS

Data storage technology has reached a level where it is feasible to archive and distribute individual photometric observations of objects over long periods of time. Such unpublished observations of variable stars have been stored and distributed by IAU Commission 27 for some years now (see Jaschek and Breger 1988). The continuing programs of APTs now operating at several observatories require the development and implementation of archiving and dissemination procedures for large numbers of observations of many objects. However, the operation of such an archiving center is not staightforward, since observations of the same objects will be submitted to the data centers in batches and will need to be combined into comprehensive files for storage and distribution. Requests for observations covering particular time periods will necessitate the extraction of data from large files, plus the centers will undoubtedly receive many requests for information pertaining to the availability of observations of individual objects over specified time periods.

An important consideration concerns the method of storage and distribution. The IAU Commission 27 archive consists of individual files of observations submitted at various times by the same or different observers. Since no processing of the data is done at the center, the files can be in a variety of formats and many files can exist for the same object. Thus, when a researcher requests all data for a particular object, it may often be required to collect a large number of files and assemble them for dissemination to the requester. At the other end, the collection of files must be combined by reformatting and merging, a task that can involve a considerable amount of time and effort before the data are even ready for analysis. A more efficient procedure, and one that can save processing time for users of the archive, is to standardize the archiving format for each photometric system and to combine all observations of each object into a single file at the data center. Each new batch of observations received would then be concatenated into the existing file for the appropriate object and system. A request for the observations of a certain object in a specified photometric system would then simply entail the retrieval of the unique file for that object and system.

Of course, the standardization of formats for the various photometric systems might present problems because observations often differ depending on the objects being observed. For example, an observation made in the UBV system usually results in determination of the quantities V, (B - V) and (U - B), but for variable stars, only differential magnitudes and colors might be measured, in which case the format could be the same, but the quantities in the file would need to be explicitly defined and declared in accompanying documentation. Observations of variables are also complicated by the fact that data for comparison stars must be archived and keyed to the variables that they were used for.

The archiving and distribution of standard-star data must also be considered. The availability of all observations of the standards that have been used to transform program-star observations to the standard system can be of great help toward determining weights for a particular set of observations when they are combined with other sets to compute homogeneous means (see Lindemann & Hauck 1973). As with comparison stars, all standard-star observations would need to be tied to the corresponding program stars that they were used for.

Although the above procedures introduce many complications into what could be a reasonably straightforward archiving task, it has happened on many occasions in the past that archived data are not usable by future researchers because not enough information was made available and documented at the time of original data archiving. It may be possible to simplify some of the methods discussed, but it is clear that a considerable amount of planning should go into the creation of an archive of photometric observations in order to ensure that their full potential is realized at all future times.

The storage of photometric files by object on partitioned (direct access) media, such as magnetic disks, would make the sorting of files and addition of new data quite simple, but the volume of data and the large numbers of objects involved may make online storage infeasible, in which case offline storage on magnetic tapes or optical disks would be necessary.

A major problem with a sequential medium such as magnetic tape is that new observations cannot be added to existing files without rewriting the files in their entirety, and this procedure would require too much time and labor to be accomplished at the data centers, at least not without the assignment of dedicated personnel to the task.

An alternative would be to store all time-ordered observations for particular objects on a collection of microdisks accessible to a personal computer, or in individual hard disk files on a dedicated PC or workstation. Particular files could then be extracted and distributed to a requester. Better yet, the host computer could be made available over computer networks so that an astronomer would log into the machine and extract data to transmit via electronic mail or anonymous FTP (file transfer protocol). Such a system would require the construction of a data base containing all information necessary for users to query and access the archive, including object designations (and alternates), coordinates, and file information. The implementation of a properly designed "self-service" system would allow users to retrieve all desired data themselves via network transfer, thus precluding the need to commit dedicated data center personnel to service requests for data on an individual basis.

4. SUMMARY AND CONCLUSIONS

The astronomical data centers currently have a rather complete collection of photometric compilation catalogs, with those covering the major systems (UBV and uvbyβ) being updated on a regular basis. However, a modern system of data storage and retrieval, including automatic network access and retrieval of data, must be developed to handle individual observations, particularly those now being obtained by existing automatic telescopes. The development of such a system for use at the data centers will allow individual astronomers to retrieve desired photoelectric data themselves, at any time, with minimum intervention by data center personnel.

REFERENCES

Ducati, J. R. 1981 A&AS 45, 119
Hauck, B. and Mermilliod, M. 1985 A&AS 60, 61
Jaschek, C. and Breger, M. 1988 Bull. Inf. CDS No. 35, p 93
Jasniewicz, G. 1982 A&AS 49, 99
Lanz, T. 1986 A&AS 65, 195
Lindemann, E. and Hauck, B. 1973 A&AS 11, 119
Magnenat, P. 1973 Centre de Données Stellaires Internal Report No. 6
Magnenat, P. 1974 Centre de Données Stellaires Internal Report No. 9
McClure, R. D. and Forrester, W. T. 1981 Publ. Dom. Astrophys. Obs., 15, 439
Mermilliod, J.-C. 1987 A&AS 71, 413
Mermilliod, J.-C. and Mermilliod, M. 1980 Bull. Inf. CDS No. 19, p 65
Morel, M. and Magnenat, P. 1978 A&AS 34, 477
Nicolet, B. 1975 A&AS 22, 239
Nicolet, B. 1978 A&AS 34, 1
Nicollier, Cl. and Hauck, B. 1978 A&AS 31, 437
Philip, A. G. D. and Egret, D. 1985 in IAU Symposium No. 111, Calibration of Fundamental Stellar Quantities, D. S. Hayes, L. E. Pasinetti, A. G. Davis Philip, eds., Reidel, Dordrecht, p. 353
Python, M. 1979 A&AS 38, 463
Rufener, F. 1981 A&AS 45, 207
Straižys, V., Kazlauskas, A., Jodinskiene, E., and Bartkevicius, A. 1989 Bull. Inf. CDS No. 37, p. 179

DISCUSSION

JANES: In your talk you emphasized situations typified by variable star observations. In such cases, there is a relatively small number of stars, but possibly many observations of each star. But there are also situations where one observes a field (i.e., a cluster). Here there will be one or two observations of a great many stars. How should one treat those data?

WARREN: They can be treated in the same way that published observations are archived now; i.e., they can be incorporated into compilation catalogs so that all data are present in one source. Of course, this is dependent upon the availability of unique identifications (or designations) for the individual stars. Lacking designations for very large numbers of stars, for example, in clusters, then the observations should be archived in separate files having coordinate designations so that any one star can be later identified unambiguously.

HAUCK: You can add another catalog to your list. We have completed our General Catalog of Photometric Data. This catalog allows one to find in which photometric systems a star has been measured. The catalog contains more than 100,000 stars and is available from the CDS in Strasbourg. An announcement will be published soon in A&AS.

WARREN: Yes, I am aware of the catalog and it will be an extremely valuable source of information. However, I only included in the table catalogs that had been received in machine-readable form as of the date that the table was prepared (28 September, 1990). Let us hope that the new catalog can be incorporated into SIMBAD without too much delay.

CRAWFORD: Are the data base catalogs making an effort to handle the problem of those who publish new data already combined with other published data? The problem is that often data then appear twice (or more often).

HAUCK: An attempt is always made to identify those observations previously published and to decouple them from newly appearing data, but it is not always easy if the old and new observations have been combined.

YOSS: If we have sent archive data to Strasbourg, can we send you new data mixed with the old, and can you then sort them out and add only the new data to the archive? We have all data (old & new) in one file, and it would be easier for us to send you one grand file periodically.

WARREN: Actually, it is easier for the data centers if the combined file is resubmitted periodically. We can then replace the old by the new, without having to separate the data, and users can obtain the comprehensive file for analysis.

CRAMER: In your table, citing compilation catalogs, you mention the 1981 edition of the Geneva photometric catalog. A new compilation containing 29,400 entries was published in 1988.

WARREN: The new Geneva catalog has been received in hard copy form, but not yet as a machine-readable file; therefore, I did not include it in the table. We exchange catalogs with the Strasbourg center on a periodic basis, so there is often a delay between the time a catalog is received in Strasbourg and at the ADC. However, if a catalog has been requested at either center, then a special effort is made to exchange it on an individual basis.

STERKEN: Archiving and computer-accessible distribution of photometric data almost exclusively covers data collected during the last couple of decades, not even half a century. But there are many good-quality data (visual estimates and early photographic and photoelectric measurements) of variable stars. These data, though of lower absolute accuracy than are the modern observations, are very important, and their greatest asset is that they can extend the available record to a time-baseline that may be longer by a factor of 5 to 10, or even more (e.g. the case of long-period variables). One should also undertake the computer-accessible archiving of these data with proper attention to the homogenization.

WARREN: I agree, but it is a large undertaking to computerize the earlier observations. The AAVSO has been working on such a project for some years now for the observations in its own archives; hence, many of the older variable-star observations are available from that organization on request.

PHILIP: Do you need positions for all the stars in a compiled catalog of photometric data? For example, on a CCD frame of cluster data could the photometric data for all the measured stars be entered into a compiled catalog?

WARREN: As mentioned above, this can be done if the stars are provided with unique identifications. If not, then special files including positional designations should be archived. We must archive the observations in a way that will ensure unambiguous identification at any future time; otherwise, the archived data will be effectively lost.

SUMMARY PAPER: PRECISION PHOTOMETRY, PAST, PRESENT, FUTURE

Kenneth A. Janes

Boston University

1. INTRODUCTION

In his welcoming remarks, Prof. Ralph Alpher of Union College commented that to a cosmologist, precision means that one is about equal to ten. In most sciences, the expectation is that one can do better; the goal of this meeting has been to find out just how much better the photometrists can do. In fact, we find that photometry can now be done with a precision undreamed of only a few years ago. Nevertheless, we have seen that in reality the meaning of the word precision can be defined rather differently, depending on the nature of the astrophysical questions one is seeking to answer. The trend however is clear - because it is now technologically possible to obtain high precision, we find that we are now asking questions that were not thought appropriate to ask in the past. In this review, I hope to highlight some of the new levels of precision, some new applications and new ways of looking at old problems in photometry. I have, somewhat arbitrarily, assigned the papers into categories of my own invention, based on first impressions from the abstracts.

2. HISTORICAL NOTES

It is appropriate to begin this review by taking note of what might be called the "Schenectady" series of meetings on photometry and related subjects, organized by our host, A. G. Davis Philip. The first of the series took place in October 1974 (in Albany, NY); many of the participants at this meeting were also there for a meeting on "Multicolor Photometry and the Theoretical HR Diagram". Since that time, Dave Philip has run or helped to run some 18 meetings on various topics; not all of them were directed to photometry or related issues, but one can very nicely trace the development of photometry and photometric techniques from a study of the proceedings of these various meetings.

The inspiration for doing precision photometry has truly come from Bengt Strömgren, who showed how, with a suitable choice of spectral pass-bands, one can obtain genuine astrophysical information. It is appropriate then, that a considerable portion of this meeting is devoted to Strömgren and to work done with the uvby, or 4-color system. In the reviews of Strömgren's life and of the 4-color photometry by Knude and Crawford, as well as in several other papers at the meeting, we have been reminded of the reasons why one does photometry - the purpose for photometry is to sample, generally at low spectral resolution, the flux from stars and galaxies in order to ascertain their physical properties. It is often easy to forget that obvious fact in the midst of worrying about the technical details of doing photometry.

3. PHOTOMETRIC TECHNIQUES AND CALIBRATIONS - uvby

It is a little surprising to see that there are still significant issues to be resolved in transformation of four-color photometry, such as those raised by Manfroid and Sterken and by Delgado and Alfaro. But on reflection one sees that in fact what has happened is that because of the improved precision, people can now dig more deeply into the system, and that even small calibration errors become noticeable. Manfroid and Sterken suggest ways to systematize observations made by different observers or equipment, and Delgado and Alfaro have shown that reddened stars may not transform properly to the standard system. In fact, there is a dependence on E(b-y).

The two poster papers by Hauck show on the one hand what can be done with 4-color photometry in calibrating the physical parameters of A and F stars, and on the other, what has been done, which is that 45,000 stars have been observed on the four-color system. The calibration of the A stars was also the topic of the paper by Figueras et al., and Garrison and Gray investigated the effects of rotation on the calibrations of B stars. The combined effect of these papers is to show that we now have a good understanding of the physical calibration of the 4-color system for stars of spectral types B, A and F.

The Twarogs' work on CCD 4-color photometry is particularly exciting. They have shown dramatically that the high precision possible with CCD's can be applied to the 4-color photometry. They have also begun to address what has become a vexing product of the success of CCD's - the problem of finding faint standards. Their efforts to produce these standards are especially important.

4. MORE TECHNIQUES - OTHER SYSTEMS

Several papers illustrate the range of photometric systems now in use. Mendoza's paper on the $\alpha(16)-\Lambda(9)$ system reminds us what we can learn from truly narrow-band photometry; Golay's poster paper describes the continuing progress with the Geneva photometry and Gray has shown how one can combine objective prism data with 4-color photometry to select a well-defined group of stars (in this case, the λ Boo stars).

Pel's paper on the intercomparison of several photometric systems provides another demonstration of the increasingly systematic way that

SUMMARY PAPER: PRECISION PHOTOMETRY, PAST, PRESENT, FUTURE

photometry is being defined. He showed that in general one can relate one system to another in a direct way, and no systematic trends are in evidence as one moves around the sky. The only discordance is with the Geneva photometry, which seems to show a slight directional dependence in the differences with other systems. Hopefully, this is an effect that can be explained or accounted for in some way.

Rich's review of infrared photometry gives a selection of tantalizing morsels from the immense possibilities of IR photometry. IR photometry may soon become as active and diverse a field as visual photometry, and Rich's "case study" of the galactic bulge gives some of the flavor of the potential of IR photometry.

A number of papers deal with various instrumental issues. Bell and Yoss have combined grism spectroscopy and a CCD camera to show that they can produce reliable synthetic photometric indices. Although the idea is old, they show that with a CCD it is now possible to get photometry in this way that is actually useful. The paper by Chromey et al. shows that with a CCD, precision photometry is even possible with a small telescope in a suburban environment. In the same vein, Sterken's poster paper shows the power of an automatic telescope for 4-color photometry.

Seitzer's careful review of photometry with HST includes an especially useful set of references to papers which discuss the best procedures to follow in doing HST photometry. But his paper also shows how much work still needs to be done to make HST a fully functional telescope.

Finally, Gilliland's extraordinary paper on low-amplitude variability among M 67 stars has re-defined the meaning of the phrase "precision photometry". He has even had to coin a new word - "micromagnitude". He showed the potential for the future of photometry; with photometric precision better than a thousandth of a magnitude, the possibilities for probing basic stellar physics begin to increase dramatically.

5. APPLICATIONS - CLUSTERS

In the papers by Demarque, by Sarajedini and by Richer, we see that there are still differences of opinion about how to interpret globular cluster photometry. As the photometry has improved, the extreme difficulty of determining absolute cluster ages becomes more apparent - the morphology of the cluster CM-Diagrams must be known extremely well in order to say anything about ages. Nevertheless, there is progress being made. There is now little uncertainty that there is a measurable range in the ages of the globular clusters. Sarajedini argues for a systematic correlation between cluster age and metallicity, whereas Richer has tried to show that at fixed metallicity, there are differences in cluster ages. These two statements are not necessarily contradictory, but on the other hand, they do show that significant uncertainties remain in determining cluster ages. Demarque's review of the theoretical side of the issue shows that at least a few of the potential problems probably are not actually important - most notably the matter of rotation, which seems not to be a significant factor in altering globular cluster CM-Diagrams. Diffusion of helium may have some small effects, but does not appear to be a major issue. People have worried about these effects for years, but if the calculations Demarque describes are confirmed, then the concerns are mostly unfounded.

Stetson's paper on M 15 shows however, that some new problems must now be considered - specifically, the possibility of radial gradients of stellar properties within the clusters. His new work confirms the existence of a color gradient within M 15. The relative populations of red horizontal-branch stars, stars at the tip of the red-giant branch and stars on the extended horizontal branch all depend on location within the cluster. That is to say, there is some kind of linkage between the processes that affect stellar evolution and those that relate to the star's position within the cluster.

If we can determine the physical properties of stars on the horizontal branch, then we will have important constraints on allowable theoretical models. For that reason, Philip's paper on CCD uvby photometry of the horizontal branch stars is especially exciting. The high-precision photometry possible using CCD's along with improved calibrations of the uvby system will ultimately have direct bearing on the age question mentioned above.

It is exciting to see, in the paper by Weller et al., that there are good possibilities for CCD photometry with Schmidt telescopes. In relatively uncrowded fields, such as the region around NGC 5822 which they studied, the quality of the photometry is competitive with that from larger scale telescopes. The way is open for precision wide-area surveys.

In the papers by Phelps and Janes and by Luginbuhl et al., we see another direction in which the improved precision of the CCD's can take us - towards studying the faint end of the luminosity function. The question of bi-modal star formation, which may or may not be just an artifact of the limitations of the older photographic photometry, should soon be answered. In the very youngest clusters such as NGC 2264, we can now study directly the pre-main-sequence phase of stellar evolution.

6. Applications - STELLAR SURVEYS AND VARIABILITY

It is now possible to consider precision photometry and large-scale surveys at the same time. Until recently, in doing large surveys one was forced to make do with rather low precision photometry, but we see here several applications where photometry of large numbers of stars or large numbers of observations of a single star are possible, even at the highest precision.

Several surveys of the stellar populations towards the galactic poles are the subject of papers: Knude, Majewski, and Drilling and Philip all made significant advances in our understanding of the complex combination of stellar populations in that direction: the entangled combination of (young and old) thin disk stars, the thick disk and the true halo. Surveys of the sort reported here are demonstrating that we can finally begin to separate these populations. Perhaps the most important result to come from these studies is the realization that the halo and the thick disk really are distinct populations with few, if any, stars in a state intermediate between the two.

Drilling's poster paper surveying the O and B stars along the galactic plane indicates what can now be accomplished in the way of determining reddening to stars, even in regions where the overall reddening is rather high.

Tyson has done a beautiful job in adapting the Washington photometry to studying the central bulge of the galaxy, although one wonders if reddening might not still be the limiting factor. We do not yet understand the stellar population of the central region of the galaxy, but the sort of work that Tyson and Rich are doing should go a long way towards that end.

Finally, the author of this review would like to direct attention to the fact that we also know little of the outermost regions of the galactic disk; with the sort of techniques discussed at this meeting, large-scale surveys are possible of the outer regions of the galaxy.

Turning from single observations of many stars to many observations of single stars, we learned from Burki's review of some recent results from the Geneva photometry database. When the photometry is both precise and systematic, an impressive variety of physical properties of stars can be probed through the fine details of their variability.

7. CATALOGS

The increased precision (not to mention the increased volume) of photometry presents problems and challenges in maintaining astronomical catalogs. In Upgren's review of the nearby stars, we see that the better photometry is now permitting a much improved picture of the stellar population of the immediate solar neighborhood and presents exciting prospects for future fundamental calibrations of stellar properties. Cramer's review of the extension of the Geneva photometric catalog shows that their group is managing to keep up with an increasing flow of data, which means that the massive collection of Geneva photometry is becoming increasingly useful to solve astrophysical problems. However, Warren cautioned that the NSSDC has difficulty in keeping up the ever greater flow of data into the center, and there are several unresolved issues of a database management sort facing the archivists. It is of course rather important that ways be found to keep up, or else the data will be of no use to anyone.

8. ASTROPHYSICAL CALIBRATIONS

The poster paper by North shows that the fundamental calibration of photometry needed to derive basic stellar properties is still not complete - the semi-empirical calibrations differ from one another, and there is no way at the moment to decide among them. There is hope, however. Kurucz has extended his spectral synthesis program to include an incredible number of atomic and molecular transitions. The fact that he can now match in some detail the uv spectrum of the Sun gives one confidence that his new synthetic colors will be a close correspondence to reality. We may finally be on the verge of being able to predict the colors of a star, given its physical properties.

If the atmospheric parameters found by Kurucz can be combined with recently derived theoretical models of HB stars described by Sweigart, then we may have another hold on the globular cluster properties and hence on the question of their ages. Sweigart's review of the status of horizontal-branch theory constitutes a virtual check-list of important observing projects, all of them within reach of modern technology.

9. FUTURE PROSPECTS - PHOTOMETRY IN 2006?

We have seen that there have been enormous advances in photometry in the 16 years that groups have been meeting in Schenectady. What might the next sixteen years hold? This is always a dangerous sort of question to ask, but it is also fun to consider what the future direction might be.

First one might ask whether people will still even be doing photometry as we now define it. Will the overall flux from stars and color indices of some sort even be interesting? The answer to that would appear to be a resounding yes - after all, even if we have thoroughly studied the nearby and bright stars, even with giant telescopes being planned, the best one could hope for in studying distant, extragalactic stars is that we will someday be able to see them no more clearly than we can now see the fainter stars in our galaxy.

In thinking of technological changes, it would seem likely that there will be a continued increase in the size of solid state detectors. There is no fundamental reason why one could not expect to have area detectors with 10,000 or so pixels on a side within just a few years from now. Obviously the storage and manipulation of such huge arrays is a problem, but that is just the sort of problem which also occurs in many other areas of information technology. The general commercial potential of managing huge databases will almost certainly encourage further development of storage and manipulation of large amounts of data.

I suspect that the most striking detector changes over the next few years will come in the infrared. IR arrays as large as 256 X 256 are already in development, and there does not seem to be any fundamental reason why one could not eventually expect IR arrays as large as those in the optical. Since in many ways the IR is a less-developed science than the optical, there should be every expectation of enormous changes in IR astronomy, including IR photometry.

Except for possible new IR systems, it does not seem likely that there will be new general purpose photometric systems developed in the next decades. Instead, it is more probable that there will be an increase in the use of special-purpose photometry - photometric indices developed to answer just one or two specific questions. There will be an increasing reliance on picking and choosing among the large number of systems now in use to define personalized sets of indices. Ultimately, however, even such systems will require a reliance on the type of photometry that has been described here, and on the calibrations of these systems. If a large enough body of stars can be put together which have precise determinations of their basic properties (t_{eff}, log g, metallicity, etc) then one might find people doing a direct transformation from the appropriate raw photometric measurements into fundamental stellar properties.

The applications of photometry will still likely be directed into two broad categories: surveys of field stars and in photometry of star clusters.

One exciting prospect is that there will soon be the "ultimate" in stellar catalogs - surveys in one or more wavelengths of the entire galaxy, to a limiting magnitude of the order of 20 or even fainter. Indeed, such surveys are being seriously discussed at the present time. Sixteen years from now, it seems likely that one will be able to retrieve from a database the magnitudes and color indices (in some system) of many hundreds of millions of stars. In the area of stellar variability, there will likely be a continued development of precise studies of various sorts of stars. The science of stellar seismology holds considerable promise of producing a fundamental increase in our understanding of stellar structure. If photometric precision can be increased to something on the order of a few micromagnitudes, then it will be possible to monitor pulsations in solar-type stars with amplitudes of the order of a kilometer. Questions relating to star spots or stellar activity in general will be similarly accessible.

Finally, perhaps we will finally be able to answer the perennial question of the ages of the globular clusters.

Section 2

Poster Session

A CCD PHOTOMETER FOR THE VASSAR COLLEGE OBSERVATORY

F. R. Chromey

Vassar College

Mary Ellen Hunt

Bryn Mawr College

Monica Edelstein and Elizabeth Bonar

Vassar College

ABSTRACT: We have installed a CCD Camera on the 0.4 meter reflector at the Vassar College Observatory. A number of astrophysically interesting problems can be addressed with such an instrument, despite the small aperture of the telescope and the daunting sky conditions in Poughkeepsie. After two months of operation, we find the system capable of differential photometry in the UBVRI system with a precision of 1% and a practical limiting magnitude of V = 16.0.

This camera is of interest to astronomers because: (1) the total cost of the installation was relatively modest (on the order of $60,000, with similar systems available for perhaps one-third that amount); (2) the system is operated and maintained by undergraduate students and two astronomers at a small liberal arts college which has only rudimentary support facilities; (3) the system can be dedicated to synoptic observations that deserve attention but cannot be supported at national or international observatories.

We describe the hardware and software components installed and present the photometric observations of standard stars used to evaluate the system.

1. THE HARDWARE SYSTEM

With the aid of an ILI grant from the NSF (#8951146) we purchased a CC200 CCD Camera system from Photometrics, Inc, of Tucson, Arizona. The PM-512 chip, supplied by Photometrics with an ultraviolet enhanced coating, has 512 x 512 pixels, each 20 microns on a side (0.7 arc sec in the focal plane of our telescope). The chip is operated at liquid nitrogen temperatures, has a readout noise of approximately 1 ADU, and negligible dark current. Sky brightness at our site is about 0.8 ADU/sec in the V band. Full well potential is 16K ADU's.

A schematic of the hardware set up appears in Fig. 1. Data are first collected into RAM (10 frame capacity), then written to the hard disk of an IBM AT microcomputer. Frames are then transferred over the campus network to a DECstation 5000, where final data reduction is done using the IRAF package supplied by the National Optical Astronomical Observatories. The DECstation was purchased with the aid of a grant by the Keck foundation. Although it greatly speeds the reduction of images, it is not absolutely essential, since most simple reductions could be done on the PC/Photometrics computers.

The 15-inch telescope, an f/15 conventional Cassegrain, was installed in the old (1865) dome in 1958. It had not been used for some time when we attached the CCD camera this spring. We discovered a number of tracking difficulties, the most serious of which was a periodic error introduced by universal joints in the drive train. We relocated the drive motor and transmission outside the pier to correct this problem. Fig. 2 shows two 40-second unguided exposures of the globular cluster NGC 6838, one taken in May of 1990, before relocation of the drive transmission, the other taken in July just after relocation.

We rewired the telescope slow motion controls, and built an offset guider in the college shop, so that, except for object and guide star acquisition, observers can do most tasks from a heated control room, rather than the telescope dome. Our attempts to use a commercial autoguider (SBIG corporation) were frustrated by the very coarse slow motion motors. Until we can replace the slow motion motors, guiding is done manually at the telescope, or in the case of bright guide stars, in the control room using a video camera and monitor.

2. STANDARD STAR OBSERVATIONS

We observed red-blue pairs from the Landolt (1983) equatorial list to evaluate the photometric potential of the camera. We find the following differential transformation coefficients (Standard Johnson/Kron system on the left, instrumental system on the right):

$\Delta(U - B) = 1.057 \, \Delta(u - b) - $ (red leak)

$\Delta(B - V) = 1.030 \, \Delta(b - v)$

$\Delta V = \Delta v + 0.017 \, \Delta(b - v)$

$\Delta(V - R) = 1.139 \, \Delta(v - r)$

$\Delta(R - I) = 0.921 \, \Delta(r - i)$

A CCD PHOTOMETER FOR THE VASSAR COLLEGE OBSERVATORY

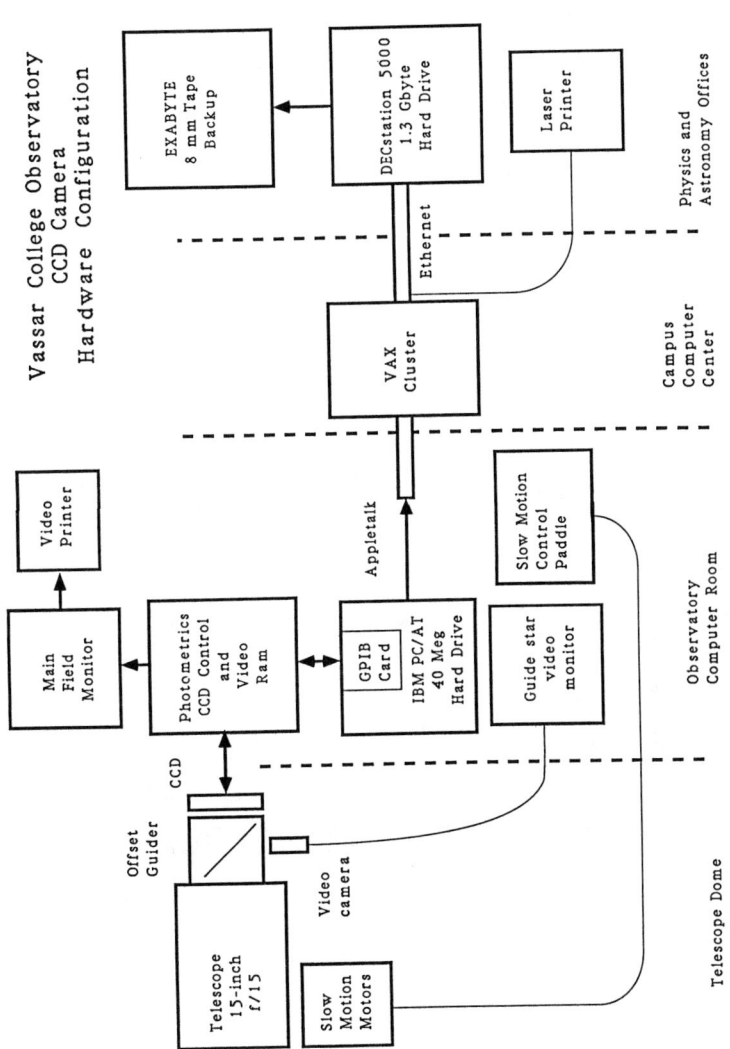

Fig. 1. A schematic of the hardware setup

Limits for stellar photometry are set by a relatively bright sky and poor seeing. For dark-of-the-moon operation, the seeing disk FWHM determines the V magnitude of the faintest star that can be measured. If we require a signal-to-noise ratio of 100 in a 15-minute exposure, the V magnitude limit is 16.3 for a FWHM of 2.5 arc sec and 15.0 for a FWHM of 4.5 arc sec (the range of typical seeing). These magnitude limits have recently been degraded by an increase in outdoor lighting on campus, and heat from the rest of the observatory building corrupts seeing to about 5 arc sec during the winter months.

REFERENCE

Landolt, A. 1983 AJ 88, 439

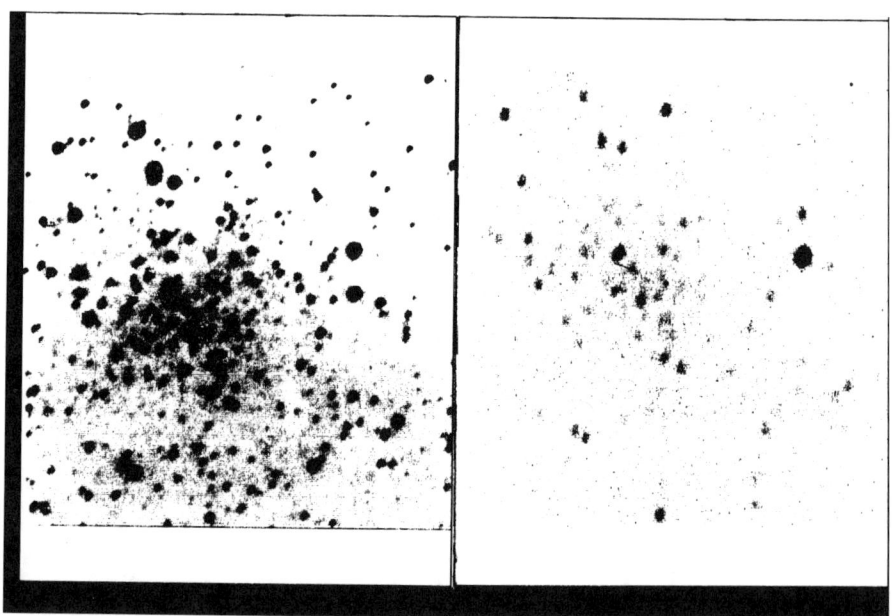

Fig. 2. 40 sec, unguided exposures of NGC 6838.
Left - July 1990, Right - May 1990.

DISCUSSION

YOSS: What are your plans for observations? How many observers are there in your group and how many nights per year can your group observe?

CHROMEY: The immediate plans for observing are to monitor supernovae light curves. We will work with seven other observatories at small colleges on the east coast, in a consortium funded by the Keck foundation on this project. We are considering a few other programs including star spots on T Tauri stars and light curves of asteroids.

We are absolutely restricted to differential work, but we can observe on approximately 50% of the nights. The schools in the consortium are scattered over two or three different weather patterns, so we hope for reasonably good coverage on the supernova project.

JANES: Is it difficult for students to learn the system?

HUNT: For an undergraduate it's not very difficult to learn how to work the system. For me it was helped by the fact that the system wasn't fully installed when I arrived. That way I was able to see how all the equipment worked and could ask specifics about each part as it came into the system rather than try to learn the whole thing at once. When I started I really didn't know anything about photometry but I found that by the end of a summer's worth of work the observing, the reduction of data really was not hard for me. So I don't think that for an undergraduate student it will be too much of a problem to learn what needs to be done.

UBV PHOTOMETRY OF OB+ STARS IN THE SOUTHERN MILKY WAY

John S. Drilling

Louisiana State University

ABSTRACT: 1226 new observations are combined with previously published results of the author to yield an internally consistent set of magnitudes and colors on the international UBV system for 666 stars classified as OB in the Stephenson-Sanduleak OB star survey. The (U-B), (B-V) diagram indicates that these stars consist primarily of O-type stars and early B-type supergiants, reddened by up to E(B-V) = 2.1 magnitudes.

The OB+ stars are hot stars for which the Balmer lines are very weak or absent at low spectroscopic resolution. These stars are interesting from the point of view of galactic structure because the O- and B-type supergiants, which define the spiral structure in other galaxies, are OB+ stars. They are also interesting from an astrophysical point of view because the following have all been classified as OB+ stars: hot hydrogen-deficient stars, very hot subluminous O-type stars, some dwarf novae, VV Cephei stars with luminous O- and B-type components, FG Sagittae, and nearly all of the optically identified X-ray sources which lie in the Milky Way.

The most complete survey of OB+ stars to date has been that of the Case and Hamburg Observatories (see Stephenson and Sanduleak 1971). This survey covers the entire Milky Way and is reasonably completely photographed to absolute magnitude V = 5. The author began a program to obtain slit spectra and UBV photometry for all stars classified as OB+ in the Case-Hamburg survey which are brighter than photographic magnitude 12.0 and for which such data have not been published. Because the UBV photometry could be obtained with small telescopes, and because it provided some advance information about the spectral types and enabled accurate exposure times to be estimated, the photometric observations were completed first. Results for 13 stars in a region in Norma and for 164 stars with 1950.0 declination greater than -15 degrees have already been published (Drilling 1972, 1975).

The present paper combines the results of 1226 new observations of stars classified as OB+ in Luminous Stars in the Southern Milky (Stephenson and Sanduleak 1971) with previously published results of the author (Drilling 1972, 1975) to yield an internally consistent set

A. G. Davis Philip, A. R.
Upgren and K. A. Janes (eds.)
PRECISION PHOTOMETRY
299 - 302 © 1991
L. Davis Press

John S. Drilling
Dept. of Physics and Astronomy
Louisiana State University
Baton Rouge, Louisiana
70803

of magnitudes and colors for 666 OB+ stars in the Southern Milky Way. The individual observations, heliocentric Julian Dates of observation, and airmasses at the time of observation have been archived at the National Space Science Data Center. Included are virtually all of the non-emission OB+ stars listed in Luminous Stars in the Southern Milky Way which are brighter than photographic magnitude 12.0, plus a number of the emission-line objects and fainter stars as well. In general, only one observation was made of the stars observed by Klare and Neckel (1977), or by Schild, Garrison and Hiltner (1983), for comparison purposes.

The new measurements were obtained with the Lowell 0.6-meter telescope and the 0.4-meter telescopes at the Cerro Tololo Inter-American Observatory during the period June 1972 through May 1976. In all cases, a standard UBV filter set was used in combination with a refrigerated 1P21 photomultiplier. Extinction coefficients, transformation coefficients, and night corrections were determined separately for each of the three instruments on each observing run using a total of 528 observations of standard stars selected from the lists of Johnson (1963) and Johnson et al. (1966). These stars were observed over the same range in color, air mass, hour angle, and declination as the program stars, and the reduction procedure described by Schulte and Crawford (1961) was followed with the modification that the extinction coefficient k4 was assumed to be zero. The mean errors of a single observation are ±0.014, ±0.011 and ±0.016 in V, (B-V) and (U-B), respectively, for the standard stars.

To check on the internal consistency of the observations, the differences between the values of V, (B-V) and (U-B) obtained with a particular instrument during a particular observing run and the mean of all of the other values obtained by the author for a particular star were plotted against (B-V), (U-B), airmass, hour angle and declination. This was done for the observations of Drilling (1972, 1975) as well as for the new observations, and in all cases, any systematic errors were found to be small compared to random mean errors of ±0.021, ±0.015, and ±0.018 in V, (B-V) and (U-B), respectively, for a single observation.

Comparison with the results of Klare and Neckel (1977) show that systematic differences are present which are not negligible with respect to the random errors. Klare and Neckel used the standard magnitudes and colors given by Gutierrez-Moreno et al. (1966), which define a different photometric system than that defined by the lists of Johnson (1963) and Johnson et al. (1966). Also, the procedures used by Klare and Neckel to handle the atmospheric extinction and transformation of the instrumental magnitudes and colors to the international UBV system are quite different from those described by Schulte and Crawford (1961). It therefore is not surprising that these systematic differences should appear for the OB+ stars, which lie on the edge of that region in the (U-B), (B-V) plane defined by the standard stars. The systematic differences are well fit by the following mean relations:

$$V - V_{kn} = +0.020(B-V)_{kn} - 0.047(B-V)_{kn}/ff/f12/fn/f0, \text{ etc.}$$
$$V - V_{kn} = +0.0298B = V9_{kn} - 0.047(B-V)_{kn}^2$$
$$(B-V) - (B-V)_{kn} = -0.079(B-V)_{kn} + 0.109(B-V)_{kn}^2$$
$$(U-B) - (U-B)_{kn} = -0.018 + 0.016(B-V)_{kn} - 0.042(B-V)_{kn}^2$$

Comparison with the results of Schild, Garrison and Hiltner (1983) show similar systematic differences. These authors used a mixture of Gutiérrez-Moreno et al. (1966) and Johnson (1963) standards, and reduced their data in a manner different both from that used in the present investigation and from that used by Klare and Neckel (1977), so it is again not surprising that these systematic differences are present. The differences are well fit by the following mean relations:

$$V - V_{sgh} = +0.021 + 0.009(B-V)_{sgh} - 0.020(B-V)_{sgh}^2$$
$$V - V_{sgh} = +0.021 + 0.009(B-V)_{sgh} - 0.020(B-V)_{sgh}^2$$
$$(B-V) - (B-V)_{sgh} = +0.007 - 0.022(B-V)_{sgh} + 0.025(B-V)_{sgh}^2$$
$$(U-B) - (U-B)_{sgh} = -0.040 + 0.008(B-V)_{sgh} - 0.039(B-V)_{sgh}^2$$

Most of the stars are seen to scatter about the reddening line for main-sequence stars of spectral class O5 - 7 in the (U-B), (B-V) diagram, as would be expected for O-type stars and early B-type supergiants. The two bluest of these are the sdO star LSS 1275 (Klare and Neckel 1977) and LSS 2018, the binary central star of the planetary nebula DS1 (Drilling 1985, Landolt and Drilling 1986). Three objects lie more than 0.4 magnitudes above the O5 - 7 reddening line: the dwarf nova LSS 2804 (Drilling 1990) and two VV Cephei stars with very luminous blue components, LSS 1916 and LSS 3371 (Drilling 1979). Stars which lie more than 0.4 magnitudes below the O5 - 7 reddening line are probably late B- and early A-type supergiants (Drilling 1975).

REFERENCES

Drilling, J. S. 1972 AJ 77, 463
Drilling, J. S. 1975 AJ 80, 128
Drilling, J. S. 1979 A&A 71, 214
Drilling, J. S. 1985 ApJL 294, L 107
Drilling, J. S. 1990, in preparation
Gutiérrez-Moreno, A., Moreno, H., Stock, J., Torres, C. and Wroblewski, H. 1966 Publ. Astr. Univ. Chile, 1, 1
Johnson, H. L. 1963 in Basic Astronomical Data, K. Aa. Strand, ed., p. 204
Johnson, H. L. 1966 ARAA 4, 193
Johnson, H. L., Mitchell, R. I., Iriarte, B. and Wisniewski, W. Z. 1966 Comm. Lunar Planet. Lab. 4, 99.
Hiltner, W. A. and Johnson, H. L. 1956, ApJ 124, 367
Klare, G. and Neckel, Th. 1977 A&AS 27, 215
Landolt, A. U. and Drilling, J. S. 1986 AJ 91, 1372
Schild, R. E., Garrison, R. F. and Hiltner, W. A. 1983 ApJS 51, 321
Schulte, D.H. and Crawford, D. L. 1961 Kitt Peak Natl. Obs. Contrib. No. 10
Stephenson, C. B. and Sanduleak, N. 1971 Publ. Warner and Swasey Obs. 1, 1

PHOTOMETRIC BOXES IN THE $UBVB_1B_2V_1G$ SYSTEM

M. Golay, B. Nicolet and N. Cramer

Geneva Observatory

The concept of "photometric boxes" was introduced by Golay et al. in 1969 and systematically applied since then by Golay et al. (1977, 1978). The exact definitions of the various types of box, the "elliptical" and "cubic" boxes, are given in an article by Nicolet (1981). Only elliptical boxes are used in the examples of Table 1. The size of the box is expressed by its radius R, which is equivalent to a spherical volume element in a n-dimensional "photometric" space.

Nicolet has developed a program (PHOTOM) which selects the stars within the volume of radius R (photometric box) centered on a given star identified by its name (Catalogue No, HD, BD, HR, name), or by its colors, or its color indices, or even by its reddening-free parameters. These various quantities are defined and given in the 1988 catalogue for each star. It is thus possible to analyze the neighborhood of a star in any space of two to six dimensions. The coordinates may be indices such as $(U-B_2)$, (B_1-B_2), (B_2-V_1) and (V_1-G), or the six normalized colors of the catalogue $(U-B)$, $(V-B)$, (B_1-B), (B_2-B), (V_1-B) and $(G-B)$, or the parameters d, Δ, g and m_2 (see definitions in 1988 catalogue) or even X, Y and Z (Cramer and Maeder 1979). Mixed spaces may also be used. We give a few examples in Table 1 with the three-dimensional d, Δ, (B_2-V_1) space. Such a space with at least one dimension a color, or color index, is sensitive to differences of interstellar reddening.

The PHOTOM program can be used with any version of the Geneva catalogue, extension or subset of data. An important subset (called PHOTHK) is that of the "Michigan Catalog of two-dimensional Spectral Types for the HD Stars", published by N. Houk, and for which we have the seven colors of 15,000 stars. Another sub-program (GGPHOT) involves the 1400 stars of the catalogue of spectral standards compiled by Goy (1991). PHOTOM can also make use of the colors computed by Kobi and North (1990) for Kurucz's grid of models. This sub-program (PHOTKUR) allows us to estimate T_{eff}, log g or [Fe/H], within a given approximation, for stars defined in the various manners mentioned above. Conversely, the program can select all the stars of the catalogue having colors or parameters within a close, and adjustable distance of a given model.

Table 1 shows a few photometric boxes for various types of stars. The sections of the table refer to the following types of stars:

1) Am star. Central star: HD 8374.   Am

2) Weak line star. Central star: HD 140283.   F7wl (Goy)

3) Horizontal-branch star. Central star: HD 161817.   A2VI H.B.

4) Peculiar stars of Mich. Cat. Central star: HD 132322.   ApSrCrEu

   Note the relatively large distance of the closest star; Ap stars are frequently loosely packed in photometric boxes.

5) λ Boo star. Central star: HD 111786. λ Boo type
   HD 6870 and HD 75654 are λ Boo (Renson, CDS Bulletin, in press).

6) Hot MK star. Central star: HD 36512.   B0 V

   A non-reddened star: $(B_2-V_1) = -0.326 = (B_2-V_1)_0$
   Hence the color excesses:

   HD 55879 $E(B_2-V_1) = 0.088$ $E(B_2-V_1) = 0.73\ E(B-V)$
   HD 265134   poor quality measurement
   BD 61°2550 $E(B_2-V1) = 0.521$
   HD 149757 $E(B_2-V_1) = 0.258$

7) T_{eff}, log g and [Fe/H] of the star HD 113139.   F2V (Goy)

REFERENCES

Cramer, N. and Maeder, A.   1979   A&A 78, 305
Golay, M.   1969   Publ. Obs. Genève, 76, série A
Golay, M., Mandwewala, N and Bartholdi, P.   1977   A&A 60, 181
Golay, M.   1978   A&A 62, 189
Goy, G.   1991, in press
Kobi, D, and North, P.   1990   A&AS 85, 999
Nicolet, B.   1981   A&A 104, 185

DISCUSSION

PHILIP: The filters and calibrations for the Geneva System can be found in articles in Problems of Calibration of Mutlicolor Photometric Systems, the proceedings of an earlier meeting held in Schenectady.

TABLE 1

Photometric Boxes for Various Types of Stars

```
Code  Rem T. Sp.    Nom      BS      HD       DM      Alpha 1950 Delta   12   b2
Bouk Sp. Qual    Vm Pv Sv    U       V        B1      B2       V1   G    Pc   Sc
     U -B2 B1-B2 B2-V1 B2-G  d       DELTA    g       X        Y    Z         dist.

10008374B    Am         47   And    395   8374 +36d 0237  1h20.7 +37d28  129.9 -24.7
                  5.591  6   4    1.505 0.618 0.979 1.397 1.338 1.742    6   5
          0.108-0.418 0.059-0.345 1.124 0.395 0.130 1.336-0.242-0.040

         Boîte elliptique sur couleurs U-B2, B1-B2, B2-V1, V1- G          Rayon 0.015

10208132D    Am+Am              8361 208132 +65d 1664 21h50.4 +65d31  105.8  +9.0
                  6.367  3   7    1.500 0.617 0.971 1.397 1.341 1.746    4   3         Part 1.
          0.103-0.426 0.056-0.349 1.138 0.393 0.124 1.342-0.227-0.039                   .011

10042968     Am                   42968 -30d 2917  6h10.5 -30d28  237.3 -21.2
AlmA5-F2    2  8.385   2   1    1.492 0.611 0.974 1.397 1.327 1.734    2   5
          0.095-0.423 0.070-0.337 1.123 0.375 0.129 1.320-0.234-0.033                   .014

10102660B    A3m                4535 102660 +17d 2402 11h46.7 +16d31  246.8 +71.9
                  6.048  3   3    1.517 0.616 0.974 1.396 1.332 1.739    3   4
          0.121-0.422 0.064-0.343 1.146 0.406 0.130 1.358-0.230-0.035                   .014


Code  Rem T. Sp.    Nom      BS      HD       DM      Alpha 1950 Delta   12   b2
Bouk Sp. Qual    Vm Pv Sv    U       V        B1      B2       V1   G    Pc   Sc
     U -B2 B1-B2 B2-V1 B2-G  d       DELTA    g       X        Y    Z         dist.

10140283     ABV    F3wf (Say)    140283 -10d 4149 15h40.4 -10446  356.4 +33.6
                  7.190  9  11    1.100 0.367 0.955 1.412 1.102 1.434   10   5
         -0.312-0.457 0.310-0.022 0.799-0.294-0.006 0.572-0.124-0.021

         Boîte elliptique sur paramètres d, Delta et g                    Rayon 0.015

00262606     (sdG1)              +26d 2606 14h46.9 +25455    36.6 +63.7
                  9.723  4   9    1.052 0.448 0.939 1.422 1.176 1.527    4   4
         -0.370-0.483 0.246-0.105 0.804-0.283-0.007 0.584-0.125-0.023                   .011   Part 2.

10200654     (G0)                200654 -50d13237 21h 3.1 -50d09  348.8 -41.9
(G)v          1  9.065   3   9    1.200 0.232 0.981 1.387 0.977 1.271    3   5
         -0.187-0.406 0.410 0.116 0.800-0.284-0.007 0.581-0.127-0.025                   .011

-00014318    (K)                      -01d 4318 22h31.9 +00-59   65.9 -47.8
                 11.058  3   4    1.133 0.339 0.968 1.407 1.074 1.390    3   7
         -0.274-0.439 0.333 0.017 0.793-0.288-0.010 0.573-0.129-0.028                   .014


10161817VR   A2VI                161817 +25d 3344 17h44.6 +25d47   50.4 +24.9
                  6.967 24  21    1.694 0.763 0.991 1.469 1.473 1.907   39   8
          0.225-0.568-0.004-0.438 1.605 0.589 0.021 1.749 0.090-0.007

         Boîte elliptique sur couleurs U-B2, B1-B2, B2-V1, V1- G          Rayon 0.015   Part 3.

10015801     AOIII                15801 -49d 0699  2h29.1 -49d36  269.0 -60.7
AO III/IV    3 10.402   3   1    1.711 0.784 0.907 1.477 1.488 1.926    3   5
          0.234-0.570-0.011-0.449 1.619 0.608 0.024 1.772 0.091-0.005                   .013


10132322     FOp                 132322 -63d 3473 14h53.3 -63432  316.6  -4.6
Ap SrCrEu     1  7.369   3   2    1.620 0.676 0.978 1.398 1.391 1.801    3   5
          0.222-0.420 0.007-0.403 1.243 0.557 0.136 1.537-0.225-0.048

         Boîte elliptique sur couleurs U-B2, B1-B2, B2-V1, V1- G          Rayon 0.030

10216823     Am        Tau 3Gru 8722 216823 -48d14364 22h51.0 -48d30  342.2 -59.5
A4mA5-F2      1  5.695   3   2    1.634 0.677 0.972 1.399 1.391 1.810    3   6
          0.235-0.427 0.008-0.411 1.273 0.577 0.142 1.570-0.213-0.039                   .019

10032035     (A3M)                32035 -72d 0334  4h55.2 -72d56  284.7 -34.4
A3mA5-F0      1  7.314   3   3    1.618 0.671 0.970 1.405 1.383 1.802    3   3          Part 4a.
          0.213-0.435 0.022-0.397 1.270 0.543 0.134 1.540-0.197-0.034                   .021

10077438     Ap SrEuCr            77438 -51d 3417  8h57.3 -52d04  271.8  -4.1
Ap EuCrSr     2 10.138   2   5    1.623 0.669 0.963 1.402 1.386 1.800    2   4
          0.221-0.439 0.016-0.398 1.288 0.552 0.123 1.556-0.182-0.039                   .024

10165040B    A7pSr    P1   Pav 6745 165040 -63d 4292 17h58.9 -63d40  330.5 -19.6
<A7p Sr           4.331  14  15    1.615 0.678 0.973 1.414 1.394 1.820   13   6
          0.201-0.441 0.020-0.406 1.273 0.539 0.137 1.538-0.196-0.028                   .028
```

```
10036060   A7m                    1827  36060 -41d 1884  5h23.9 -41d02   246.0 -32.7
A5mA5-F2    1  5.860   4   1  1.618 0.674 0.966 1.409 1.386 1.807    4    3
          0.209-0.443 0.023-0.398 1.285 0.540 0.128 1.545-0.182-0.031         .028

10011346   (A2)                  11346 -74d 0132  1h46.4 -74d35   297.7 -42.5
Ap SrEuCr   1  9.926   2  11  1.645 0.663 0.977 1.402 1.376 1.784    3    5
          0.243-0.425 0.026-0.382 1.276 0.561 0.129 1.557-0.198-0.043         .028

10189832   (F0p)                189832 -39d13583 19h56.9 -39d08   1.7 -30.0
Ap SrCrEu   1  6.902   3   8  1.597 0.654 0.980 1.391 1.367 1.782    3    3
          0.206-0.411 0.024-0.391 1.205 0.531 0.152 1.496-0.250-0.041         .030
```

Part 4a. Cont.

```
10132322   F0p                  132322 -63d 3473 14h53.3 -63d32   316.6  -4.6
Ap SrCrEu   1  7.369   3   2  1.620 0.676 0.978 1.398 1.391 1.801    3    5
          0.222-0.420 0.007-0.403 1.243 0.557 0.136 1.537-0.225-0.048         .028

         Boite elliptique sur parametres d, Delta et g             Rayon 0.030

10125555   (A0)                 125555 -74d 1168 14h15.0 -74d35   309.0 -13.3
Ap SrEuCr [< 1  9.397   3   1  1.695 0.555 1.002 1.384 1.274 1.659    3    3
          0.311-0.382 0.110-0.275 1.239 0.540 0.140 1.521-0.222-0.040         .022

10011346   (A2)                  11346 -74d 0132  1h46.4 -74d35   297.7 -42.5
Ap SrEuCr   1  9.926   2  11  1.645 0.663 0.977 1.402 1.376 1.784    3    5
          0.243-0.425 0.026-0.382 1.276 0.561 0.129 1.557-0.198-0.043         .027

10216823   Am        Tau 3Gru 8722 216823 -48d14364 22h51.0 -48d30  342.2 -59.5
A4mA5-F2    1  5.695   3   2  1.634 0.677 0.972 1.399 1.391 1.810    3    6
          0.235-0.427 0.008-0.411 1.273 0.577 0.142 1.570-0.213-0.039         .029

10132322   F0p                  132322 -63d 3473 14h53.3 -63d32   316.6  -4.6
Ap SrCrEu   1  7.369   3   2  1.620 0.676 0.978 1.398 1.391 1.801    3    5
          0.222-0.420 0.007-0.403 1.243 0.557 0.136 1.537-0.225-0.048         .028
```

Part 4b.

```
         Boite elliptique sur parametres d, Delta et B2-V1          Rayon 0.030

10032035   (A3M)                 32035 -72d 0334  4h55.2 -72d56   284.7 -34.4
A3mA5-F0    1  7.314   3   3  1.618 0.671 0.970 1.405 1.383 1.802    3    3
          0.213-0.435 0.022-0.397 1.270 0.543 0.134 1.540-0.197-0.034         .020

10216823   Am        Tau 3Gru 8722 216823 -48d14364 22h51.0 -48d30  342.2 -59.5
A4mA5-F2    1  5.695   3   2  1.634 0.677 0.972 1.399 1.391 1.810    3    6
          0.235-0.427 0.008-0.411 1.273 0.577 0.142 1.570-0.213-0.039         .022

10165040B  A7pSr        Pi  Pav 6745 165040 -63d 4292 17h58.9 -63d40 330.5 -19.6
<A7p Sr     4.331  14  15  1.615 0.678 0.973 1.414 1.394 1.820   13    6
          0.201-0.441 0.020-0.406 1.273 0.539 0.137 1.538-0.196-0.028         .025

10077438   Ap SrEuCr             77438 -51d 3417  8h57.3 -52d04   271.8  -4.1
Ap EuCrSr   2 10.138   2   5  1.623 0.669 0.963 1.402 1.386 1.800    2    4
          0.221-0.439 0.016-0.398 1.288 0.552 0.123 1.556-0.182-0.039         .029
```

Part 4c.

```
Code  Rem T. Sp.     Nom       BS     HD      DM     Alpha 1950 Delta   12   b2
Bouk Sp. Qual    Vm  Pv  Sv    U     V     B1    B2    V1    G      Pc   Sc
      U -B2 B1-B2 B2-V1 B2-G    d   DELTA    g     X    Y    Z              dist.

10111786   (A0)                  4881 111786 -26d 9369 12h49.3 -26d28  303.1 +36.1
A0 III      1  6.141   3   9  1.409 0.681 0.929 1.446 1.392 1.815    3    3
         -0.037-0.517 0.054-0.369 1.219 0.270 0.057 1.277-0.081-0.013

         Boite elliptique sur couleurs U-B2, B1-B2, B2-V1, V1- G    Rayon 0.020

10015287   (A0)                 15287 -29d 0884  2h24.7 -29d38   224.8 -68.9
A2 III      2  9.930   4   2  1.403 0.681 0.928 1.448 1.392 1.816    4    4
         -0.045-0.520 0.056-0.368 1.219 0.261 0.055 1.269-0.077-0.011         .005

10006870VR A2IIp    BS    Tuc    6870 -62d 0090  1h 6.1 -62d08   299.5 -55.1
A2 IIVp     2  7.470   6   6  1.401 0.671 0.928 1.443 1.384 1.802    8    4
         -0.042-0.515 0.059-0.359 1.209 0.257 0.052 1.260-0.080-0.016         .011

10075654   A5III                 3517  75654 -38d 4925  8h48.0 -38d58  260.5  +3.0
A2/3 II/III 1  6.373   3   7  1.434 0.672 0.934 1.446 1.387 1.807    3    3
         -0.012-0.512 0.059-0.361 1.232 0.288 0.058 1.299-0.080-0.014         .019

10111786   (A0)                  4881 111786 -26d 9369 12h49.3 -26d28  303.1 +36.1
A0 III      1  6.141   3   9  1.409 0.681 0.929 1.446 1.392 1.815    3    3
         -0.037-0.517 0.054-0.369 1.219 0.270 0.057 1.277-0.081-0.013
```

Part 5a.

```
         Boite elliptique sur parametres d, Delta et B2-V1          Rayon 0.020

10015287   (A0)                 15287 -29d 0884  2h24.7 -29d38   224.8 -68.9
A2 III      2  9.930   4   2  1.403 0.681 0.928 1.448 1.392 1.816    4    4
         -0.045-0.520 0.056-0.368 1.219 0.261 0.055 1.269-0.077-0.011         .006

10006870VR A2IIp    BS    Tuc    6870 -62d 0090  1h 6.1 -62d08   299.5 -55.1
A2 IIVp     2  7.470   6   6  1.401 0.671 0.928 1.443 1.384 1.802    8    4
         -0.042-0.515 0.059-0.359 1.209 0.257 0.052 1.260-0.080-0.016         .013
```

Part 5b.

PHOTOMETRIC BOXES IN THE $UBVB_1B_2V_1G$ SYSTEM

```
10111786     (A0)            4881 111786 -26d 9369 12h49.3 -26d28  303.1 +36.1
A0 III       1  6.141   3   9  1.409 0.681 0.929 1.446 1.392 1.815   3    3
            -0.037-0.517 0.054-0.369 1.219 0.270 0.057 1.277-0.081-0.013

     Boite elliptique sur parametres d, Delta et g              Rayon 0.015

10015287     (A0)                 15287 -29d 0884  2h24.7 -29d38  224.8 -68.9
A2 III       2  9.930   4   2  1.403 0.681 0.928 1.448 1.392 1.816   4    4
            -0.045-0.520 0.056-0.368 1.219 0.261 0.055 1.269-0.077-0.011        .008

10006870VR A2IIp   BS  Tuc        6870 -62d 0090  1h 6.1 -62d08  299.5 -55.1
A2 IIVp      2  7.470   6   6  1.401 0.671 0.928 1.443 1.384 1.802   8    4
            -0.042-0.515 0.059-0.359 1.209 0.257 0.052 1.260-0.080-0.016        .013

 Code Rem T. Sp.   Nom      BS     HD       Alpha 1950 Delta    12    b2
 Houk Sp. Qual    Vm  Pv Sv    U     V    B1    B2   V1    G    Pc  Sc
      U -B2  B1-B2  B2-V1  B2-G    d     DELTA   g     X    Y    Z         dist.

10036512   B0V     Ups Ori 1855  36512 -07d 1104  5h29.5 -07d21  210.4 -21.0
             4.604    5   7  0.131 1.281 0.758 1.644 1.970 2.532   5    8
            -1.513-0.886-0.326-0.888 0.640-0.774-0.123 0.073 0.019 0.003

     Boite elliptique sur parametres d, Delta et g              Rayon 0.007

10055879   O9.5II              2739 55879 -10d 1933  7h12.1 -10d13  224.7  +0.4
             6.003    4   5  0.196 1.166 0.768 1.613 1.851 2.384     4    3
            -1.417-0.845-0.238-0.771 0.636-0.776-0.122 0.071 0.016 0.003         .003

10265134   O9.0I              265134 +13d 1440  6h48.7 +13d41  200.8  +6.2
             9.166   1ᵖ  0ᵐ  0.280 1.042 0.788 1.593 1.742 2.242   0ᵐ  0ᵐ
            -1.313-0.805-0.149-0.649 0.643-0.773-0.127 0.076 0.023 0.001         .005

00612550   O9.5III                  +61d 2350 23h49.8 +61d51  116.1  +0.0
             9.277    2   7  0.376 0.602 0.862 1.513 1.318 1.704    2    4
            -0.937-0.651 0.195-0.193 0.645-0.776-0.124 0.074 0.024 0.005         .005

10149757   O9.5V   Zet Oph 6175 149757 -10d 4350 16h34.5 -10d28   6.3 +23.6
             2.358   10  15  0.356 0.955 0.818 1.588 1.656 2.135   10    6
            -1.232-0.770-0.068-0.547 0.639-0.777-0.120 0.071 0.018 0.006         .007

 Teff Log(g) [Fe/H] Mod      U     V    B1    B2    V1    G
 Houk Sp. Qual    U -B1 B1-B2 B2-V1 B2-G    d    DELTA   g     X      Y      Z dist.

10113139D    4.928   7   3  1.355 0.509 0.988 1.398 1.238 1.610    7    6
            -0.043-0.410 0.160-0.212 0.953 0.133 0.095 1.022-0.241-0.038

     Boite elliptique sur couleurs U-B2, B1-B2, B2-V1, V1- G      Rayon 0.040

7000.  4.00 0.50 Nouv Conv 1.395 0.523 0.988 1.394 1.248 1.625
            0.407-0.406 0.146-0.231 0.988 0.193 0.105 1.092-0.245-0.038  .035


10113139D    4.928   7   3  1.355 0.509 0.988 1.398 1.238 1.610    7    6
            -0.043-0.410 0.160-0.212 0.953 0.133 0.095 1.022-0.241-0.038

     Boite elliptique sur parametres d, Delta et g              Rayon 0.040

6500.  3.50 0.00 Nouv Conv 1.375 0.464 0.988 1.392 1.193 1.553
            0.387-0.404 0.199-0.161 0.965 0.117 0.085 1.014-0.220-0.036  .003

7000.  4.00 0.00 Nouv Conv 1.343 0.559 0.963 1.415 1.281 1.669
            0.380-0.452 0.134-0.254 1.026 0.139 0.074 1.064-0.179-0.027  .004
```

Part 5c.

Part 6.

Part 7a.

Part 7b.

Remarks:

Pv, Sv   Weight, and standard deviation for V in units of 0.001 m

Pc, Sc   Weight, and standard deviation for colors, units 0.001 m

dist.   Distance in magnitudes from the central star, or model of
           the box

A METHOD FOR SELECTING LAMBDA BOOTIS CANDIDATES FROM STROMGREN PHOTOMETRY AND OBJECTIVE PRISM SPECTRA OF FIELD STARS

Richard O. Gray

Appalachian State University

The λ Bootis stars are a rare class of chemically peculiar A-type stars. They are characterized (cf. Morgan et al. 1943, Hauck and Slettebak 1983, Abt 1984, Gray 1988) by a weak Mg II > λ 4481 I line, a K-line type of A0 or slightly later, and hydrogen-line types between A0 and F0. For their hydrogen-line types, the metallic-line spectrum is weak. It is generally believed from their space velocities that λ Bootis stars are Population I. Gray (1988) has pointed out that most λ Bootis stars show broad, although often shallow, hydrogen-line wings, and thus are dwarfs.

The λ Bootis stars constitute approximately 1% of the field A-type stars, and their discovery until now has been mostly a matter of chance. As a consequence, only about 18 stars of this class are known. While the use of photometric criteria can help to improve the probability of finding a λ Bootis star in a sample of field stars (cf. Hauck 1986, who derived valuable photometric criteria in the Geneva system), photometric criteria alone are not sufficient to narrow the search enough to make it efficient. This is made clear in Fig. 1. Nor can λ Bootis stars usually be distinguished on objective-prism plates. This is unfortunate, as we need a larger sample of known λ Bootis stars for statistical studies.

I have found that by combining photometric (uvby) data with objective-prism data (especially of the quality found in the work of Nancy Houk) it is possible to select a group of stars which have a much higher probability (about 25%) of being true λ Bootis stars. The procedure is as follows: Fig. 1 shows a box in the (b-y), m_1 plane which includes all of the known λ Bootis stars with Strömgren photometry. This box contains many other stars, especially in the blue end. These stars (cf. Gray and Garrison 1989), besides being apparently normal A-type stars, also include reddened B-type stars, rapidly rotating A-type stars, shell stars, Ap stars, evolved (supergiant) A-type stars, horizontal-branch stars and λ Bootis stars. The photometric discrimination can be somewhat improved by imposing the condition (satisfied by all the λ Bootis stars) that $0 < \delta c_1 < 0.30$.

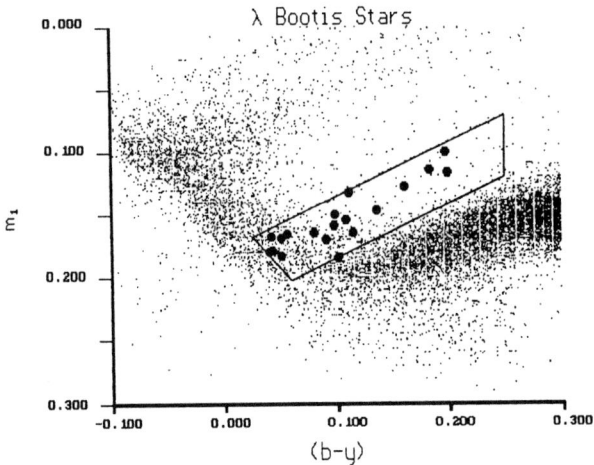

Fig. 1. m_1 versus (b-y) showing location of λ Bootis stars.

However, objective-prism spectral types can lead to an order of magnitude better discrimination. With objective prism types, B-type stars can be eliminated (the earliest type for λ Bootis stars is A0), and some of the evolved stars, especially the supergiant stars can be eliminated. The Ca II K-type for λ Bootis stars lies between A0 and A3, while the hydrogen-line type, if mentioned, may be as late as F0. The real discrimination comes, however, in the following observation. λ Bootis stars in the redward 3/4 of the box in Fig. 1 are classified, almost without exception, in the Michigan Spectral Catalogue (1975, 1978, 1982, 1988) with luminosity types of II, II/III and more rarely III, despite the fact that when their slit spectra are examined, they turn out to be dwarfs. The origin of this difference is that the broad, but shallow hydrogen-line wings of the λ Bootis stars cannot be detected on objective prism plates. Thus, with little loss of candidates, stars with luminosity types other than III, II/III and II can be eliminated. λ Bootis stars in the blue end of the box are almost impossible to distinguish from ordinary field stars by any method except slit spectroscopy. During my last trip to Las Campanas, I used these criteria to prepare a list of 17 λ Bootis candidates. One on this list was already a known λ Bootis star. Because of lack of time, only seven stars on this list were observed with the Garrison spectrograph, with a resolution of 1 Å. Two of these stars (HD 107233 and HD 142703) turned out to be λ Bootis stars. Two are shell stars, one a rapidly rotating star, and the other a possible field horizontal-branch star. The descriptions of these spectra can be found in Table 1.

I estimate that with existing photometric data, and the present volumes of the Michigan Spectral Catalogue, the number of known and confirmed λ Bootis stars could be at least doubled. When the Michigan Catalogue is extended to the northern hemisphere, the number could double again.

REFERENCES

Gray, R. O. 1988 AJ 95, 220
Gray, R. O. and Garrison, R. F. 1989 ApJS 70, 623
Hauck, B. 1986 A&A 154, 349
Hauck, B. and Slettebak, A. 1983 A&A 127, 231
Houk, N. 1978 Michigan Catalogue of Two-Dimensional Spectral Types for the HD Stars, Vol 2, University of Michigan
Houk, N. 1982 Michigan Catalogue of Two-Dimensional Spectral Types for the HD Stars, Vol 3, University of Michigan
Houk, N. 1988 Michigan Catalogue of Two-Dimensional Spectral Types for the HD Stars, Vol 4, University of Michigan
Houk, N. and Cowley, A. P. 1975 University of Michigan Catalogue of Two-Dimensional Spectral Types for the HD Stars, Vol 1, University of Michigan.
Morgan, W. W., Keenan, P. C. and Kellman, E. 1943 An Atlas of Stellar Spectra, University of Chicago Press, Chicago

DISCUSSION

DEMARQUE: How does the vsini distribution of the λ Bootis stars compare with that of the normal A stars?

GRAY: As far as I can tell from the very small sample of λ Bootis stars available to us (only 18 - 20 are confirmed members of this class) there is no distinction between the vsini distributions of normal A stars and λ Bootis stars. They do not seem to be particularly rapid rotators nor slow rotators.

PHILIP: Do λ Bootis stars exist among the Population II stars?

GRAY: Chris Corbally and I have observed spectroscopically a number of your high latitude A-stars. Among these stars we have found a handful of stars which look like λ Bootis stars, but we are not absolutely certain of these identifications because the S/N of our spectra was not quite sufficient to distinguish the spectral characteristics of this group with confidence. We plan to re-observe these stars at higher S/N.

STETSON: I would like to point out that high latitude does not necessarily make them Population II.

GRAY: Granted, many of these A-stars appear spectroscopically to have normal solar compositions. The fact that they are at high galactic latitudes and have high space velocities is very interesting indeed.

TABLE 1

SLIT SPECTRUM CLASSIFICATIONS OF λ BOOTIS CANDIDATES

HD	Sp T.	Notes
HD 34799	kA3hF0mA3 FHB?	Mg II 4481 is weak but visible Sr II 4077 is prominent, and probably indicates low gravity, hence FHB designation. Spectrum is generally similar to that of HD 161817, a FHB star, only slightly later.
HD 38616	A2 Ib Shell	Shell star.
HD 83277	A2 II Shell	Hydrogen lines show sweak hell cores, Mg I 4481 is very weak, Sr II 4077 and 4216 are both prominent, although not strong.
HD 94326	A3 IV nn	Normal, high vsini subgiant.
HD 97411	A0 IV	Mg II 4481 is slightly weak, but not abnormally so. Normal
HD 107233	kA1hF0mA1 Va λ Boo	Hydrogen lines show weak cores and broad and shallow wings. Mg II 4481 is very weak.
HD 142703	kA1hF0mA1 Va λ Boo	Hydrogen lines show cores similar to F0 stars, but with very broad and shallow wings. Mg II 4481 is very weak. Very similar to HD 107233.

PHOTOMETRIC ABUNDANCES FOR A AND F STARS

B. Hauck and S. Berthet

Institute of Astronomy of the University of Lausanne

1. INTRODUCTION

Photometry is undoubtedly one of the best ways to obtain a rapid estimation of the metallicity of stars. We can use photometric data, for example, to pick out Population II stars from Population I stars. We also need to know the metallicity in order to discuss the many problems related to stellar evolution. A. Maeder (1990) recently pointed out this aspect for the massive stars by presenting tables for assive star evolution at various metallicities. CCD photometry will play an important role in the study of star clusters, not only in our Galaxy but also in external galaxies. Knowledge of the metallicity will be an important factor for a correct interpretation of the data. Many photoelectric photometric systems are suitable for determining an estimation of the metallicity or at least segregating Population I from Population II stars. However we need good calibrations of the photometric parameters in terms of [Fe/H]. The Geneva and Strömgren systems are both suitable for this purpose.

2. CALIBRATION OF BLANKETING PARAMETERS IN TERMS OF [Fe/H]

Strömgren (1964) proposed the first calibration of the δm_1 parameter in terms of [Fe/H] for F8 - G2 stars. Many other relations have been established since, among which we would mention those of Crawford and Perry (1976), Nissen (1981) for F-type stars and Olsen (1984) for early G- and K-type stars. Hauck (1978) derived a relation between Δm_2 and [Fe/H] for F- and early G-type stars and pointed out that this type of relation cannot be linear. An improved relation was then published for the F-type stars (Hauck 1985). All the above-mentioned relations were established with main-sequence stars, but Berthet (1990a) has recently shown that on the basis of abundance determinations for F giant stars, this type of calibration can be extended to giant stars. On this basis, Berthet (1990b) has more recently undertaken to derive new calibrations of Δm_2 and δm_1 in terms of [Fe/H]. These new calibrations are based on a sample of 164 A - F main-sequence and giant stars for which [Fe/H] values are from the

Cayrel et al. (1988) catalogue. As the range in [Fe/H] is large ($-2.13 \leq$ [Fe/H] ≤ 0.5), quadratic relations have to be derived. The following were obtained:

$$[Fe/H] = -16.079 \, (\Delta m_2)^2 + 5.935 \, \Delta m_2 + 0.081$$
$$[Fe/H] = -35.139 \, (\delta m_1)^2 - 6.515 \, \delta m_1 + 0.081$$

3. APPLICATION TO THE WEAK-LINED F-TYPE STARS

In their survey of F-type stars, Jaschek et al. (1989) found 90 weak-lined F- and early G-type stars. On this basis they show that F-type stars with $\delta m_1 > 0.030$ are weak-lined stars and find that 25% of F-type stars are weak-lined stars. This sample has allowed Hauck et al. (1990) to compare the capability of the Geneva and Strömgren systems to detect weak-lined F-type stars. They first found a very good correlation for the whole sample between Δm_2 and δm_1 ($\Delta m_2 = -1.185 \, \delta m_1 + 0.001$, $r = 0.88$) and then considered all the F-type bright stars measured in the Geneva system. The sample consists of 442 stars of luminosity classes V, V-IV and IV, unresolved visual binaries being excluded. Considering stars with a Δm_2 value less than -0.034, they found that 24% of the sample consists of weak-lined stars. Jaschek et al. have found that 25% of their sample consists of weak-lined stars. They have also considered separately the early F stars (F0 - F4) and the late F stars (F5 - F9) and found that respectively 13% and 34% of each sample consists of weak-lined stars. For the Geneva system, Hauck et al. have found 16% and 32% respectively. These results lead us to conclude that the ability to pick out weak-lined F stars is very similar in both systems.

4. CONCLUSIONS

We repeat the conclusion of Hauck et al. (1990) that the Geneva and Strömgren systems are both very suitable for deriving estimations of [Fe/H] and that both are efficient for the detection of metal-weak stars.

REFERENCES

Berthet, S. 1990a A&A 227, 156
Berthet, S. 1990b A&A 236, 440
Cayrel de Strobel, G., Hauck, B., Thevenin, F., Frangois, F. and Mermilliod, M. 1988 A Catalogue of [Fe/H] Determinations, preprint
Crawford, D. L. and Perry, C. L. 1976 PASP 88, 454
Hauck, B. 1978 A&A 63, 273
Hauck, B., Foy, R. and Proust, D. 1985 A&A 149, 167
Hauck B., Jaschek, C., Jaschek, M. and Andrillat, Y., in preparation
Jaschek, M., Andrillat, Y. and Jaschek, C. 1989 A&A 218, 180
Maeder, A. 1990 A&AS 84, 139
Olsen, E. H. 1984 A&AS 57, 443
Strömgren, B. 1964 Astrophys. Norveg. 9, 333

TWENTY-FIVE YEARS OF UVBY-BETA DATA

B. Hauck and M. Mermilliod

Institute of Astronomy of the University of Lausanne

1. INTRODUCTION

The uvbyβ system is one of the most important photometric systems thanks to its numerous applications for the study of the B, A and F stars. It is also the second system, after the UBV system, in the number of stars measured.

The data are dispersed in the literature and are obtained with various types of equipment, thus the need to gather them into a single catalogue was expressed by many astronomers. We started to collect all the published data and in 1973 we were ready to publish the first version of the uvbyβ catalogue (Lindemann and Hauck, 1973). Since that time we have regularly updated this catalogue and the latest version, the fifth, will soon be published (Hauck and Mermilliod, 1990).

In fact we publish only a part of our work, that containing the homogeneous data obtained after a critical evaluation (cf. Lindemann and Hauck, 1973). The other part is our main file containing all the data found in the literature. As this part, as well as the first one, is available in machine-readable form from the Stellar Data Center of Strasbourg, it could also be useful for various purposes, as for example, the retrieval of all data published by one of a group of authors.

2. DATA GROWTH

Table 1 shows the growth of data since the publication of the first sets of data by Strömgren and Perry (1965) and Crawford et al. (1965), respectively.

TABLE 1.

Growth of data in the uvbyβ system

Year of publication	No. of Stars	No. of Statements	Publications
1965	1217	1217	76
1973	7600	10500	106
1975	9407	12900	223
1980	19884	31161	336
1985	40484	60016	414
1989	44896	69190	

3. WHICH STARS ARE MEASURED?

We give in this section some information based on the latest version of the catalogue. The distribution in magnitude is given in Fig. 1. 6008 stars (13%) belong to the Bright Star Catalogue, while 39441 (79%) stars are in the HD catalogue. If we consider the distribution of the bright stars by spectral type class, we see that nearly all bright O, B, A and F stars are measured in the Strömgren system.

REFERENCES

Crawford D. L., Barnes J. V., Faure B. Q. and Golson J. C. 1966
    AJ 71, 709
Hauck B. and Mermilliod M. 1990 A&AS, in press
Lindemann E. and Hauck B. 1973 A&AS 11, 119
Strömgren B. and Perry C. L. 1965 Institute for Advanced Study,
    Princeton, N. J., unpublished

DISCUSSION

DEMARQUE: Are the Geneva and Strömgren systems complementary or do they have overlap?

HAUCK: I think we can answer that both systems are complementary and have also some overlaps! For example the Geneva system is more suitable for cool stars. In the range of A - F stars we have overlap, however the Geneva system offers more possibilities for Ap stars and for other peculiar A and F stars due to the larger number of filters. The purpose of this study with M. Jaschek, C. Jaschek and Y. Andrillat was to consider the ability of both systems to detect weak-lined F stars. Our conclusion was that both systmes are well suited for that and offer similar ability. The catalogue is available on tape from the CDS and it will appear on microfiche in a next issue of A&AS.

TWENTY-FIVE YEARS OF uvby-BETA DATA

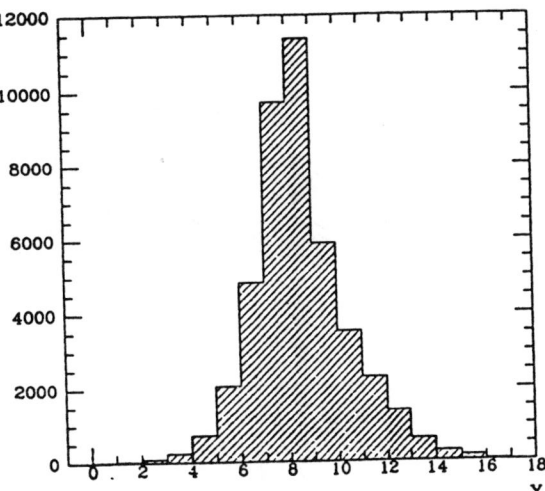

Fig. 1 Distribution in magnitudes

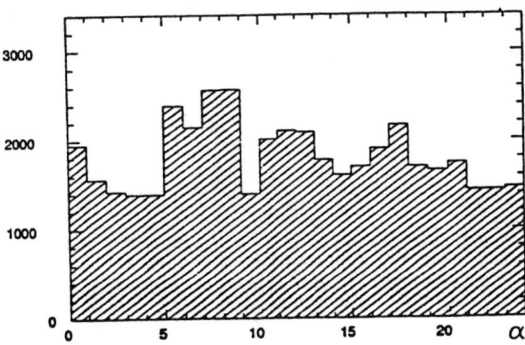

Fig. 2a Distribution in right ascension

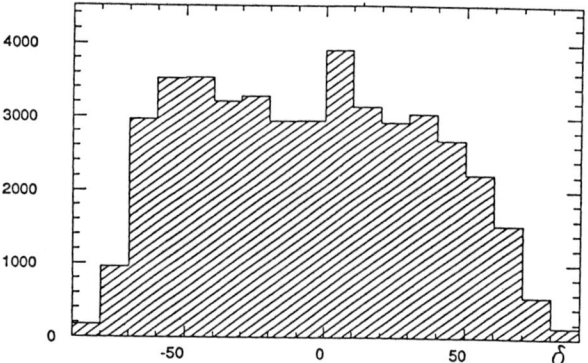

Fig. 2b Distribution in declination

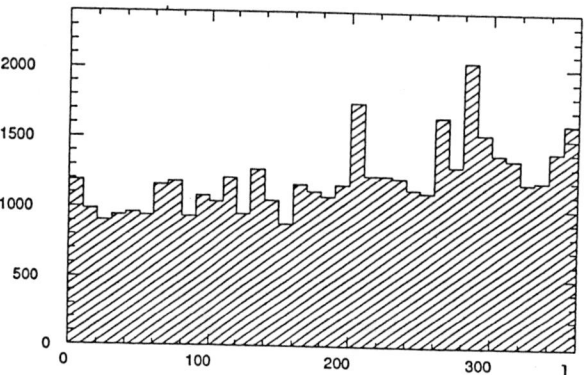

Fig. 2c Distribution in galactic longitude

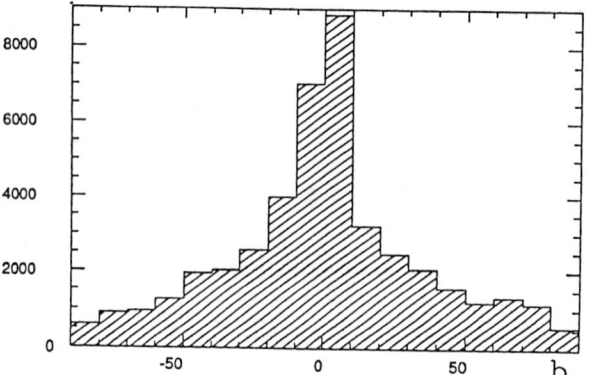

Fig. 2d Distribution in galactic latitude

DISTANCES AND AGES FOR A V STARS FROM uvby AND Hβ PHOTOMETRY

F. Figueras, J. Torra and C. Jordi

University of Barcelona

ABSTRACT: Several calibrations and methods to determine the physical parameters of main-sequence A-type stars from uvby and Hβ photometry are reviewed. The new algorithm elaborated allows an accurate determination of individual ages and distances, which will be used to analyze the kinematic properties of these stars.

1. INTRODUCTION

With the aim of analyzing kinematic properties of main-sequence A-type stars and the possible variation of kinematics with age (Gómez et al. 1990), we have undertaken a program, in collaboration with Meudon Observatory, devoted to obtain accurate individual photometric distances and ages from uvby and Hβ observations. Our sample contains main-sequence A-type stars up to tenth magnitude having accurate proper motions in the SIMBAD data base, spectral types compiled by Jaschek (1978) and measured radial velocities. To obtain uvby and Hβ photometry for those stars not included in the Hauck and Mermilliod (1990) compilation, an observing program has been undertaken at C.A.H.A. and O.A.N. Observatories (Spain). A first list of the photometric results has been published in Figueras et al. (1990).

Since the work of Strömgren (1966), several procedures and calibrations to determine the physical parameters through uvbyβ photometry have been elaborated. Before choosing one we have made a quasi-exhaustive check of the different methods and calibrations leading to an updated algorithm for the determination of absolute magnitudes, effective temperatures and surface gravities for main-sequence A stars. All this study has been extensively described in Figueras el al. (1990) and a brief summary is presented here. Also we present a preliminary comparison between the ages obtained using Vandenberg (1985) and Maeder and Meynet (1988) isochrones, which will allow us to analyze the influence of the stellar evolutionary models on the study of age-kinematic relations.

2. DETERMINATION OF PHYSICAL PARAMETERS

As indicated by Strömgren (1966), to obtain the physical

parameters of main-sequence A-type stars from uvbyβ photometry, it is necessary to make a correct assignment of the stars to the three categories: earlier than the Balmer maximum, near, and later than the Balmer maximum. Our classification is performed both, following the Strömgren (1966) criteria when using uvbyβ data and the Moon (1985a) criteria when using spectral type information. Furthermore, a procedure similar to that used by Philip et al. (1976) has been applied for discordant cases.

The complete procedure adopted for the determination of absolute magnitude, effective temperature and surface gravity, in each one of the three above mentioned regions, is presented in Table 1. The choice of the Moon and Dworetsky (1985) grids for the derivation of T_{eff} and log g instead of the ones of Relyea and Kurucz (1978), has recently been discussed in Torra et al. (1990). In that paper we applied our procedure to some stars belonging to the Hauck and Mermilliod (1990) compilation and having empirical determinations of T_{eff} and log g. The mean differences are -60 $\pm$180 K for T_{eff} (49 stars) and -0.03 $\pm$0.12 dex for log g (42 stars). We tested M_v determinations using stars from Gliese's catalog. Mean differences are 0.09 $\pm$0.22 mag having only 9 stars in common. We think it will be interesting to repeat this work with the new version of the Yale Parallax catalog, when available.

A first application of our procedure to the observations published in Figueras et al. (1990) shows that the agreement between the photometric values and the spectral information compiled by Jaschek (1978) has been excellent for more than ninety per cent of the sample. As an example, we show in Fig. 1 the $(b-y)_o$ vs. m_o intrinsic colors obtained for the subsample of 24 metallic stars classified spectroscopically. Only three stars fall above the Hyades standard relation from Crawford (1979), suggesting a normal metal content when compared with the Hyades. As expected, no peculiar Ap star (E(b - y) < -0.040) has been found in our sample and only few program stars show a negative color excess never exceeding -0.025. Nearly all the stars observed up to now satisfy $-0.010 < \delta m_o < 0.025$, indicating that they belong to Population I.

We have derived individual stellar ages using both, the Maeder and Meynet (1988) and Vandenberg (1985) evolutionary tracks. For each model two different third-order polynomials are fitted to obtain log (age) as a function of log T_{eff} and log g. These relations fit the evolutionary tracks to within 0.02 - 0.03 dex for Maeder and Meynet (1988) models in the interval 1.5 - 4 M_o, and 0.01 - 0.03 dex for Vandenberg (1985) models in the interval 1.5 - 3 M_o. When propagating this error and the error on the input parameters, our preliminary results indicate that the errors in log (age) are on the range 0.03 to 0.24 dex, which translates into a 25% mean uncertainty in the age of a given star. Although further analysis are required, the comparison of the ages derived from the two models indicates that a systematic

DISTANCES AND AGES FOR A V STARS FROM uvby AND Hβ PHOTOMETRY 321

TABLE 1

The procedure used to compute the astrophysical parameters

	Early Group	Intermediate Group	Late Group
Deredenned Indices - Reference Lines	Crawford (1978)	Hilditch et al. (1983) Moon (1985a,b)	Crawford (1979)
M_v, δM_{bol}	Balona and Shobbrook (1984)	Strömgren (1966) (a,r) grid	Guthrie (1987)
T_{eff}, log g	Moon and Dworetsky (1985) (c_o,β) grid	Moon and Dworetsky (1985) (a,r) grid	Moon and Dworetsky (1985) (β,c_o) grid Dworetsky and Moon (1986)

Fig. 1. m_o vs $(b - y)_o$ for 24 Am stars. The fully drawn line is the Hyades standard relation from Crawford (1979).

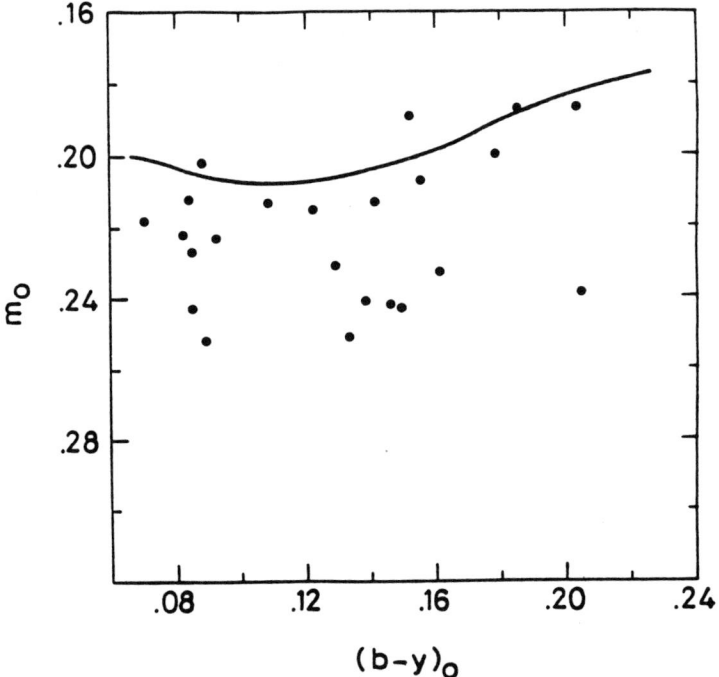

Fig. 1. m_o vs $(b - y)_o$ for 24 Am stars. The fully drawn line is the Hyades standard relation from Crawford (1979).

difference of 0.20 dex in log(age) - in the sense Maeder and Meynet minus Vandenberg - is found in all the range of our program stars.

ACKNOWLEDGEMENT

This work has been supported by the CICYT under contract ESP88-0731 and by the program of Acciones Integradas Hispano-Francesas.

REFERENCES

Balona, L. A., Shobbrook, R. R. 1984 MNRAS 211, 375
Crawford, D. L. 1978 AJ 83, 48
Crawford, D. L. 1979 AJ 84, 1858
Dworetsky, M. M. and Moon, T. T. 1986 MNRAS 220, 787
Figueras, F., Torra, J., Jordi, C. 1990 A&AS, in press
Gómez, A., Delhaye, J., Grenier, S., Jaschek, C. and Jaschek, M. 1990 A&AL 236, 95
Guthrie, B. N. G. 1987 MNRAS 226, 361
Hauck, B. and Mermilliod, M. 1990 A&AS 86, 107
Hilditch, R. W., Hill, G. and Barnes, J. V. 1983 MNRAS 204, 241
Jaschek, M. 1978 Inf Bull Strasbourg Data Center 15, 124
Maeder, A. and Meynet, G. 1988 A&AS 76, 411
Moon, T. T. 1985a Comm Univ of London Obs No. 78
Moon, T. T. 1985b, private communication
Moon, T. T. and Dworetsky, M. M. 1985 MNRAS 217, 305
Philip, A. G. D., Miller, T. M. and Relyea, L. J. 1976 Dudley Obs Rpt No 12
Relyea, L. J. and Kurucz, R. L. 1978 ApJS 37, 45
Strömgren, B. 1966 ARAA 4, 433
Torra, J, Figueras, F., Jordi, C. and Rossell, G. 1990 AP&SS 170, 251
Vandenberg, D. A. 1985 ApJS 58, 711

V, R, I, Hα PHOTOMETRY OF NGC 2264

C. B. Luginbuhl and F. J. Vrba

U. S. Naval Observatory
Flagstaff Station

S. E. Strom and K. M. Strom

University of Massachusetts

1. INTRODUCTION

NGC 2264 has long been recognized as a relatively nearby (approx. 800 pc) site of recent low-mass star formation. Walker's (1956) pioneering photometric and spectroscopic study, limited to $V \leq 17$, revealed that many stars below about two solar masses are still contracting to the main sequence. Adams, Strom and Strom (1983) (hereinafter AS^2) used primarily photographic and video camera photometry to extend the study of NGC 2264 to approx. $V = 22$. They found an additional 300 candidate pre-main sequence (PMS) stars on the bases of ultraviolet excesses, light variability, or Hα emission. While this database provides an extremely valuable tool for understanding low mass stellar evolution in NGC 2264, it is limited by the usual large corrections which need to be applied at all brightnesses to the photographic data to be placed on the standard photometric system and by video camera errors which exceed 0.10 mag below $V = 20$ mag. Due to these limitations, we chose to re-examine many of the AS^2 fields via a CCD-based V, R, I, Hα photometric survey with the aims of improving our understanding of the low mass star formation history and the evolution of Hα emission and variability in NGC 2264.

2. OBSERVATIONS

Observations were obtained using a TI 800^2 thinned CCD at the USNO, Flagstaff Station 1.0-m telescope, providing a field of view of 5.7 x 5.7 arcminutes. Ten fields in the NGC 2264 area were selected to avoid the brightest stars and to provide overlaps with the Walker (1956) and AS^2 databases. The VRI observations were made using HST WF/PC photometric filters with 2 and 15 minute exposures on each of two nights separated by a few days and a single set of 45 minute VRI exposures obtained approximately two years later. Each night's observations included typically 15 Landolt standards to allow transformation to the Kron-Cousins VRI system. The VRI observations were reduced using the DAOPHOT profile fitting routine, NSTAR. The nightly 2 and 15 minute observation pairs were combined into a single nightly VRI set leading, in most cases, to three VRI measures for each star on three different nights. The three nightly VRI sets were compared using χ^2 analysis to detect variability.

The Hα observations were obtained using a 30 Å FWHM filter with single 40 minute exposures for each field and not concurrent with any of the broadband observations. The instrumental Hα magnitudes were combined with the mean broadband R value for each star, forming a preliminary R-(Hα) instrumental color. The mode of these preliminary colors (due to the difference in bandwidth and average transmission between the R and Hα filters) was then subtracted to form the R-(Hα) colors. We estimate that the detection limit for Hα emission to be a Wλ of ≈ 5 Å.

The decreasing sensitivity of the CCD to the blue and the narrowness of the Hα filter meant that observations were not as deep in Hα and V as in R and I. The following table summarizes the final data set obtained with $\sigma \leq 0.20$ mag:

TABLE 1

Data Obtained With CCD

I, (R-I)	for 1671 stars
I, (R-I), (V-I)	for 1002 stars
I, (R-I), (V-I), (R-(Hα))	for 574 stars

3. RESULTS

All stars with measured VRI and Hα are plotted in a (R-I) versus (V-I) diagram in Fig. 1. Symbol size is proportional to (R-(Hα)). Here, as in all figures, the position in the diagram reflects the weighted average of all measures. The main sequence (MS), represented by dwarf colors, and reddened by E(V-I) = 0.102 mag, is shown for comparison. The spread of points lying to the left of the MS is consistent with contamination of R by H-α emission. Though the Hα equivalent widths in these stars can be expected to be highly variable, we find that our (R - (Hα)) values correlate reasonably well with the equivalent widths listed in Herbig and Bell (1988) over the range of $0.0 \leq (R - (Hα)) \leq 1.50$, $0 \leq Wλ \leq 125$. Our values of (R - (Hα)) of up to 2.6 then imply Wλ of approx. 225 Å. We note that an emission line with an equivalent width of 225 Å observed through our R filter (947 FWHM) would be expected to result in a blueward displacement in (R-I) of 2.5log[(225 + 947)/947] = 0.23 mag, which is typical of displacements seen in Fig. 1. Though in individual cases we cannot demonstrate that the blueward displacement in the color-color diagram is the result solely of Hα emission, since our Hα measures were not taken concurrently with the broadband R measures, it is consistent with the sometimes very strong emission. Figure 2a shows the (V-I) versus I

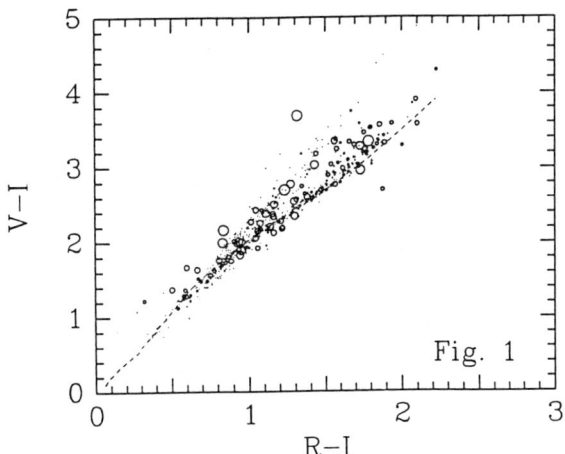

Fig. 1. (R-I) vs (V-I) with symbol size proportional to (R-Hα).

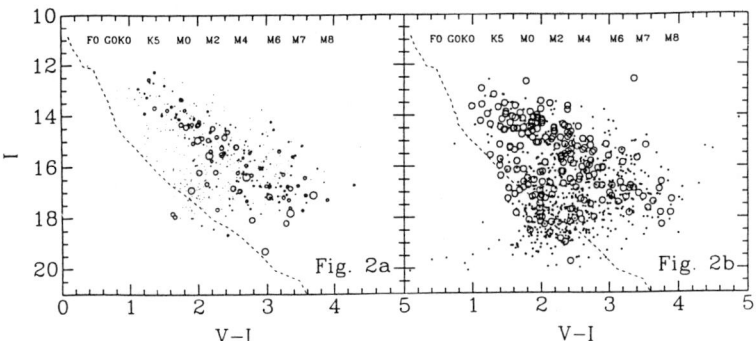

Fig. 2a. (V-I) vs I symbols as in Fig. 1.

Fig. 2b. (V-I) vs I for stars with more than one measure. Circles represent variable stars, dots the others.

color-magnitude diagram for stars and a reddened MS [$(I-M_I) = 9.85$] with symbols as in Fig. 1. The (V-I) color index is uncontaminated by emission at Hα and so provides a better comparison with the MS. Here a well populated locus of stars lying about three magnitudes above the MS is clearly seen, as is a more thinly-populated region of older PMS stars approaching the MS, and some presumably background stars lying below the MS. The majority of the stars lying in the upper group show weak or strong Hα emission. A number of the older stars approaching the MS, redder than (V-I) approximately 2, show strong emission, though the proportion is small. The CM-diagram for all stars with more than one measure (and therefore a variability estimate) in I is plotted in Fig. 2b. Circles represent the clearly variable stars ($\chi^2 \geq 10.0$), while dots are plotted for the remaining stars ($\chi^2 < 10.0$). Here many more stars appear than in Fig. 2a since many stars were not detected at Hα. The fainter limit of I approximately 19 at (V-I) = 2 reflects a V limit of about 21. The younger PMS group is clearly discernable here, as is a narrow gap on the fainter edge before the older PMS group begins to appear. This gap suggests that NGC 2264 has undergone episodic star formation. Variable stars spread across both the younger and older PMS groups. Star by star comparison of Figs. 2a and 2b shows that the youngest group of stars in this cluster is characterized by stars showing strong Hα emission, by stars exhibiting variability at I, and by stars showing both characteristics. As the stars descend toward the MS, the proportion showing strong Hα emission decreases dramatically, while the proportion showing variability does not change obviously; many more become non-emitting variables. The variability shows no preference for particular masses over the range observed here (about 2.0 - 0.1 solar masses), though strong Hα emission is undetected in the older PMS stars for (V-I) < 2 (M > 0.5 solar masses). We conclude that the characteristic light variability of PMS stars persists longer than strong line emission for a broad range of stellar masses.

REFERENCES

Adams, M. T., Strom, K. M. and Strom, S. E. 1983 ApJS 53, 893
Herbig, G. H. and Bell, K. R. 1988 Lick Obs. Bull. No. 1111
Walker, M. F. 1956 ApJS 2, 365

DISCUSSION

DEMARQUE: What kind of information do you learn about the luminosity function?

LUGINBUHL: This is one of the principal aims of our study, but one that is awaiting a more complete analysis of reddening and the collection of a consistent set of pre-main-sequence evolutionary tracks. When we have achieved these we will examine the mass function as a function of age and the changes in pre-main-sequence characteristics as stars approach the main sequence.

COMPARISON OF TWO SEMI-EMPIRICAL CALIBRATIONS OF THE uvbyβ SYSTEM

P. North and D. Kobi

Institute of Astronomy, University of Lausanne

ABSTRACT: We compare the calibration of the uvbyβ system in terms of T_{eff} and log g proposed by Moon and Dworetsky (1985) with that proposed by Lester et al. (1986). We show that systematic differences exist, and stress the importance of using numerous and reliable reference stars to correct the synthetic colors computed from model atmospheres.

1. INTRODUCTION

The uvbyβ system is among the most widely used photometric system for obtaining the physical parameters of stars. The synthetic colors obtained from the atmosphere models of Kurucz (1979a,b) have to be transformed to the standard system using standard stars with known T_{eff} and log g, and this has been attempted especially by Moon and Dworetsky (1985, hereafter MD) and by Lester et al. (1986, hereafter LGK). The results of these two papers are compared in order to highlight the limits of the method and to show how it can be improved.

2. COMPARISON OF THE TWO CALIBRATIONS

The diagrams c_1 vs β (Fig. 1 and 3) and r^* vs a_0 (Fig. 2) show in superposition the grids of each work, for cool, intermediate and hot stars. The agreement between the grids of MD and of LGK is good (i.e. $\Delta T_{eff}/T_{eff}$ < 1% and Δlog g < 0.1 dex) in the range T_{eff} = 7500 to 12,000 K and log g = 3.5 to 4.5. However, it becomes very unsatisfactory outside this range: at T_{eff} = 18,000 K, we have Δlog g = 0.25 dex for log g = 4.5, and 0.4 dex for log g = 3; at T_{eff} = 6000 K, we have ΔT_{eff} = 250 K, i.e. 4% and Δ log g = 0.2 to 0.25 dex around log g = 4.0. These differences are larger than the observational uncertainties, so the following question naturally arises: which grid is closer to reality, and how can the method of semi-empirical calibration be improved?

3. DISCUSSION

LGK use only 5 standard stars for fitting 3 parameters; MD use 19 standard stars (although some of them do not have a fundamental value of either log g or T_{eff}), so that their calibration appears more reliable a priori.

The calibrations are tested using the method proposed by North and Nicolet (1990) and by Kobi and North (1990), i.e. by means of cluster stars. Members of the Pleiades are used for the "cool" part of the calibrations (Fig. 1) and members of the Orion OB1 association for the "hot" part (Fig. 3). Since these clusters are young, the stars are essentially unevolved and are very good surface gravity standards, as far as we can trust the internal structure models of Maeder and Meynet (1988): near the ZAMS, log g is a very slow function of mass (or equivalently of T_{eff}) and remains close to 4.3, as indicated on Figs. 1 - 3 under the tick marks. The ZAMS defined by Crawford 1975, 1978 and 1979 on the basis of field stars agrees fairly well (within about 0.1 dex in log g) with the cluster data, while that of Hilditch et al. 1983 (Fig. 2) is less satisfactory.

Philip and Relyea (1979) applied roughly the same idea to the c_0 vs $(b - y)_0$ grid for $T_{eff} < 8500$ K, but using Crawford's ZAMS for F stars instead of individual cluster members.

4. CONCLUSION

• Both calibrations tend to overestimate log g for $T_{eff} < 6500$ K and to underestimate it for $T_{eff} > 12,000$ K, but that of MD is better in the latter range.

• It is possible to improve the calibration, at least concerning log g, by using cluster members as standard stars and relying on the internal structure theory. This work is underway (Kobi and North, in preparation).

• The range of log g of the standard stars is not yet large enough to allow very good surface gravity determinations for evolved stars.

REFERENCES

Crawford, D. L. 1975 AJ 80, 955
Crawford, D. L. 1978 AJ 83, 48
Crawford, D. L. 1979 AJ 84, 1858
Hilditch, R. W., Hill, G. and Barnes, J. V. 1983 MNRAS 204, 241
Kobi, D. and North, P. 1990 A&AS, in press
Kurucz, R. L. 1979a ApJS 40, 1
Kurucz, R. L. 1979b in Problems of Calibration of Multicolor Photometric Systems, A. G. D. Philip, ed., Dudley Obs Rept No 14, p. 271
Lester, J. B., Gray, R. O. and Kurucz, R. L. 1986 ApJS 61, 509
Maeder, A. and Meynet, G. 1988 A&AS 76, 411
Moon, T. T. and Dworetsky, M. M. 1985 MNRAS 204, 241
North, P. and Nicolet, B. 1990 A&A 228, 78
Philip, A. G. D. and Relyea, L. J. 1979 AJ 84, 1743

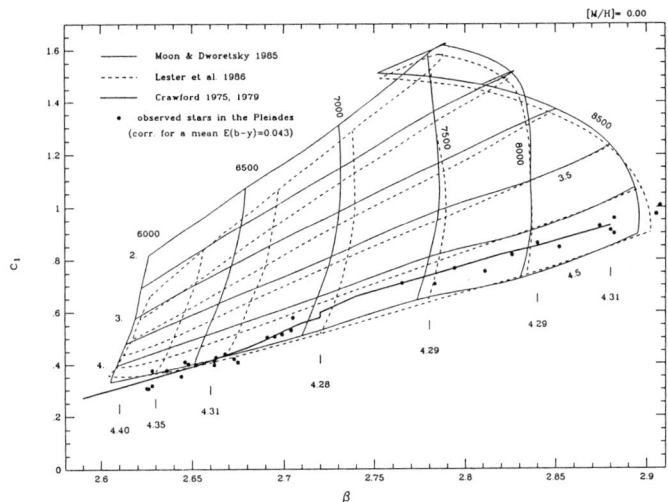

Fig. 1. Comparison of the MD and LGK calibrations for cool stars of solar metallicity. The T_{eff} and log g values are given near the corresponding lines. The ZAMS defined by Crawford is shown for comparison, as well as the Pleiades' stars, whose log g values given by stellar structure theory, are written under the tick marks. Notice that the T_{eff} = 5500 K line of LGK coincides with the T_{eff} = 6000 K line of MD.

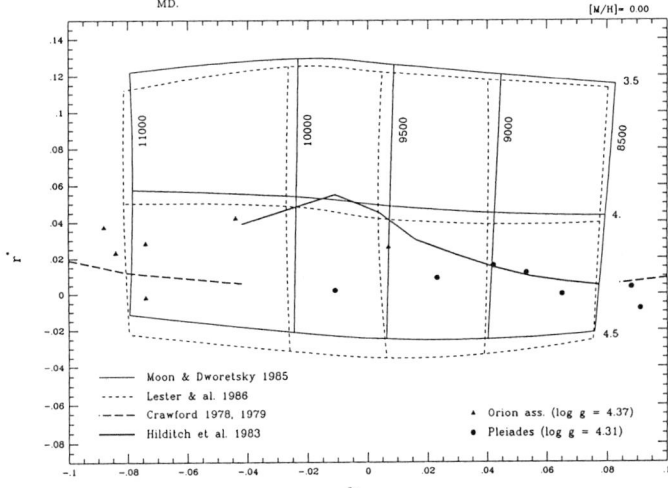

Fig. 2. Same as Fig. 1, but for intermediate (i.e. B9 - A3) stars. A few (individually dereddened) stars of the Orion OB1 association are shown as well. Notice the bad fit between the ZAMS of Hilditch et al. (1983) and those of Crawford.

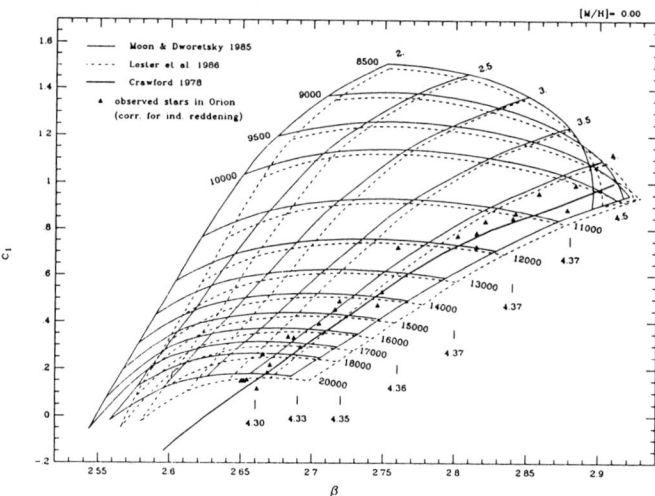

Fig. 3. Same as Fig. 1, but for hot stars and members of the Orion association.

PHOTOMETRY OF THE YOUNG OPEN CLUSTER NGC 581

Randy L. Phelps and Kenneth A. Janes

Boston University

ABSTRACT: We have begun a systematic photometric investigation of a group of young open clusters in the Perseus spiral arm. This study will yield information on the time scale of cluster formation in this region of the galaxy, and whether or not cluster formation occurred randomly or sequentially along a preferred direction. In addition we expect to gain insight into the star formation process itself by examining the differences in each cluster's luminosity function (Zinnecker 1987). We report here the preliminary results of our investigation of one of the clusters in our program, NGC 581.

1. DATA

The data were obtained with the Kitt Peak National Observatory's No. 1 0.9-m telescope, operating at f/7.5 and using the TEK 1 CCD, on observing runs in November, 1988 and September, 1989. This setup gives a scale of 0.77 arc-seconds/pixel and a field of view of 6.6 arc-minutes. The KPNO Harris B and V filters, a standard KPNO $CuSO_4$ + UG2 U filter and a KPNO Mould interference I filter were used.

Due to the relatively young age of NGC 581, there typically was at least one bright (V ~ 9) B-type star in the field of view. Multiple short exposures (20 - 300 seconds), with exposure times depending upon the presence of brighter stars and the filter being used, were co-added when necessary in order to achieve a total integration time of 10 minutes for the B, V and I filters and 15 minutes in the U filter. Landolt standard stars were observed for calibration.

2. REDUCTIONS

Bias subtraction and flat-fielding were done with the NOAO IRAF package. Sky flats were used exclusively for the U and B images and for the V and I frames when they were available. When sky flats for the V and I frames were not obtained, dome flats for these filters were used. The photometry was done using a point-spread function photometry package written by one of us (KAJ) along with James Heasley of the University of Hawaii.

3. RESULTS

Our photometry shows NGC 581 to have a mean reddening of $E(B-V)$ = 0.44, using the reddened ZAMS line of FitzGerald (1970) (solid line in Fig. I). The ZAMS stars were reddened according to the reddening lines of Crawford and Mandwewala (1976).

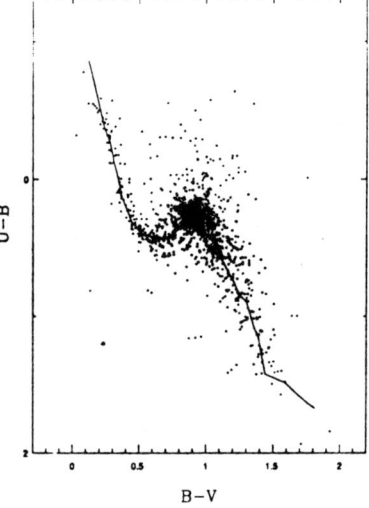

Fig. 1. Two-color diagram of NGC 581. See text for details.

The CM-Diagram of Fig. 2a consists of our photometry and Hoag's (1961) photometry of the 6 brightest stars, which were saturated on our images. No systematic difference in the unsaturated stars in our data, which were also observed by Hoag (1961), were apparent. A fit of the Schmidt-Kaler (1982) ZAMS, assuming $E(B-V) = 0.44$ and $(m-M)_o = 13.55$ is also shown as the solid line. Thus the cluster is at a distance of ~2700 ±100 pc, or $(m-M)_v$ = 12.15 ±0.1. A value for the ratio of total to selective absorption, A_v, of 3.1 was assumed. These values are slightly different than those reported in Phelps and Janes (1990a) due to the new data on the cluster, with improved calibrations, and the use of a different ZAMS.

The CM-Diagram of four regions (Fig. 2b) away from the cluster is used to correct for field star contamination in the construction of the luminosity function. The derived luminosity function (Fig. 3) shows no clear evidence of a turnover at lower luminosities, to the level where the photometry is reasonably complete. The possible existence of such turnovers has been used to argue for the concept of Bimodal Star Formation, where high mass stars are formed in different regions than low mass stars (cf. Larson 1986).

PHOTOMETRY OF THE YOUNG OPEN CLUSTER NGC 581

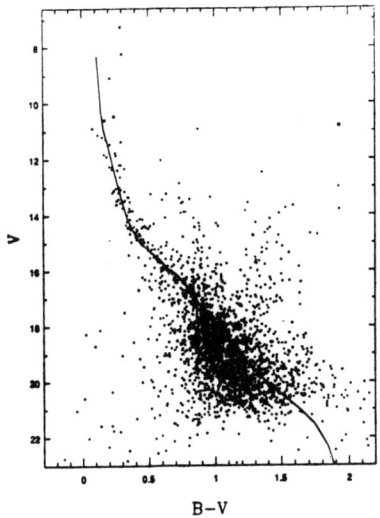

Fig. 2a. The CM-Diagram for NGC 581.

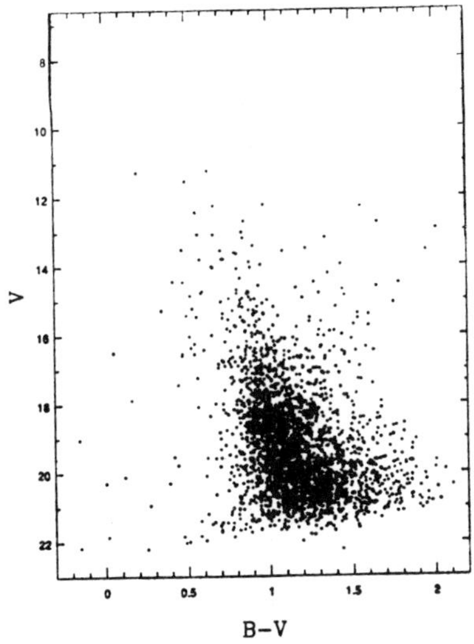

Fig. 2b. The CM-Diagram for four regions. See text for details.

The turnover observed in Fig. 3 can be explained by incompleteness of the photometry at the fainter magnitudes. A more detailed analysis of completeness will be undertaken in the future with a yet to be implemented fake star routine in the photometry software.

The chosen normalization of the Scalo (1986) luminosity function shown in Fig. 3 does, however, indicate that the cluster contains ~solar mass stars in numbers consistent with a field star luminosity function. The excess number of stars at higher luminosities is due, in part, to evolutionary effects, as well as to a true excess of such stars.

REFERENCES

Crawford, D. L, and Mandwewala, N. 1976 PASP 88,917
FitzGerald, M. P. 1970 A&A 4, 234
Hoag, A. 1961 Pub US Naval Obs 17,349
Larson, R. B. 1986 MNRAS 218,409
Phelps, R. L. and Janes, K. A. 1990a in CCDs in Astronomy. II.
    A. G. D. Philip, D. S. Hayes and S. J. Adelman, eds., L. Davis Press, Schenectady, p. 257
Scalo, J. 1986 Fund Cos Phys 11, 144
Schmidt-Kaler, T. 1982 in Landolt-Bornstein VI, Vol. 2b, Springer-Verlag, Berlin
Zinnecker, H. 1987 Pub. Astr. Inst. Czech Acad. Sci. 69, 77

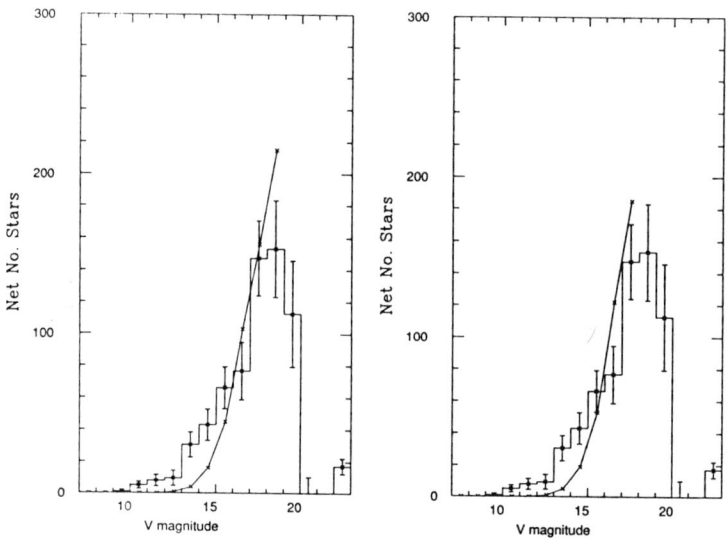

Fig. 3a. Luminosity function for NGC 581 (histogram), from V magnitudes binned in one magnitude intervals, and subtracting the similarly binned V magnitudes of the field region frames, corrected for area differences. Error bars are one σ errors from Poisson statistics. The solid line is the Scalo (1986) luminosity function, normalized to the net number of presumed cluster stars observed with V < 19, shifted according to the derived cluster distance and reddening, $(m-M)_0$ = 13.55.

Fig. 3b. Same as in Fig. 3a except that the normalization for the Scalo (1986) luminosity function is for the net number of presumed cluster stars observed with V < 18, which is likely to be a more complete sample.

DISCUSSION

YOSS: Do the dips in $\Phi(M)$ correspond to $M_v \sim 8$ as is found in the general luminosity function? And could that be used as a distance indicator?

PHELPS: No, the possible dip at magnitude 16 - 17 corresponds to an absolute magnitude at $M_v \sim 3$. Dips at similar absolute magnitudes have been observed in several clusters. I am not sure that I would trust such a dip as a distance indicator due to the limited statistics included.

AUTOMATIC PHOTOMETRIC TELESCOPES AND PHOTOMETRIC ACCURACY

C. Sterken,

Astrophysical Institute, University of Brussels

J. Manfroid

Institute of Astrophysics, University of Liège

ABSTRACT: By means of the mean standard deviations of differential uvby magnitudes of constant stars, we discuss some effects which may negatively influence the precision of automatically performed measurements.

1. AUTOMATIC TELESCOPES AND DIFFERENTIAL PHOTOMETRY

Automatic telescopes represent a novel concept leading to a radically new way of planning and conducting observations. This is best illustrated in photoelectric photometry where the human factor is one cause of errors. Man, with his slow reaction time and high susceptibility to fatigue, certainly cannot compete with a computer and with ultrafast equipment. In manually conducted photometric observations, most of the time is spent with the photometer in idle status, when the observer moves the telescope to the next star, or when the observer is identifying or centering the object, or when he is planning the rest of the night. Above all there is the problem of manpower: for each telescope in operation a skilled observer is needed all year round, and this is a major limitation on the total number of measurements that can be made.

Short integration time and short time intervals between successive measurements are essential for high accuracy photometry: this is the only way to eliminate the effect of short term variations in the atmosphere. This is thoroughly discussed by Young et al. (1990) who investigated the feasibility of reaching millimagnitude accuracy in photometry. They suggest 10 sec as an adequate timescale (for every filter). Each band should be measured separately under the same conditions. Fast measurements also mean that a lot of measurements can be made each night, and this means that it is much easier to incorporate many more standard and constant star measurements. This in turn leads to more consistent reductions and higher accuracy and homogeneity in the data.

To benefit from the greater accuracy of the observations, the comparison stars have to be very stable. Little is known about variability at a level of a few millimagnitudes, but the general feeling is that many - most - stars are variable. Consequently Young et al. correctly suggest that one should observe many comparison stars. After a long series of observations the constancy of some of them could be established, and in the long term this would lead to the establishment of extensive lists of reliable comparison stars. In this case only one or two of them would be necessary for differential photometry. But for the time being one should consider large sequences of comparison stars in each field, and accept the penalty of long sequences. If all comparisons in one field do vary, some kind of average value should be adopted for the microvariables with the smallest amplitudes in order to obtain a lower noise level, and the long sequences should be kept forever.

Sequences of variable and comparison stars should have a rapid turnaround time. Hence each star should ideally fall in the right magnitude interval, and if this is not possible, the faintest star sets the number of integrations for the whole sequence. But this may mean that the time interval between the program star and most comparison stars (or between any comparison pairs) would be excessively long when compared to the suggested 10 seconds.

Automatic telescopes, specifically built for observing without human assistance, will always have an edge over conventional telescopes, even over those which are computer controlled, since they hop more quickly from star to star. The ratio of the time spent outside integrations, to the time during integrations, is a good assessment of the efficiency of the equipment. Dedicated telescopes, such as the APTs, are optimal in this respect since they take only a few seconds for pointing and centering. Automated telescopes on the other hand suffer from their large inertial momentum and need tens of seconds to find and center a star. Roughly speaking, half of the possible observing time is wasted, and the interval between successive star pointings is also roughly doubled, so that larger changes in the atmospheric conditions will intervene. We do not say that renovating existing telescopes with automatic control is not interesting (we describe below such a system), but there are many arguments against it. The costs should be compared with those of buying cheap, reliable APTs which are commercially available. The advantage of the APTs in terms of efficiency but also of floor space (several 1-m APTs use as little space as one 50-cm conventional telescope) and maintenance should be considered. Moreover, running automatic telescopes requires a huge effort to develop the programming language. Very good and versatile software exists for the APTs, with ingenious communication and networking systems. The cost of re-inventing the wheel for one's own telescope is certainly prohibitive.

2. OUR EXPERIENCE WITH AN AUTOMATIZED MANUAL TELESCOPE.

The SAT (Strömgren Automatic Telescope, Florentin Nielsen et al. 1987) is the name given to the ESO Danish 50-cm telescope located at the European Southern Observatory at the La Silla site in Chile, after it was refurbished and provided with full computer control. The SAT has now been used for several years with considerable success. It is essentially a mission instrument where as a rule each observer gets a few weeks observing time per run. A rather flexible programming language was developed, and it is the responsibility of the user to code the observing sequences for the night. There is complete freedom in that respect, so that each observer will program the telescope in his own way. The result is that the SAT is functioning essentially in the same way as before automation, but it is faster and has a larger output. However, when compared to dedicated APTs the typical setting time of 30 seconds is pretty long.

A big advantage of the SAT (over any existing APT) is its four-channel photometer which allows the measurement in the four Strömgren bands at the same time. Moreover, $H\beta$ photometry can be performed by simply turning a lever to enter the $H\beta$ mode which yields simultaneous measurements in the two β bands. Hence the slowness of the telescope motion is largely compensated by the simultaneity of the measurements in the different colors.

Programming the SAT in an efficient way requires a thorough knowledge of the language, and an evaluation of all possible situations that can be expected during the night. Loops and IF's conditions should be included wherever needed. Since observing runs are of relatively short duration, few astronomers make the effort. They either construct short programmed sequences and monitor the telescope all night, or they hastily write a poor program, and leave the telescope unattended for many hours. This frequently leads to inferior results because standard stars are sometimes observed at airmass as high as 4, or because a critical phase of an eclipsing binary has been missed. Our experience shows that, for a similar observing program, the average airmass is larger in automatic mode by systematically 0.05 to 0.15.

In the framework of the Long Term Photometry of Variable program at ESO, a lot of observing time has been attributed to the SAT. Several observers have carried out the observations with varying degrees of success. Each observer has about one month at the SAT, and we call this period a "run". The SAT telescope has a diameter of only 50-cm and many stars have more than 5 mmag photon noise, and the policy of course is to increase the number of integrations for the fainter stars. This is a rule which is easily forgotten when programming the sequences in automatic mode, especially if the telescope is used by people with limited observing experience in photometry. About 50 independent differential measurements are made every night. All in all, the rms values are of about 5 millimag in uvby. A few stars in every 1000 give identification problems.

The graph shown in Fig. 1 gives the rms value of the differential results obtained for comparison star pairs having more than 10 measurements. This is probably the best estimate of the overall accuracy of each run. All observations occurred according to the same instructions, except for the first and last run, which are the only runs carried out by the same observer (C.S., but with a very different observing program). The run indicated with a cross is an observing run carried out at the ESO 50-cm telescope, and is given for reference only. "Automatic" operation started in December 1987. It is clear that rather large variations occur between the DAN50 and SAT block of runs (in spite of completely comparable observing missions).

Though we cannot rule out a hardware effect, or incorrect centering (which in the case of a spectrophotometer of this type would introduce larger scatter, we think a factor affecting the overall accuracy of the result may probably be found in the programming of the SAT. The worst cases were obtained by inexperienced astronomers who wanted to write long programs and leave the telescope alone during a major part of the night. It is absolutely necessary to do such programming very carefully and test the code exhaustively to avoid disastrous errors. This shows the importance of the software in developing APT systems. A lot of planning has to be done before efficient observations are carried out. We conclude that automatic telescopes are an improvement only when they are being programmed by observers who have extensive experience in manually conducted observations.

3. CONCLUSION

The experience we got on the SAT is certainly positive, but we noticed several problems with this setup. Most or all of these problems would not appear in the APT environment.

REFERENCES

Florentin Nielsen, R., Norregaard, P. and Olsen, E. H. 1987 The Messenger, 50, 45
Sterken, C. 1983 The Messenger, 33, 10
Sterken, C. 1986 in The Study of Variable Stars Using Small Telescopes, J. R. Percy, ed., Cambridge University Press, p. 165
Young, A. T., Genet, R. M., Bodys, L. J., Borucki, W. J. and Lockwood, G. W., Henry, G. W., Hall, D. S., Smith, D. P., Baliunas, S. L., Donahue, R. and Epand, D. H. 1990 PASP, in press

DISCUSSION

KNUDE: Did you calibrate the photometer before each run?

STERKEN: Our observers (and other visiting astronomers) do not perform the calibrations, but calibration is regularly verified by a Danish observer. But a change in the instrumental system due to poor calibration would not cause those systematically high differences between comparison stars.

AUTOMATIC PHOTOMETRIC TELESCOPES AND PHOTOMETRIC ACCURACY

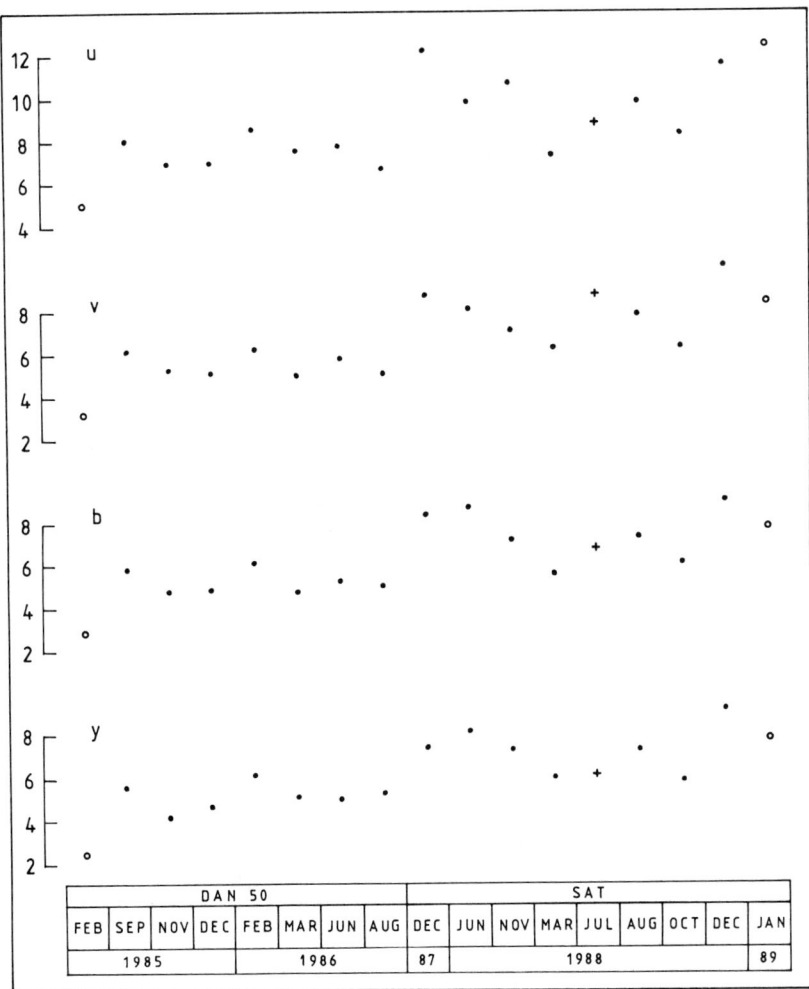

Fig. 1. Systematic difference between mean standard deviation of one differential measurement for the same telescope and photometer operated manually (DAN50), or automatically (SAT). Open circles denote measurements obtained by the same observer, and the + sign is a reference point obtained at the ESO 50-cm telescope.

THE PRESENT STATUS OF INVESTIGATIONS USING THE VILNIUS PHOTOMETRIC SYSTEM

V. Straižys

Institute of Theoretical Physics and Astronomy

The Vilnius photometric system, consisting of seven bandpassses of medium halfwidth, has been developed for completely photometric classification of stars of all spectral types in two or three dimensions in the conditions of different amounts of interstellar reddening. The system also gives identification of stars with different peculiarities, including Be stars, Herbig Ae/Be stars, metallic-line stars, peculiar stars of Si and Cr-Sr-Eu types, T Tauri-type stars, F-G-K subdwarfs, metal-deficient giants, carbon and barium stars, and white dwarfs. The most recent and most extensive description of the system and the methods of photometric classification and quantification is given in my monograph (Straizys 1990).

During the last few years the use of the Vilnius system has been concentrated mainly in three programs. One of them is an investigation of dark clouds and their vicinities. Another program is an investigation of interstellar extinction in the directions of globular clusters. And the third program is a study of stellar populations at different Galactic latitudes with the intention of improving the model of the Galaxy. Additional details about all these programs are given below.

The Vilnius system, which allows one to make two-dimensional classifications of stars even in regions where there is interstellar reddening, is an ideal tool for determination of the "extinction versus distance" dependences in different directions of the Galaxy. The method is very well suited for determination of distances and extinction values of dark clouds in the solar neighborhood. At the distance of the dark cloud the extinction values show a steep rise. The distance of the cloud is determined taking into account the dispersion of points due to distance errors. The method is applied to make a distance survey of a number of nearby dark clouds. In most cases the photoelectric photometry of stars down to $V = 12$ or 13, projected on a dark cloud, is used. So far we have investigated a chain of dark clouds: Lynds 1495, 1521, 1528 and 1538 in Taurus

(Straižys and Meištas 1980), another chain, Lynds 1543, 1546 and 1551 in Taurus (Meištas and Straižys 1981), the Merope dark cloud in the Pleiades (Černis 1987), the cloud in front of the open cluster NGC 1746 in Taurus (Straižys et al. 1990b), the cloud surrounding the reflection nebula NGC 1333 and the open cluster IC 348 in Perseus (Černis, 1990a, b), the cloud around the reflection nebula NGC 7023 in Cepheus (Straižys et al. 1990a), the cloud dividing the North America and Pelican nebulae (Straižys et al. 1989), and the Rho Ophiuchi dark cloud (Straižys 1984). The results are summarized in a paper presented to the Nordic-Baltic Astronomy meeting at Uppsala in June, 1990 (Meištas et al. 1990).

The second program, concerned the investigation of interstellar extinction in the vicinities of globular clusters, is based on photoelectric photometry of stars down to V = 12 or 13 and on CCD photometry of stars down to V = 17. The following areas have been investigated: the M 71 area in Janulis and Straižys (1984), Janulis (1984) and Boyle et al. (1990), the M 56 area in Janulis (1986), Boyle et al. (1989) and Smriglio et al. (1990), the NGC 6712 area in Janulis and Straizys (1990), and the M 15 area in Janulis and Straižys (1990).

The third program partly overlaps the above mentioned investigations of globular cluster areas and has the aim of determining stellar distribution functions down to 17 or 18 made at different Galactic latitudes and longitudes. This work is based on CCD photometry and photoelectric standards (see Boyle et al. 1989, 1990). To achieve photometric accuracy of $\pm$ 1 - 2% at V = 17 mag, we need exposure times in ultraviolet filters of the order of 1 hour working with the 90-cm telescope at Kitt Peak. Thus, there is no problem to collect CCD observations in all seven filters for one field during one night. One must consider that the possibility of three-dimensional classification of hundreds of stars at this magnitude limit has appeared for the first time by using an effective photometric system and light detectors of the highest sensitivity.

REFERENCES

Boyle, R. P., Smriglio, F., Nandy, K. and Straizys, V. 1989
    A&AS, in press.
Boyle, R. P., Smriglio, F., Nandy, K. and Straizys, V. 1990
    A&AS, in press
Černis, K. 1987 ApSS 133, 355
Černis, K. 1990a ApSS 166, 315
Černis, K. 1990b ApSS, in preparation
Janulis, R. 1984 Bull Vilnius Obs, No. 67, 18
Janulis, R. 1986 Bull Vilnius Obs, No. 75, 8
Janulis, R. and Straižys, V. 1984 ApS, 100, 95
Janulis, R. and Straižys, V. 1990 Uppsala Obs Reports, in press
Meištas, E., Černis, K. and Straižys, V. 1990 Uppsala Obs
    Reports, in press
Meištas, E. and Straižys, V. 1981 Acta Astronomica 31, 85

Smriglio, F., Nandy, K., Boyle, R. P., Dasgupta, A. K., Straižys, V. and Janulis, R. 1990 A&A, in press
Straižys, V. 1984 Bull. Vilnius Obs., No. 67, 3
Straižys, V. 1990 Multicolor Photometry, Pachart Publ, Tucson
Straižys, V., Goldberg, E. P., Meištas, E. and Vansevicius, V. 1989 A&A 222, 82
Straižys, V. and Meištas, E. 1980 Acta Astronomica 30, 541
Straižys, V., Meištas, E., Černis, K. and Kazlauskas, A. 1990a, in preparation
Straižys, V., Meištas, E., Černis, K. and Paupers, O. 1990b, in preparation

CCD SCHMIDT PHOTOMETRY OF NGC 5822

W. Weller

Cerro Tololo Inter-American Observatory

Barbara J. Anthony-Twarog and Bruce A. Twarog

University of Kansas

ABSTRACT: Though CCD photometry has become the technique of choice for cluster research, the small field of most systems remains unsuitable for some applications, particularly nearby open clusters. As an alternative to using a larger format CCD, we have been exploring the option of using a Schmidt telescope with a standard CCD to improve the field size. VBR photometry of the open cluster NGC 5822 (diameter ~ 35') shows that simple aperture photometry of the frames provides accuracy in uncrowded fields equal to that of more conventional CCD data. The cluster cmd exhibits a large degree of scatter, consistent with an earlier, reliable photographic survey. Main-sequence fitting under the assumption of solar abundance and E(B-V) = 0.15 leads to an apparent modulus close to 10.0, significantly larger than earlier estimates from photometry. The CM-Diagram scatter and low photometric distances are consistent with a large population of binaries, in agreement with the radial velocity studies of the red giant branch.

I. INTRODUCTION

The study of intermediate-age open clusters has grown in significance as a test of the need for convective overshoot in stellar models of intermediate mass. This need is heightened by the fact that most of the clusters of this type lie near the Galactic plane where field star confusion is a problem, that the higher mass turnoffs are often poorly populated due to the declining luminosity function with higher mass, and that many of the more distant clusters suffer from variable reddening. The obvious solution of analyzing nearby clusters is weakened by the large area covered by the clusters. To overcome this constraint, we have been testing the feasibility of using a standard CCD on the Schmidt telescope at CTIO. Our first example of the value of this approach is with the intermediate-age open cluster, NGC 5822.

2. OBSERVATIONS AND REDUCTION

For NGC 5822, four frames, 20' on a side, were made into a mosaic with 50% overlap to cover a square area 30' on a side. For each field, a series of three exposure sets was obtained in each color. The exposure sets were timed to produce an increase in limiting magnitude between 2 and 2.5 magnitudes at each step. Exposures were obtained in B, V and R for the cluster and a number of E-region standard fields with typically five standards per field. Fig. 1 illustrates the standard deviation in V as a function of magnitude for a complete set of frames for one field. The average error in V remains at or below 0.01 mag down to almost V = 16; the limit is fainter for R and brighter for B. The photometry was then transformed to the Kron-Cousins system.

3. RESULTS AND DISCUSSION

Fig. 2 illustrates the CM-Diagram for one complete field in the cluster in BV; an analogous result occurs in VR. In both cases, the CM-Diagram contains a red giant branch distributed in two distinct clumps and a broad main-sequence turnoff, merging into the field star distribution at fainter magnitudes. The scatter in the main sequence is two to three times larger than predicted from the photometric errors, a result corroborated by the photographic analysis by Hirshfeld et al. (1978). The cluster exhibits no evidence for variable reddening. Values have been estimated from DDO (Claria et al. 1989, Dawson 1978) and uvbyβ (Stetson 1981) photometry. The current best estimate appears to be E(B-V) = 0.15 ±0.02. Note how sharply defined the blue edge of the CM-Diagram appears for (V-R), an unlikely result if reddening is the problem. A potential explanation of importance is that the dispersion is caused by binaries. Support for this idea comes from three sources: a) the uvbyβ distance moduli are distributed bimodally with a peak near (m-M) = 10.0 and a second near (m-M) = 9.2; the second peak would be dominated by binaries with mass ratios near 1.0. b) the radial velocity survey of the giants in the cluster by Mermilliod and Mayor (1990) shows that NGC 5822 is anomalously rich in binaries. c) One can attempt to fit the cluster cmd to a standard ZAMS to obtain the distance modulus. For (B-V) we have adopted the solar relation of Vandenberg and Poll (1989), adjusted to a solar color of (B-V) = 0.65. The resulting ZAMS adjusted for the reddening of E(B-V) = 0.15 and an apparent modulus of 10.0 is shown as a solid line in Fig.2, significantly larger than the photometric means unless the lower moduli are deleted as binaries.

REFERENCES

Claria, J. J., Lapasset, E. and Minniti, D. 1989 A&AS 78, 363
Dawson, D. W. 1981 AJ 86, 237
Hirshfeld, A., McClure, R. D. and Twarog, B. A. 1978 in IAU Symposium No. 80, The HR Diagram, A. G. D. Philip and D. S. Hayes, eds., Reidel, Dordrecht, p. 163

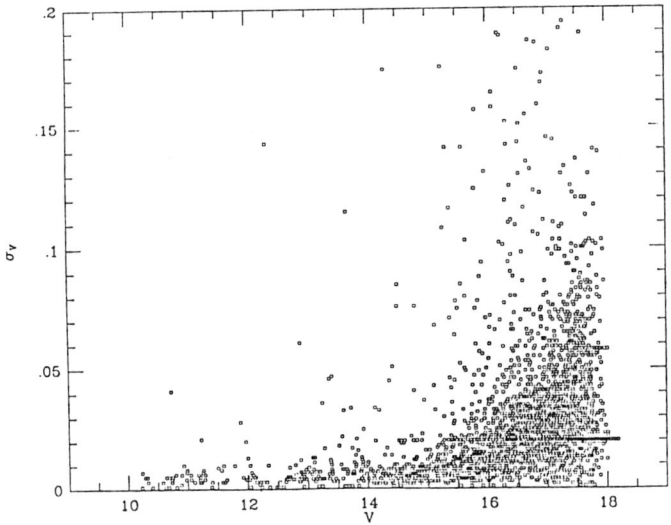

Fig. 1. Standard Deviation in V as a function of V.

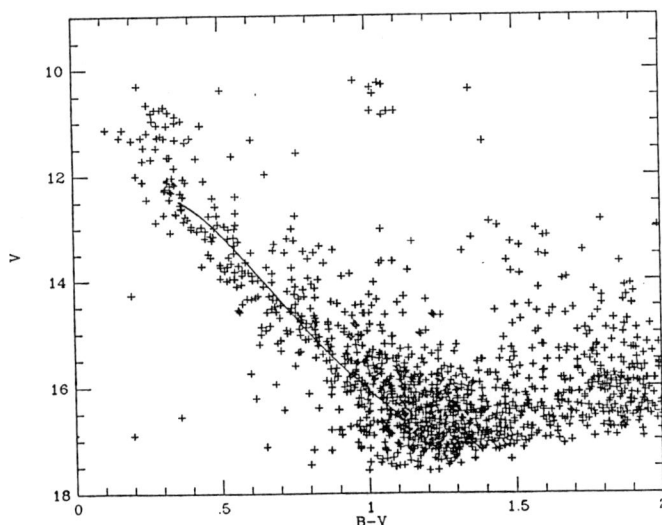

Fig. 2. CM-Diagram compared to the solar ZAMS with (m-M) = 10.0 and E(B-V) = 0.15.

Mermilliod, J.-C. and Mayor, M.   1990, preprint
Stetson, P. B.   1981   AJ 86, 1500
VandenBerg, D. A. and Poll, H.   1989   AJ 98, 1451

DISCUSSION

TWAROG: The photometry was reduced using simple aperture photometry rather than fitting a point-spread function. Attempts at using a point-spread function proved unreliable, probably due to the small number of pixels sampling each star. The frame-to-frame comparisons showed excellent agreement with the simple approach.

Section 3

Banquet Talk and Photos

ITERATED FUNCTION SYSTEMS AND THE DYNAMICAL REPRESENTATION OF NATURAL STRUCTURES

Michael Frame

Union College

Look for a moment at Fig. 1, a picture of a tree, complicated in appearance though obviously computer-generated. Storing or transmitting this image pixel-by-pixel requires a noticeable amount of information, and yet we shall see this picture can be (almost) entirely described by 36 3-digit numbers. ("Almost" because we must specify also starting point, number of iterations, and scale of the picture.) The underlying notion is to describe the dynamical system which draws this picture.

To begin we consider a simpler picture, the "Sierpinski Gasket," G, of Fig. 2. Observe that the gasket can be decomposed into three smaller copies of the gasket: the lower left labeled G_1, the lower right G_2, and the upper G_3. It is easy to find functions taking G to G_1, G_2 and G_3. With the scale shown in Fig. 2, these functions are $T_1(x,y) = (x/2, y/2)$, $T_2(x,y) = (x/2 + 1, y/2)$, and $T_3(x,y) = (x/2 + 1/2, y/2 + \sqrt{3}/2)$. We shall see there is a sense in which these three functions uniquely determine the gasket.

Instead of applying each function separately to the gasket G, suppose we apply all three simultaneously. That is, from each point (x,y) of G we generate three points $T_1(x,y)$, $T_2(x,y)$, and $T_3(x,y)$. Then since $T_1(G) = G_1$, $T_2(G) = G_2$, and $T_3(G) = G_3$, we see that applying all three functions together to G returns $G_1 \cup G_2 \cup G_3 = G$. If we denote the simultaneous application of these three functions by $T_1 \cup T_2 \cup T_3$ - that is, $(T_1 \cup T_2 \cup T_3)(x,y) = T_1(x,y) \cup T_2(x,y) \cup T_3(x,y)$ - then the previous comment becomes

Observation 1   G is left invariant by $T_1 \cup T_2 \cup T_3$.

Having noticed this, a natural question is

Question 1   What happens if we apply these transformations to another shape?

Fig. 1. A tree. Is this picture simple or complicated?

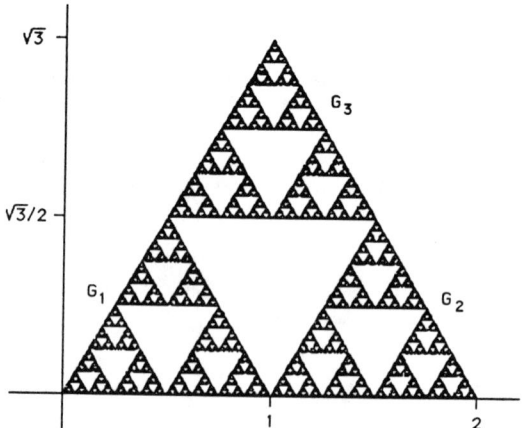

Fig. 2. The Sierpinski Gasket, together with its decomposition into three smaller copies of itself, - G_1, G_2 and G_3.

The first picture of Fig. 3 shows a square, while the second picture shows the result of applying $T_1 \cup T_2 \cup T_3$ to this square: three smaller squares - S_1, S_2 and S_3 - arranged roughly in the locations of G_1 G_2 and G_3. This is not the original square, so right away we suspect the gasket is some sort of "special shape" for $T_1 \cup T_2 \cup T_3$. This notion is reinforced by the second picture of Fig. 4, which shows the result of applying $T_1 \cup T_2 \cup T_3$ to a flower. Comparing the first two pictures of Figs. 3 and 4 with Fig. 2, we may begin to see why the gasket is the only shape exactly covered by its image under $T_1 \cup T_2 \cup T_3$.

Even more is true, though. Suppose we now apply $T_1 \cup T_2 \cup T_3$ to S_1, S_2 and S_3. The third picture of Fig. 3 shows the result: nine still smaller squares. Applying $T_1 \cup T_2 \cup T_3$ again and again generates the other pictures of Fig. 3, a sequence of pictures converging to the gasket. (This rough notion of convergence can be made precise using the "Hausdorff metric.") Fig. 4 presents evidence that these pictures converge to the gasket, regardless of the starting shape. So we have

Observation 2 The transformations T_1, T_2 and T_3 characterize the gasket G in this sense:

$$T_1 \cup T_2 \cup T_3(G) = G \text{ and}$$

$$(T_1 \cup T_2 \cup T_3)^{on}(S) \text{ converges to G for any shape S.}$$

Here $(T_1 \cup T_2 \cup T_3)^{on}$ means apply $(T_1 \cup T_2 \cup T_3)$ n times in succession, and the implication is to observe limiting behavior as n becomes large.

A collection of functions like $T_1 \cup T_2 \cup T_3$ generating a picture by repeated application is called an Iterated Function System (IFS). Building on the work of Dekking (1982) and Hutchinson (1981), Barnsley is responsible for most of the work popularizing this approach. His text (Barnsley 1988) develops the properties of iterated function systems and gives many examples of realizing fractal modeling through IFS. Chapter nine of (Falconer 1990) and the recent Scientific American article (Jurgens et al. 1990) contain other good expositions of IFS. In essence, the IFS method is an efficient way to encode images of many natural objects because as Mandelbrot recognized (Mandelbrot 1983), natural objects often exhibit the same structures on many scales of length. IFS is a functional way of unpacking this scale independence, a way of recursively finding smaller pictures within larger and hence of elucidating the fractal nature of objects.

Magnified sufficiently, the last picture of Fig. 4 really is 3^7 tiny copies of the flower, but the limited resolution of the computer screen renders it indistinguishable from the gasket. This "loss of

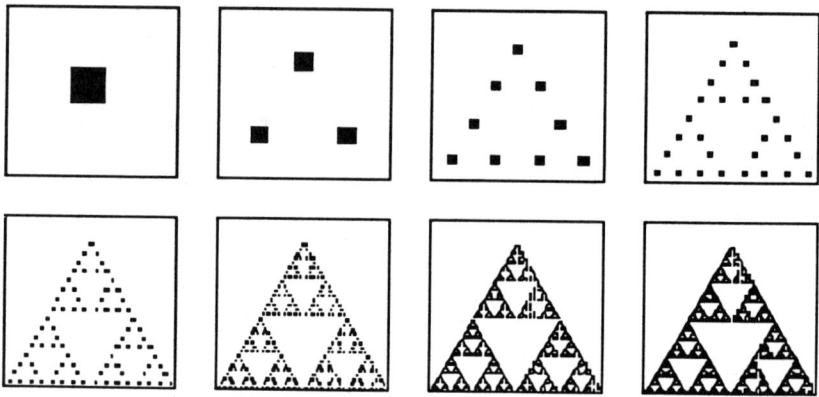

Fig. 3. The Sierpinski Gasket, obtained as the limit of applying $T_1 \cup T_2 \cup T_3$ to a square.

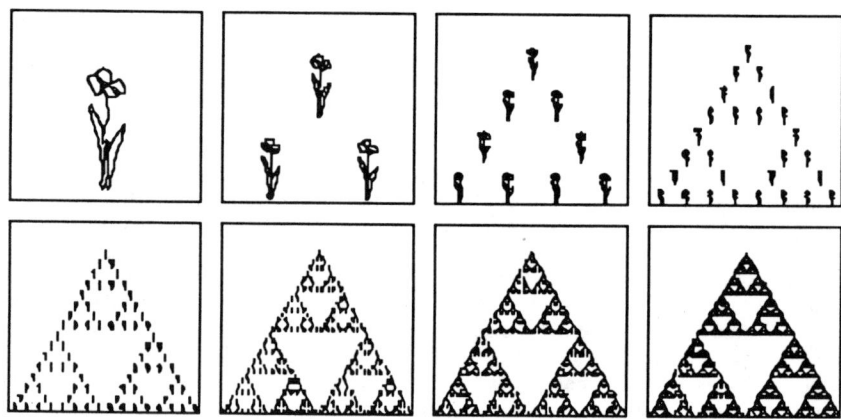

Fig. 4. The Sierpinski Gasket, obtained as the limit of applying $T_1 \cup T_2 \cup T_3$ to a picture of a flower. Along with Fig. 3, this illustrates that the resulting shape does not depend on the starting shape, at least if the pictures are viewed in limited resolution.

Fig. 5. A "transformation" of the Hubble Space Telescope into a 6" backyard telescope. The trick is the identification of the eighth and ninth pictures.

information" is not surprising, since any starting picture produces a sequence with limit the gasket. Together with a touch of whimsy, this convergence can be used to produce some interesting fake transformations of one picture into another. Fig. 5 gives an example.

The idea to take so far is that at least for some shapes, S, there are sets of transformations which, when applied together, leave S invariant and when repeatedly applied together to any other shape yield a sequence of shapes converging to S. Questions which arise naturally now are

Question 2 If we change the transformations a small amount, how will the limiting picture change?

Question 3 Given a shape S, how can we find the transformations which have S as limiting picture?

A "yes" answer to Question 2 gives some reason to be hopeful of a finding an answer to Question 3: if small changes in the functions produce small changes in the picture, then perhaps we need not find the right functions exactly, but only get close to them. So we are lucky that the answer to the first question is yes, for in a sense that can be made precise; the limiting process depends continuously on the transformations. Consequently, a small change in the transformations gives rise to a small change in the limiting shape. This should not suggest that small changes are uninteresting. For example, Fig. 6 shows a sequence of nine modifications of the gasket, each adding a 40° rotation to T_1 of the previous picture. Of course, many other modifications are possible, and exploration usually produces surprises.

As for Question 3, the Inverse Problem of finding the transformations to produce a given picture, consider for a moment the reason behind Observation 1. We recognized G could be decomposed into three pieces - G_1, G_2 and G_3 - and for each of these it was a simple matter to find the transformations T_1, T_2 and T_3 taking G to each of G_1, G_2 and G3. So a rough guideline for finding the transformations to encode a shape, S, is to split S into pieces "looking like" S in some sense, and then find the transformations which realize this similarity. This method of finding decompositions is called the Collage Theorem by Barnsley. Fig. 7 gives an example of this process for another fractal, H, the "Hangman". Transformations T_1 and T_2, producing H1 and H_2, are clear: $T_1(x,y)$ = (x/2, y/2) and $T_2(x,y)$ = (x/2 + 1, y/2). Fig. 7 illustrates how T_3 is found.

Of course, the Sierpinski gasket and the Hangman only look complicated for a moment, until we see the exact self-similarity, the precise replication of the shape at all magnifications. Viewed in this fashion, it is not such a surprise that these shapes can be described using only a few functions. Can we do a similar encoding for a more complicated, natural shape, a shape like the tree in Fig. 1?

The answer, of course, is "yes". Fig. 8 outlines the decomposition of this shape into the relevant pieces. A moment's study of the tree reveals most of the decomposition: the four main branches are shrunken, rotated, and translated copies of the entire tree. The trunk posed more of a challenge: the solution shown in Fig. 8 uses two copies of the tree (reduced significantly in the x-direction and by a factor of one-half in the y-direction) to cover the trunk. Each of these functions is determined by six parameters: shrinking in the x- and y-directions, rotation of the x- and y-axes (different rotations of the axes allow a skewing of the whole picture to form the pieces), and translation in the x- and y-directions. Since only six functions are used to generate Fig. 1, we can say the tree picture is described by 36 numbers. Viewed in this way, the information content of Fig. 1 is quite small. Again, there is a rigidity to the picture, each piece exactly reflecting the whole, but now the image is so complicated that

ITERATED FUNCTION SYSTEMS 357

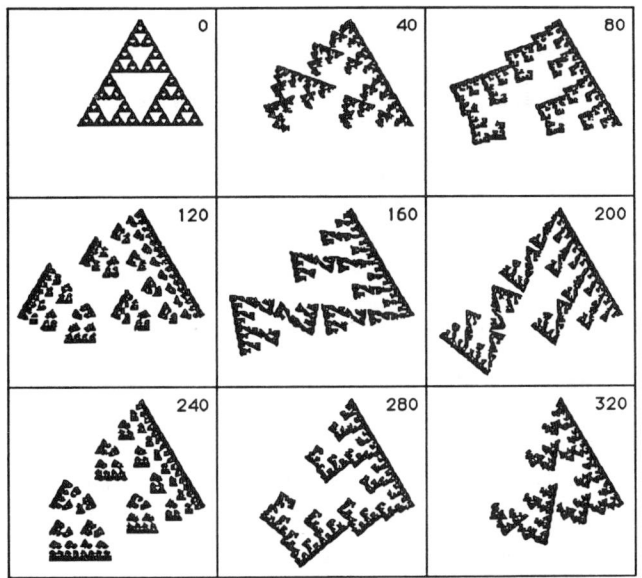

Fig. 6. Changes of the Sierpinski Gasket by including the rotations listed to the transformation T_1. Note that the lower right triangle of the lower right triangle rotates twice in the 360°.

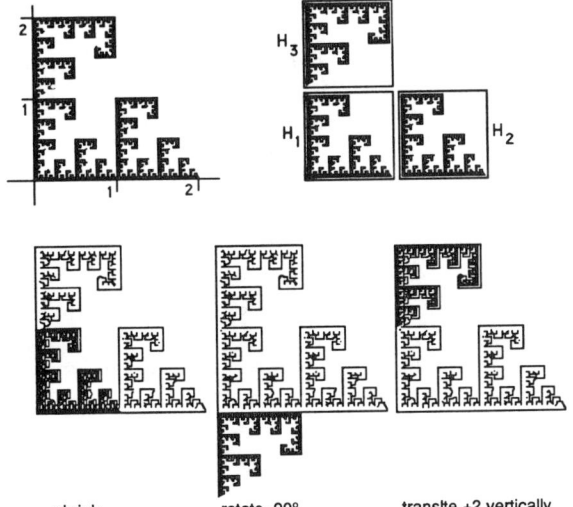

Fig. 7. Upper left, the "Hangman". Upper right, decomposition into three pieces. Bottom, shrink, rotate -90°, translate +2 vertically; the transformation giving H_3.

one may have some difficulty perceiving this rigidity. Of course, this picture is too regular to fool anyone who has carefully studied trees. For example, a botanist criticized the picture for the fact that the branches do not exhibit heliotropy. This could be remedied, but at the expense of including more functions.

Having produced a reasonable forgery of natural structure, two more questions now are suggested.

Question 4 Can these methods be extended to produce three- dimensional pictures?

Question 5 Can we produce color images in a similar fashion?

The answer to Question 4 is immediate: all the arguments and results for two-dimensional pictures transport rigorously to three dimensions, though of course more parameters are needed to specify a transformation. Question 5 has (at least) two answers. In chapter eight of Barnsley (1988), Barnsley describes a method of assigning a measure (probability distribution) to the picture and determining colors by this measure. The mathematical techniques guaranteeing the IFS method will produce a picture can be adapted directly for measures, and the color Inverse Problem can be solved by extending the Collage Theorem to measures. This approach is very versatile, though visualizing colors in terms of measures requires some practice.

Another way to color IFS images is described in Frame and Erdman (1990). As an illustration, consider the tree again. IFS images can be rendered as we have described, by applying all the functions simultaneously (the Deterministic Algorithm), or by following the path of a single point as the functions are applied one at a time in random order. This Random Iteration Algorithm is guaranteed to produce the same picture as the Deterministic Algorithm, but we shall not pursue the proof of this here (see Barnsley 1988 or Frame and Erdman 1990). The coloring method of (Frame and Erdman 1990) is described using the Random Iteration Algorithm. Denote by $T_1, \ldots, T_6$ the IFS transformations used to generate the tree. In particular, let T_1 and T_2 cover the trunk, T_3 the lower left branch, T_4 the upper left branch, T_5 the upper right branch, and T_6 the lower right branch. In the left side of Fig. 8 we have labeled a point x of the trunk. We know $T_3(x)$ will lie in the lower left branch, but in fact we know more about its position: $T_3(x)$ lies in the image in the lower left branch of the trunk. Similarly, $T_5(T_3(x))$ lies in the image in the upper right branch of the image in the lower left branch of the trunk. So an elementary way to color the picture is to keep track of the last several transformations applied and paint the point brown if any of these transformations has been T_1 or T_2, otherwise, paint the point green. An illustration of this coloring method is shown in Frame and Erdman (1990).

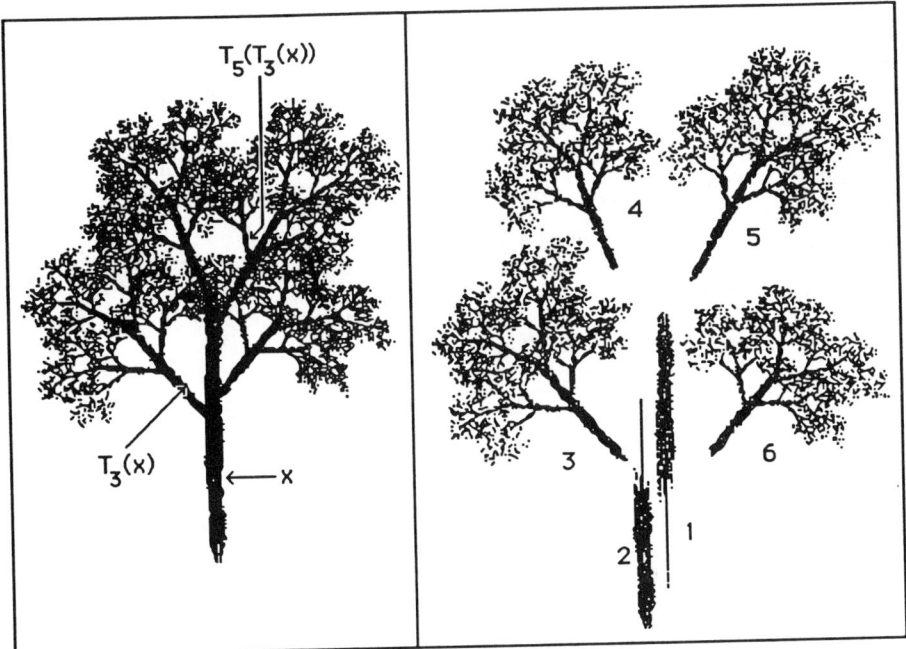

Fig. 8. The tree of Fig. 1, together with its decomposition into six pieces. The transformations taking the whole tree to each of the pieces determines the functions of the Iterated Function System generating the tree. The left panel exhibits a point, x, of the trunk, its image $T_3(x)$ in the lower left branch, and the image $T_5(T_3(x))$ in the upper right branch.

So what does this have to do with astronomy? Almost since their origin, fractals have been used for representing the large-scale structure of the universe - for example, as models of galaxy cluster distributions (Marcus 1987, Peebles 1989 and Jones et al. 1988). Chaos, often a dynamical manifestation of a fractal in phase space, has been used to describe how meteoroids are injected into earth-encountering orbits (Wisdom 1985), the mechanism by which the 3/1 Kirkwood gap arises in the asteroid belt (Wisdom 1983), the irregular rotation of Hyperion (Binzel 1986), the long-term orbit of Pluto (Murray 1989), the persistence of Jupiter's Great Red Spot (Marcus 1987), and variability in the solar cycle (Spiegal and Wolf 1987). An excellent summary of many of these results is given in (Wisdom 1987), and there are doubtless more instances where fractals and chaos arise in astronomy. We suggest there may be other applications in which a physical fractal (not a phase space fractal, but some visible structure) may be used to interpret the underlying dynamics.

Natural objects like trees, ferns, and flowers manifest fractal structures over a range of length scales, and over these scales their formation is governed by a complex nonlinear dynamical process. The rigidity of cell walls, the balance of gravitation and hydrodynamical forces, the self-organization of living tissue - these effects argue for complex forms and interconnected dynamical explanations. Large-scale structures in the universe probably are determined by forces just as complex, and so a fractal interpretation of these structures might be appropriate. As the differences between randomness and chaos are better understood, there is more hope of reading dynamical features from fractal appearances. For example, wispy, interwoven ribbons now can be viewed (sometimes) as the underlying stretching and folding characterizing chaos. By looking for the functional dependences generating a shape, rather than simply considering the shape as frozen in time, we open up a new way to understand the world. Fractals certainly cannot describe everything, but they provide a new perspective, a more direct linking of form and process. No tool should be overlooked in our efforts to understand the universe.

REFERENCES

Barnsley, M, 1988 Fractals Everywhere, Academic Press, Boston
Binzel, R., Green, J. and Opal, C. 1986 Nature 320, 511
Dekking, F. 1982 Advances in Mathematics, 44, 78
Falconer, K. 1990 Fractal Geometry: Mathematical Foundations and Applications, John Wiley & Sons, Chichester
Frame, M. and Erdman, L. 1990 Computers in Physics, 500
Hutchinson, J. 1981 Indiana University Mathematics Journal, 30, 713
Jones, B., Martinez, V., Saar, E. and Einasto, J. 1988 ApJ 332, L 1
Jurgens, H., Peitgen, H. and Saupe, D. 1990 Scientific American 263, 60
Mandelbrot, B. 1983 The Fractal Geometry of Nature, W. H. Freeman, New York
Marcus, P. 1987 Nuclear Physics B 2, 127
Murray, C. 1989 New Scientist, 25 Nov, 1989, 60
Peebles, P. 1989 Physica D 38, 273
Spiegel, E. and Wolf, A. 1987 in Chaotic Phenomena in Astrophysics, J. Buchler and H. Eichhorn, eds., New York Academy of Sciences, New York, p. 55
Wisdom, J. 1983 Icarus 56, 51
Wisdom, J. 1985 Nature 315, 731
Wisdom, J. 1987 Icarus 72, 241

PHOTOS

Bengt Strömgren working at his office at the University Observatory in Copenhagen sometime in the middle 1930's. The calculator, which is electro-mechanical, was the leading edge at the time. Note the handle at the top of the picture. The operator used this in connection with exponential calculations. The sound of the gears was quite satisfying, and gave the impression of work being done.

Entrance to Union College

Campus Center and Nott Memorial

Entrance to
Jackson Garden

Nott Memorial

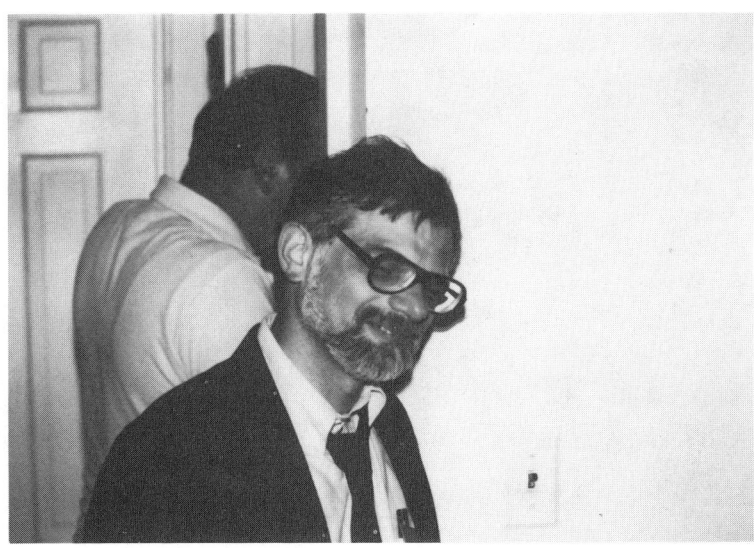
Frederick Chromey

Mary Ellen Hunt

Noel Cramer

Francesa Figueras

Harvey Richer   Arthur Upgren

David Crawford

Kavan Ratnatunga

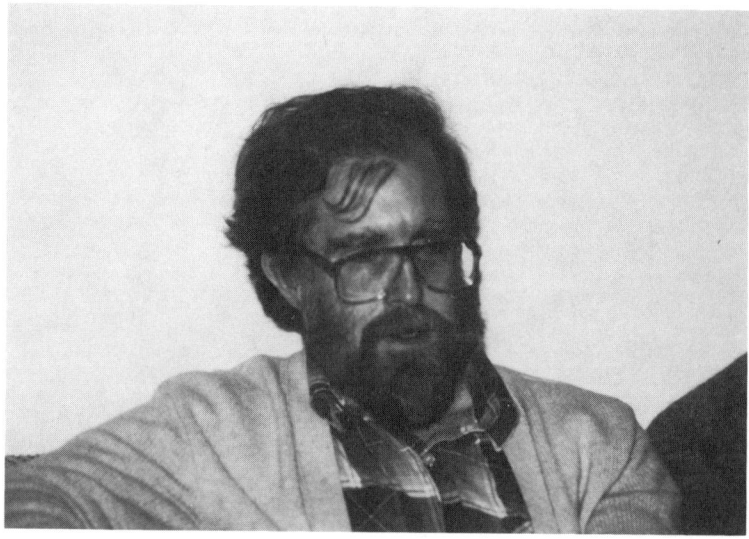

Peter Stetson

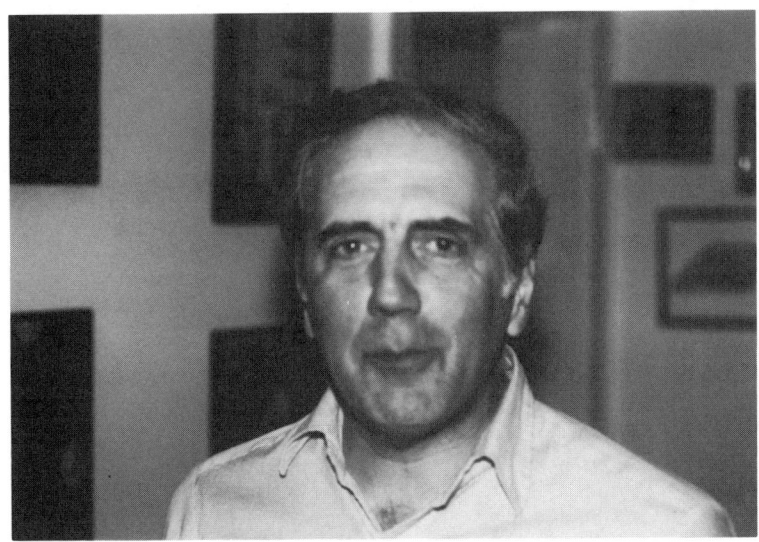

Kenneth Janes

Steven Majewski

Ata Sarajedini

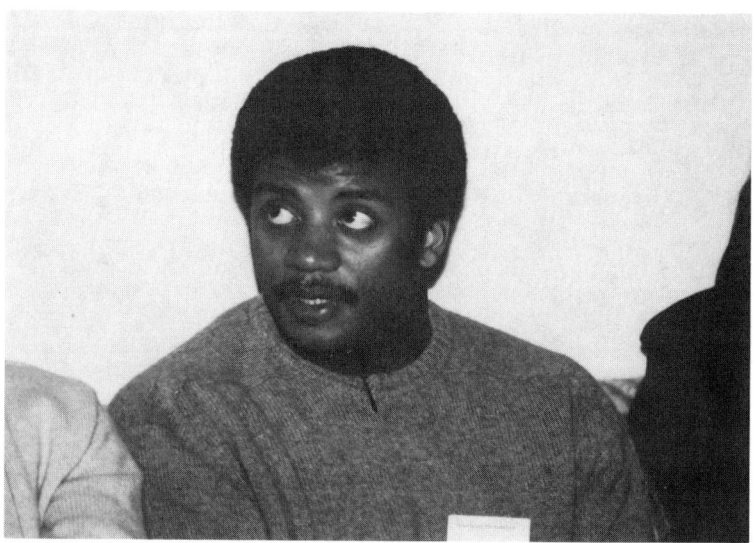

Neil Tyson

Christian Sterken

Bruce Twarog

1125 Oxford Place, Wednesday Reception

Kristina and Davis Philip

INDICES

Name Index   375

Object Index   378

Subject Index   380

If a page number is underlined in the Name Index it indicates the name of an author of a paper. An underlined page number in the Object or Subject Indices indicates that the object or subject was mentioned in the title of a paper. Underlined page numbers of institutions are the institutions of authors of papers. Subjects that are mentioned in chapter headings may be listed in the index but the page numbers are not underlined.

NAME INDEX

A

Aaronson, M.	246
Aghostino	117
Alfaro, E. J.	<u>163</u>, 284
Alpher, R.	283
Ambartsumian, V. A.	3
Andrillat, Y.	316
Anthony-Twarog, B.	<u>127</u>, 284, <u>345</u>
Arp, H. C.	159
Avrett, E. H.	43

B

Baade, W.	50, 184, 196 200, 203-4, 206 251, 253-4
Bailyn, C.	70, 72, 80, 84 87
Balmer, J.	31, 39-40, 56 122-3, 173, 277 299, 320
Barbier, D.	173
Barnes, J.	151
Bell, B.	29
Bell, D. J.	<u>111</u>, 285
Berthet, S.	<u>313</u>
Bessell, M.	247
Bingham, E. A.	72, 74, 82
Biretta, J.	254
Blanco, V.	253, 255
Bolte, M.	91
Bonar, E.	<u>293</u>
Brown, T.	273
Buonanno, R.	72, 74, 82
Burki, G.	12, 146, 172 <u>183</u>, 192
Burrel	116

C

Carney, B.	221
Cayrel, G.	160
Cederbloom, S.	87
Chalonge, D.	173
Chandrasekhar, S.	12
Cheboyer, B.	65
Chiu, L.	211
Chromey, F.	285, <u>293</u>, 297 365
Clark, J. P. A.	241
Cohn, H.	672, 87
Collins	232
Corbally, C.	311
Cote, P.	96
Cousins, A.	196, 200, 211
Cramer, N.	<u>173</u>, 282, 287 <u>303</u>, 366
Crawford, D. L.	1, 7, <u>121</u>, 137 145, 281, 284 328-9, 366
Cray, S.	30-1

D

Darwin, C.	4, 8
Delgado, A. J.	<u>163</u>, 284
Deliyannis, C.	65
Demarque, P.	24-5, <u>45</u>, 65 87, 97, 138, 192 206, 274, 285-6 311, 316, 326
Dirac, P.	30
Divan, L.	173
Djorgovski, G.	87
Doppler	266
Doubleday, E.	221
Dragon, J.	29
Drilling, J. S.	1, <u>225</u>, 230 287, <u>299</u>

E

Eddington, A.	12
Edelstein, M.	<u>293</u>
Eggen, O.	117

F

Fabry	174
Fahlman, G.	<u>89</u>
Fermi, E.	30
Figueras, F.	<u>319</u>, 366
Fitzgerald, P.	237-9, 243
Fourier	187
Fouts, G. A.	223
Frame, M.	<u>351</u>
Friel, E.	243

G

Garrison R.	109, 116, 137 145, 160 <u>231</u> 232, 243, 259 284
Gauss, W.	202, 215
Geisler, D.	205
Gilliland, R.	<u>265</u>, 274, 285
Golay, M.	151, 173, 284 <u>303</u>
Gray, R.	137, <u>231</u>, 232 284, <u>309</u>, 311
Grindlay, J.	72, 87
Guenther, D. B.	274
Gunn, J.	254

H

Harris	331
Hauck, B.	1, 43, 232, 276 281, 284, <u>313</u> <u>315</u>, 316
Hayes, D. S.	156
Heasley, J.	331
Henning, P.	243
Herbig, G. H.	99, 103, 107-8
Hesser, J.	72, 87

Name Index

Hilditch, R. W.	12, 151
Hill, P. W.	151
Hipparcos	177
Houk, N.	303, 309
Hubble, E.	52, 77, 251-2 262-4, 285, 355
Hunt, M. E.	293, 297, 365

J

Jacoby, G.	273
Janes, K. A.	1, 12, 26, 88, 160 206, 233, 243 255, 262, 274 281, 283, 286 297, 331, 369
Jaschek, C.	316
Jaschek, M.	316
Johnson, H.	84-5, 137, 196 211, 245
Jordi, C.	319

K

Keck	294
Kerr, F.	243
King, I.	93
Kirkwood, D.	359
Klare, G.	300
Knude, J.	1, 3, 12, 121 145, 147, 151 284, 287
Kobi, D.	327, 238
Koopmann, R.	154
Kron, R.	154, 211, 277
Kurucz, R.	27, 29, 43, 88 160, 288

L

Landolt, A.	324
Larson, R.	234
Leyden, A.	65
Lick	6
Lugger, P.	72, 87
Luginbuhl, C. B.	286, 323, 326
Lutz, T.	88

M

Maeder, A.	322
Majewski, S. R.	151, 209, 223-4 262, 287, 369
Manfroid, J.	12, 139, 284 335
Marsakov	148
Mateo, M.	205
McClure, R.	72, 87, 241
Mendoza, E. E.	1, 99, 109, 284
Mermilliod, M.	276, 315
Meynet, G.	322
Michaud, G.	28
Morgan, W. W.	231
Morrison, H.	221
Mould, J.	255, 331

N

Neckel, H.	300
Nemec, J.	96
Nicolet, B.	303
Nissen, P.	161
North, P.	288, 327, 328

O

Olson, E.	147, 171
Oosterhoff	46

P

Paschen, F.	173
Patterson, J.	205
Payne-Gaposhkin, C.	12
Pel, J. W.	1, 165, 172 224, 284
Perry, C.	12
Petro, L.	273
Peytremann, E.	29
Phelps, R. L.	286, 331, 334
Philip, A. G. D.	1, 12, 25-6, 65 153, 154, 160 182, 225, 282-3 286-7, 304, 311 372
Philip, K. D.	372
Poisson, S. D.	77, 334
Probst, R.	273

R

Ratnatunga, K.	67, 137, 224 262, 368
Rayet	99, 102
Reed, C.	243
Richard, C.	182
Richer, H.	89, 97, 138 145,160, 206 230, 285, 366
Rich, R. M.	43, 67, 136 193, 201, 204-5 245, 259, 285
Rose, J. A.	71-2, 87, 117
Rosseland	43
Rossi, L.	29
Rufener, F. G.	173
Russell, H. N.	12

S

Saha	215
Sandage, A. R.	74, 221, 223
Sarajedini, A.	55, 67, 224 274, 285, 370
Sawyer, B.	46, 137, 210 286, 345
Schwab, J. J.	182
Schmidt, B.	210, 345, 345
Searl, L.	219, 234
Seitzer, P.	243, 261, 262 285

NAME INDEX

Shavelov	148
Sierpinski, W.	351-2, 354, 356, 358
Slettebak, A.	232
Stark, J.	30
Sterken, C.	1, 12, <u>139</u>, 145-6, 172, 282, 284, <u>335</u>, 371
Stetson, P.	1, <u>69</u>, 88, 91, 97, 136-7, 146, 153-4, 232, 243, 262, 273-4, 286, 311, 368
Stobie, R. S.	210-11
Straizys, V.	<u>341</u>
Stromgren, B.	3-6, 8, 10-12, 108, 121-3, 125, 127, 138-9, 145, 163, 166, 283-4, 309, 314, 316, 337, 361
Strom, K. M.	<u>323</u>
Strom, S. E.	<u>323</u>
Sweigart, A. V.	<u>13</u>, 24-6, 87, 288

T

Thomas, R.	30
Thompson, L.	111
Tinsley, B.	234
Tsuji	248
Tombough, C.	233, 236-7, 243
Torra, J.	<u>319</u>
Twarog, B.	1, 88, 116 <u>127</u>, 136-8, 205, 235, 284, <u>345</u>, 348, 371
Tyson, N. D.	<u>193</u>, 204, 206, 243, 287, 370

U

Umar, S.	205
Upgren, A. R.	1, <u>117</u>, 223, 287, 366

V

Van der Waalsn	30
VandenBerg, D.	87, 91, 322
Vrba, F. J.	<u>323</u>
Varner, T.	29

W

Waelkens, C.	185
Wallerstein, G.	193
Walraven, Th.	165-6
Warren, W.	138, 145, 230, 232, 262, <u>275</u>, 281-2
Weis, E.	<u>117</u>
Weller, W.	286, <u>345</u>
Wesselink,	50, 184, 251
Wolf, M.	99, 102

Y

Yerkes	121
Yoss, K. M.	<u>111</u>, 116, 151, 160, 206, 223, 281, 285, 297, 334

Z

Zinn, R.	216, 219, 234

Object Index

OBJECT INDEX

A

AREAS
- 1 HLF 4   227
- 3 HLF 4   227-8
- 4 HLF 4   225-30
- Baade's window   203-4, 207, 254
- Cepheus   342
- NGP   116, 149, 151, 209-12, 214, 225, 227, 229-30
- North American Nebula   342
- Perseus   342
- Puppis window   238, 240, 243
- S 458   151
- SA 57   211
- SA 98   200
- SA 110   200
- SA 114   200
- SA 127   217
- SA 141   217
- SA 189   217
- SGP   217, 227-9
- Solar neighborhood   247, 341
- Taurus   342

C

CLOUDS, dark
- in front of NGC 1746   342
- in Merope   342
- in North American Neb.   342
- Lynds 1495   342
- 1521   342
- 1528   342
- 1538   342
- 1543   342
- 1546   342
- 1551   342
- Rho Oph   342
- surrounding NGC 1746   342

CLUSTERS
- associations
  - Orion OB 1   329-30
  - Puppis OB 1   239
- globular
  - NGC 104, 47 Tuc   59, 63-4, 197
  - 121, SMC   249
  - 288   25, 46, 48-50, 59-61
  - 362   25, 46, 48-50, 59-61, 91, 154
  - 1261   64
  - 2808   25, 64
  - 4147   70-1
  - 4590, M 68   95
  - 4833   154
  - 5053   93-96
  - 5139, Omega Cen   251, 127, 131-2, 136, 138
  - 5272, M 3   64
  - 5904, M 5   46, 64, 153-9, 198
  - 5927   197, 199
  - 6121, M 4   153, 251
  - 6293   71
  - 6341, M 92   64, 153
  - 6397   64, 70-1, 78, 82, 127, 129, 130-2, 136
  - 6496   197, 199
  - 6624   70-1
  - 6656, M 22   153
  - 6712   342
  - 6752   64, 91-2
  - 6779, M 56   342
  - 6809, M 55   153
  - 6838, M 71   95, 294, 296, 342
  - 7078, M 15   21, <u>69</u>, 70-80, 83-8, 286, 342
    - Stars AC 214   82
    - AC 215   82
    - AC 216   82
  - 7099, M 30   70-1, 74, 78, 82, 84, 86, 90-2, 96, 97
  - 7492   59, 91
  - Pal 5   46, 59-60
  - Pal 12   46
  - Ruprecht 106   46
- CLUSTERS open
  - Berkeley 21   236-7
  - Hyades   166, 170-1, 212, 320-1
  - IC 348   342
  - 4651   127-8
  - NGC 188   235
  - 581   <u>331</u>, 332, 334
  - 752   129
  - 1502   163
  - 1746   342
  - 2169   163
  - 2264   <u>323</u>, 325-6
  - 2682, M 67   127-9, 266, 269-70, 272, 285
  - 3680   127-9, 199, 200
  - 5822   <u>345</u>, 346
  - 6205, M 11   136
  - 6791   235
  - Ori-OB1   170-1
  - Persei, h   184
  - X   184
  - Pleiades   166, 170-1, 329, 342
  - Tombaugh 2   233, 236-7 243

G

GALAXIES
- Fitzgerald 7   239
- Local
  - Carina Dwarf   253

OBJECT INDEX

```
Draco                           198
Fornax                     198, 205
Ursa Minor                      198
Magellanic Clouds     246-7, 251-3
    Large                 166, 252
    Small                 166, 249
Milky Way          89, 148, 150
                       200, 205, 210
                     213, 217-8, 221
                       235, 238, 246
                            253, 341
    Bulge              44, 201-2
                          207, 245-7
                     251, 253, 255-8
    Southern MW          299-300
NGC  224, M 31, Andromeda   193
                            198, 246
                            253, 255-6
     598, M 33                  251
    1399             197, 200, 205
    4486, M 87                  200
```

N

```
Nebula
  Reflecting
    NGC 1333                    342
        7023                    342
```

Q

```
Quasar
  3C 273                        184
```

S

```
Solar System
  Sun         32, 265-6, 274, 288
Stars
  BD 18 5550                    134
     61 2550                    304
  CD -38 245           133-4, 137
  Eri, RZ = HD 30050   183, 185-6
                                188
  HD  34799                     312
      38618                     312
      55879                     304
      61421 = HR 2943 = Alp
        CMa = Procyon            41
      74560 = HR 3467 =
        HY Vel          183, 187-8
                                192
      83277                     312
      94326                     312
      97411                     312
     107233                     312
     111786                     304
     132322                     304
     140283                     304
     142703                     312
     149757                     304
     161817                     304
     172167 = HR 7121 = Alp
        Lyr = Vega      31, 39, 42
                                232
```

```
     265134                     304
  Supernova 1987a    183-4, 189-91
  Table of          101, 103, 305-7
  Tau, Eta                      100
```

Subject Index

A

ABUNDANCE vs ROTATIONAL VEL   217
AGE
  Calibration, current issues   49
  Metallicity relation   46
  Relative   59
  Spread   <u>45</u>, 48

B

BAADE'S WINDOW   196, 203
BAADE-WESSELINK METHOD   49, 184
BALMER
  Discontinuity   86, 122-3
  Jump   173
  Line   5, 39, 299
BLANKETING PARAMETERS
  Calibration   313
BREATHING PULSES   20-1

C

CATALOGS
  Cayrel Fe/H   160, 165
  General, of phot. data   281
  Geneva   282, 303
    1988   175-6
    extension   173
  Gliese   117, 119
  LHS   117
  Michigan Spectral   303-4, 310
  Photometric   276
  Table of catalogs   277
CLUSTERS
  Globular
    age spread   <u>45</u>
    ages   89
      relative   55, 59
    CCD detectors, applied   89
    color gradients   <u>69</u>, 70
    cusp   <u>69</u>
    luminosity function   91
    M 15   <u>69</u>, 77
      surface photometry   78
    mass functions   91
    population gradients   <u>69</u>
    research   <u>89</u>
    turn-off   <u>55</u>
  Open
    NGC 581   <u>331</u>
    NGC 5822   343
CM-DIAGRAM   70
  Frame in M 5   154
COMPUTERS
  Cray   31
  YMP   30
  VAX   31
CORE HELIUM EXHAUSTION   20

D

DARWIN LECTURE   4, 8
DELTA(B-V)   58-9, 61-3, 67

DIFFUSION
  Helium   62
DOPPLER SHIFTS   256

E

EVOLUTION
  Age determination   3
  Chemical comp. parameter   4
  Cluster ages   13
  Convective overshooting   13
  Core mass   14
  Gravity parameter   5
  Helium abundance   13
  Hydrogen burning shell   14
  Oosterhoff effect   13
  Period shift, Sandage   13
  Pre horizontal branch   14
  RGB phase   14
  Second parameter   13, 45, 60, 65
  Semiconvection   13
  Zero-age line   4

F

FIGURES
  A tree   317
  Abundance distr.   204
    three models   202
  Age as the second parameter   60
  Age vs metallicity   64
  Alpha Lambda photom.   104, 107
  Ave dist. from plane
    function of color excess   149
  (b-r) profile for M 15   83
  Binary fraction   132
  c vs (b-y)
    4 HLF 4   226
    metal-poor giants   135
    new models   43
  CCD image of NGC 6836   296
  CCD spectra of M giants   254
  Changes in Sierpinski
    Gasket   357
  CM-Diagram
    M 5 (four color)   157
      BHB stars   158
    M 15   73, 76
    NGC 581   333
    NGC 5822   <u>345</u>
    NGC 6397   130
    NGC 7492   90
    Omega Cen   132
    SN 1987A   190
    Washington System
      Galactic bulge   202
  Comparison of MD and LGK
    calibrations   329-30
  Comparison of residuals in
    phot. systems   167-9
  Core helium distribution   22
  Core mass vs shell thickness   15
  DDO passbands   113
  Del c vs distance   149
  Del m and Del c vs E(b-y)   149

SUBJECT INDEX

Del mass vs log z 15
Difference, photoelectric
  minus grism 114
Distribution by spectral
  types, Geneva Catalog 176
Distribution by transverse
  velocity 118
(Fe+Mg) vs (J-K), bulge K III 254
Fe/H vs Z 217
Finding chart for M 5 155
Four-color
  distribution in
    dec. 318
    Gal. Lat. 318
    Gal. Long. 318
    mag. 317
    RA 317
Flux distribution, new model 40-1
Hangman 357
HB tracks
  for He abundance 19
  for Z abundance 19
Histograms of Geneva Cat. 178-9
HR diagrams compared
  NGC 7492 and 6752 92
HST transformed to a 6" 355
I vs (V-I) for NGC 2264 325
Light curve of RZ Eri 188
Luminosity function
  M 31 bulge 256
  NGC 581 334
  NGC 5053 94
m vs (b-y)
  4 HLF 4 226
  Am stars 321
  Lambda Boo stars 310
(m-51) vs T1T2 204
Mass function for NGC 5053 94
Mean errors for trans. coef. 142
Metallicity calibration 124
Mosaic image, M 31 bulge 257-8
Mv vs del(B-V) 57
Number of stars in Geneva Cat. 174
Observed and predicted
  spectrophotometry 42
Passbands
  DDO 113
  UBV 113
Random photometric errors in
  pm survey 214
(R-I) vs (V-I) for NGC 2264 325
Schematic of CCD system 295
Sierpinski Gasket 352, 354, 357
Sigma V vs V for SN 1987A 190
Solar irradiance 32
Standard deviation in V 345
Systematic diff., manual
  vs automatic 339
Theoretical isochrones for
  globular clusters 90
Three not-radial modes of
  HD 74560 188
Time dependence of mass 22
Time series for a star on
  a good night 271

Tree 317
Tree of Fig. 1 359
Two-color diagram
  M 15 85
  NGC 581 332
  Puppis window 230
UBV passbands 113
UV excess as function of Z 214
V brightness profiles
  M 15 79
  versus radial dist. 79
Variation
  del(B-V) with age 57
  del(B-V) with Fe/H 60
  RZ Eri component 186
Washington system
  separation of III and V 195
FOURIER ANALYSIS 187
FRACTALS
  Astronomical examples 359

G

GALACTIC MODELS
  ELS 45, 48
GALAXIES
  Dwarf spheroidal gal. 253
  Fitzgerald 07, reddening 239
  Integrated light 252
  Milky Way
    blue stellar sample 213
    bulge 193, 200-1
      abundance distr. 193
      CCD survey 201
      chemical evolution 235
      collapse models 218
    disk
      formation 233
      stellar population 335
      thick 209, 210, 213
        215, 221
    halo population 209
    high galactic lat. 225, 229
    OB+ stars 299
    open clusters, in outer
      disk 236
    reddening and stellar pop-
      ulation, line of sight 238
    star clusters 233
    southern MW 299
    UV excess 209
    windows in galactic plane 233
  Magellanic Clouds 252
GAUSSIAN DISTRIBUTION 215

H

HORIZONTAL BRANCH 62
  Models 13
  Vertical structure 18
HR DIAGRAM 3-4, 6

I

INSTITUTIONS

Subject Index

Acciones Integradas Hispano-Francesas		322
ADC		275
Appalachian State Univ.	231,	309
Astrophys. Inst., Univ. of Brussels	139,	335
AURA	121,	265
Boston University	233, 238,	283
Bryn Mawr		293
Carnegie Inst. of Washington		209
Center for Astrophysics	27,	29
Center for Solar and Space Research		45
CICYT		322
Columbia University	193,	245
Copenhagen Univ. Obs.	3, 147,	361
CTIO	137,	343
DAO	69,	154
David Dunlap Obs.		231
ESO	140, 165,	174
	183-4,	337
Swiss Station		184
Geneva Observatory	146, 173,	183
		303
Goddard Space Flight Center	13,	275
Herzberg Inst. of Astrophysics		69
IAU		278
Inst. of Astrofisica Spaziale		29
Astron., UNAM		99
Astron., University of Lausanne	313, 315,	327
Astrophys of Andalucia		163
Theo. Physics and Astron., Vilnius		341
Kapteyn Astron. Inst.		165
Kiso Observatory		210
KPNO	116-7,	121
	154, 156,	265
	268-9, 273,	331
Lab for Astron. and Solar Physics, NASA		13
Las Campanas Obs.	145,	310
Leiden Observatory		165
Southern Station, South Africa		165
Lick Observatory		6
Los Alamos		44
Louisiana State Univ.	225,	299
Lowell Observatory		300
MacDonald Observatory		121
NASA	13, 23, 39,	52
	265,	275
National Research Council of Canada		69
National Sciences and Engin. Council, Canada		96
NOAO	121,	331
NSF	39, 121,	265
		293
NSSDC	275,	300

OHP		174
Palomar Observatory		223
Photometrics Inc.		293
San Diego Supercomputing Center	30,	39
Space Tel. Sci. Inst.	261,	262
		265
Sphinx Obs, Jungfraujoch		174
Union College	153, 225,	351
University of		
Aarhus		11
Barcelona		319
British Columbia		89
Cal. Tech.		248
Hawaii	71, 111,	331
Illinois		111
Kansas	127,	343
Massachesetts		323
Toronto	145,	231
Washington		193
US Naval Observatory Flagstaff Station		323
Van Vleck Observatory	117,	153
		225
Vassar College Obs.		293
Yale University	45,	155
Observatory		55
Yerkes Observatory		131
INSTRUMENTS		
Schmidt, Kiso Obs.		210
0.4 m tel.		
CTIO		300
Gornergat		174
Sphinx Obs.		174
0.5 m tel.		
ESO	140,	337
Strömgren Automatic		337-8
0.6 m tel.		
Las Campanas		145
Lowell Observatory		300
0.61 m tel.		
ESO, Bochum		140
0.9 m tel.		
CTIO		238
ESO, Dutch	165,	172
1 m tel.		
Flagstaff Station		323
KPNO	268,	331
Schmidt		116
Mt. Laguna Obs.		111-2
Obs. of Lyon		174
OHP		174
2.1 m tel.		
KPNO	269,	273
2.2 m tel.		
Univ. of Hawaii		71
2.4 m tel.		
Hubble Space Tel.	251, 259,	261
	263-4,	355
3.6 m tel.		
CFHT	71,	93
4 m tel.		
KPNO	211,	223
5 m tel.		
Palomar		223

SUBJECT INDEX

CCD
  detectors   45-6, 49, 55
                                        71, 89, 91, 93
                                      95-6, 111-2, 129
                                         145, 154, 194
                                                200, 238
  photometer   298
COSMOS   210
HIPPARCOS   148
IRAS   148
ITERATED FUNCTION SYSTEMS   351, 353

L

LUMINOSITY
  Red giant branch   16

M

METHODS
  M and D (Geneva System)   173
  MgH INDICES   241
  MICROTURBULENT VELOCITY   28
MODELS
  ELS   218-9, 221
  New lines, models, colors   27

O

OPACITIES
  Los Alamos   44
OXYGEN
  Abundance   61

P

PERIOD SHIFT EFFECT   21
PHOTOMETRY
  13 COLOR   108
  Abundances, PHOTOMETRIC   313
  Accuracy
    Automatic photoelectric
      telescope   335
    photometric systems   165
  Ages, A stars   319
  Alpha Lambda system   99, 108
  Applications
    clusters   285
    surveys   286
    variability   286
  Archiving photometric data   275
  Astrophysical calibrations   288
  A-type stars   225
  Automatic photometric
    telescopes   335
  Catalogs   287
  CCD
    advantages   268
    disadvantages   268
    ensemble   265
    techniques   267
    four color   127, 153
    photometer for Vassar   293
    standards   136
  Cousins system   200
  DDO   115, 238
  Four Color   3, 5-8. 31
                             108, 121, 122
                           124, 127, 128-9
                           131, 133, 137
                           140, 147, 148
                           150, 170, 225
                           230, 231, 309
                        319-20, 314, 316
  25 years of data   315
  B-type stars   163
  calibrations   327
  CCD   127, 128, 153
  comparison - CCD and
    photoelectric   128
  globular cluster stars   153
  techniques and calibrations   284
  definitions
    (b-y)   7
    c   7
    m   7
  Geneva system   170-1, 173, 175
                           182, 184, 189
                           232, 314, 316
    CCD   175
    high precision   183
    stellar boxes   303
  Globular cluster   55, 153
  H alpha   323
  H beta   3-5, 9-10
                           127-8, 147, 232
                           315, 319-20, 323
  25 years of data   315
  definition, Beta   8
  Infrared   245, 248, 251
    applications   252
    precision physics   245
  lc system   5
  MgH band indices   241
  NGC 581   331
  NGC 2264   323
  Other systems   284
  Photometric abundances   313
  Precision
    future needs   52
    high   183
    past, present, future   283
  Schmidt CCD phot. of NGC 5822 343
  Stars in source catalogs   117
  Stellar, with HST   261
  Strömgren's work   3
  Time resolved CCD ensemble   265
  UBV   31, 58, 71, 84-6

S

SECOND PARAMETER   13, 45, 60, 65
SIERPINSKI GASKET   352, 354-5, 357
SPECTRA
  Grism   111, 113
    DDO colors   111
    UBV colors   111
  Objective, thin prism   309
STARS
  A   313
    abundance   313
    distances and ages   319
    four-color phot   225

B	231
F	<u>313</u>
Fundamental stellar param.	246
HD	
30050 – RZ Eri	185
74560 – HY Vel	187
Lambda Boo	<u>309</u>
Metallic lined	107
OB+ in S Milky Way	<u>299</u>
Population II	3
SN 1987A	189
T Tauri	107
Wolf Rayet	102
STROMGREN	3
SUMMARY PAPER	<u>283</u>
SURVEYS	
Along galactic disk	237
Basel	210-1
Chiu	211
High galactic lat.	<u>225</u>
Majewski	211
Stobie	211
Puppis window	238

T

TABLES	
Alpha lambda phot.	
of MK stds	101
of non MK stars	103
Bandpasses for Cal. Tech	
IR system	246
BHB stars in M 5	159
Data obtained with CCD	326
Fields in high galactic	
latitudes	227
Filter passbands	
four color and H beta	7
lc system	6
Globular clusters and	
delta t	64
Growth of data in four	
color and H beta syst.	316
Mean errors from four color	9
Observational material on M 15	72
Open clusters, Be 21, To 3	236
Photometric boxes in Geneva	
system	305-7
RMS values of mj	143
Sky brightness determinations	81
Slit spectra of Lambda Boo	
candidates	312
Star counts in M 15	75
UBV and four-color colors	
for new models	33
Variation of Fe/H with age	9
Washington photometric syst.	196
standard fields	200
TELESCOPES	
See INSTRUMENTS	
Automatic photoelectric	335
Network of 4 m	270

Other Publications in Astronomy

THE EVOLUTION OF POPULATION II STARS (1972), A. G. D. Philip, ed., Dudley Observatory Report No. 4.

MULTICOLOR PHOTOMETRY AND THE THEORETICAL HR DIAGRAM (1975), A. G. D. Philip and D. S. Hayes, eds., Dudley Observatory Report No. 9.

UBV COLOR-MAGNITUDE DIAGRAMS OF GALACTIC GLOBULAR CLUSTERS (1976), A. G. D. Philip, M. F. Cullen and R. E. White, Dudley Observatory Report No. 11.

AN ANALYSIS OF THE HAUCK-MERMILLIOD CATALOGUE OF HOMOGENEOUS FOUR-COLOR DATA (1976), A. G. D. Philip, T. M. Miller and L. J. Relyea, Dudley Observatory Report No. 12.

GALACTIC STRUCTURE IN THE DIRECTION OF THE POLAR CAPS (1977), M. F. McCarthy and A. G. D. Philip, eds., in Highlights of Astronomy, Vol 4, Reidel, Dordrecht.

IN MEMORY OF HENRY NORRIS RUSSELL (1977), A. G. D. Philip and D. H. DeVorkin, eds., Dudley Observatory Report No. 13.

IAU SYMPOSIUM NO. 80, THE HR DIAGRAM (1978), A. G. D. Philip and D. S. Hayes, eds., Reidel, Dordrecht.

IAU COLLOQUIUM NO. 47, SPECTRAL CLASSIFICATION OF THE FUTURE (1979), M. F. McCarthy, A. G. D. Philip and G. V. Coyne, eds., Vatican Observatory.

PROBLEMS OF CALIBRATION OF MULTICOLOR PHOTOMETRIC SYSTEMS (1979), A. G. D. Philip, ed., Dudley Observatory Report No. 14.

X-RAY SYMPOSIUM 1981 (1981), A. G. D. Philip, ed., L. Davis Press.

IAU COLLOQUIUM NO. 68, ASTROPHYSICAL PARAMETERS FOR GLOBULAR CLUSTERS (1981), A. G. D. Philip and D. S. Hayes, eds., L. Davis Press.

A DEEP OBJECTIVE PRISM SURVEY OF THE LARGE MAGELLANIC CLOUD FOR OB AND SUPERGIANT STARS. PART I. (1983), A. G. D. Philip and N. Sanduleak, L. Davis Press.

IAU COLLOQUIUM NO. 76, THE NEARBY STARS AND THE STELLAR LUMINOSITY FUNCTION (1983), A. G. D. Philip and A. R. Upgren, eds., L. Davis Press.

IAU SYMPOSIUM NO. 111, CALIBRATION OF FUNDAMENTAL STELLAR QUANTITIES (1985), D. S. Hayes, L. Pasinetti and A. G. D. Philip, eds., Reidel, Dordrecht.

IAU COLLOQUIUM NO. 88, STELLAR RADIAL VELOCITIES (1985), A. G. D. Philip and D. W. Latham, eds., L. Davis Press.

HORIZONTAL-BRANCH AND UV-BRIGHT STARS (1985), A. G. D. Philip, ed., L. Davis Press.

SPECTROSCOPIC AND PHOTOMETRIC CLASSIFICATION OF POPULATION II STARS (1986), A. G. D. Philip, ed., L. Davis Press.

STANDARD STARS (1986), A. G. D. Philip, ed., in Highlights of Astronomy, Vol 7.

IAU SYMPOSIUM NO. 126, GLOBULAR CLUSTER SYSTEMS IN GALAXIES (1987), J. E. Grindlay and A. G. D. Philip, eds., Reidel, Dordrecht.

IAU COLLOQUIUM NO. 95, THE SECOND CONFERENCE ON FAINT BLUE STARS (1987), A. G. D. Philip, D. S. Hayes and J. W. Liebert, eds., L. Davis Press.

NEW DIRECTIONS IN SPECTROPHOTOMETRY (1988), A. G. D. Philip, D. S. Hayes and S. J. Adelman, eds, L. Davis Press.

CALIBRATION OF STELLAR AGES (1988), A. G. D. Philip, ed., L. Davis Press.

STAR CATALOGUES: A CENTENNIAL TRIBUTE TO A. N. VYSSOTSKY (1989), A. G. D. Philip and A. R. Upgren, eds., L. Davis Press.

THE GRAVITATIONAL FORCE PERPENDICULAR TO THE GALACTIC PLANE (1989), A. G. D. Philip and P. K. Lu, eds., L. Davis Press.

CCDs IN ASTRONOMY. II. (1990) A. G. D. Philip, D. S. Hayes and S. J. Adelman, eds., L. Davis Press.